T0191906

Lagrangian Oceanography

Physics of Earth and Space Environments

The series *Physics of Earth and Space Environments* is devoted to monograph texts dealing with all aspects of atmospheric, hydrospheric and space science research and advanced teaching. The presentations will be both qualitative as well as quantitative, with strong emphasis on the underlying (geo)physical sciences. Of particular interest are

- contributions which relate fundamental research in the aforementioned fields to present and developing environmental issues viewed broadly

- concise accounts of newly emerging important topics that are embedded in a broader framework in order to provide quick but readable access of new material to a larger audience

The books forming this collection will be of importance for graduate students and active researchers alike.

Series Editors:

Rodolfo Guzzi
Responsabile di Scienze della Terra
Head of Earth Sciences
Via di Villa Grazioli, 23
00198 Roma, Italy

Louis J. Lanzerotti
Bell Laboratories, Lucent Technologies
700 Mountain Avenue
Murray Hill, NJ 07974, USA

Ulrich Platt
Ruprecht-Karls-Universität Heidelberg
Institut für Umweltphysik
Im Neuenheimer Feld 229
69120 Heidelberg, Germany

More information about this series at http://www.springer.com/series/5117

Sergey V. Prants • Michael Yu. Uleysky
Maxim V. Budyansky

Lagrangian Oceanography

Large-scale Transport and Mixing
in the Ocean

 Springer

Sergey V. Prants
Pacific Oceanological Institute
  of the Russian Academy of Sciences
Laboratory of Nonlinear Dynamical Systems
Vladivostok, Russia

Michael Yu. Uleysky
Pacific Oceanological Institute
  of the Russian Academy of Sciences
Laboratory of Nonlinear Dynamical Systems
Vladivostok, Russia

Maxim V. Budyansky
Pacific Oceanological Institute
  of the Russian Academy of Sciences
Laboratory of Nonlinear Dynamical Systems
Vladivostok, Russia

ISSN 1610-1677   ISSN 1865-0678   (electronic)
Physics of Earth and Space Environments
ISBN 978-3-319-85041-2   ISBN 978-3-319-53022-2   (eBook)
DOI 10.1007/978-3-319-53022-2

Printed on acid-free paper

This Springer imprint is published by Springer Nature
The registered company is Springer International Publishing AG
The registered company address is: Gewerbestrasse 11, 6330 Cham, Switzerland

Preface

The great Tohoku earthquake followed by the catastrophic tsunami inflicted heavy damage on the Fukushima Nuclear Power Plant in Japan. Large amount of radioactive water leaked directly into the ocean. The radioactive pollution was also caused by atmospheric deposition on the ocean surface. Fukushima-derived radionuclides, advected by oceanic currents and eddies, propagated over a broad area in the North Pacific. It was very important to know by which transport pathways they did that and which mesoscale eddies gained and retained radioactive water and for how long a time. Could the Fukushima-derived radionuclides cross the strong Kuroshio Extension current and propagate to the south?

In April 2010, the explosion at the Blue Horizon mobile drilling rig in the Gulf of Mexico caused the catastrophic offshore oil spill. It was not, of course, a problem to monitor the propagation of oil on the sea surface. The problem was a short-term prediction for a few days of the shape, deformation, and metamorphoses of the oil plume in an unsteady velocity field.

Is it possible to find more or less robust material (Lagrangian) structures in chaotic flows governing mixing and transport of tracers? Could some of them create transport barriers preventing diffusive-like propagation of a contaminant? If we could identify such structures in the ocean, we would predict for a short and medium time where a contaminant will move even without a precise solution of the Navier–Stokes equations. The standard approach is to run global or regional numerical models of circulation to simulate propagation of pollutants and try to forecast their trajectories. It has been, of course, made after each of those events. The outcomes provide the so-called spaghettilike plots of individual trajectories that are hard to interpret. Moreover, majority of trajectories in a chaotic environment are very sensitive to small and inevitable variations in initial conditions. Those trajectories are practically unpredictable even for a comparatively short time [see Eq. (1.2)].

Hydrological fronts in the ocean, which are boundaries between waters with strong horizontal gradients of temperature, salinity, and density, have long been recognized by fishermen to attract squid, fish, mammals, and other marine organisms. These fronts are not stationary features, but they might change the form, intensity, and location and could even disappear in the course of time. In order to know the origin and history of frontal water masses to estimate their productivity and fishery conditions, it is necessary to simulate advection of virtual particles in a given velocity field backward in time. It is also important to estimate how robust and strong the fronts could be and try to predict their location for a short and medium time.

We list a few but rather dramatic events and practically important issues which require not the Eulerian but Lagrangian approach to deal with it. In the Eulerian framework, the flow is described in terms of the velocity field, while in the Lagrangian one, it is characterized by trajectories of a large number of fluid tracers which are tagged and tracked individually [2]. While in the Eulerian framework we get frozen snapshots of data, Lagrangian diagnostics enable to quantify spatiotemporal variability of the velocity field. Lagrangian structures are difficult to extract from Eulerian data, as they do not show up in a Eulerian velocity field.

The development of the Lagrangian methods in oceanography in the last years was advanced due to several factors: (1) The impressive progress in satellite monitoring has provided us continuous, near-real-time and global data at high space resolution for many oceanic and atmospheric parameters and the global altimetric velocity field. (2) The recent advance in satellite technology has revolutionized measurements taken by buoys drifting in the ocean which provide real-time information about ocean circulation in a high-frequency manner. (3) The development of high-resolution numerical models of ocean circulation has opened up new opportunities in simulating mesoscale and submesoscale processes. The new branch of oceanography, Lagrangian oceanography, is developing rapidly [3, 14]. Our personal scientific interests have been inspired by the penetration of ideas and methods of dynamical systems and chaos theory in a geophysical context in recent decades [4–7, 9–13, 15].

The ocean is a highly turbulent medium and presents a variety of dynamical phenomena with different space and time scales ranging from millimeters to a few thousand of kilometers and from milliseconds to thousands of years. The ocean is subjected to a variety of small- and large-scale random perturbations. Some of them produce water movement that is supposed to be inherently unpredictable. The others generate well-ordered and long-lived coherent structures due to the Earth's rotation, density stratification, wind stress, and bottom topography. How to find some order in this chaos?

In this book, we focus on large-scale Lagrangian transport and mixing of water masses in the ocean. We mean by large scale the motions affected by planetary rotation when the effects of planetary rotation are large. The Rossby number, $\text{Ro} = U/Lf$, should be small for such a motion as compared to unity. Here, U is a characteristic velocity, and f is the Coriolis parameter. The horizontal scale of large-scale motion, L, can vary and exceeds a few kilometers in temperate and high latitudes where we work in this book.

The starting point in the Lagrangian approach is a velocity field which is supposed to be derived analytically or given as an output of a numerical model of circulation or estimated from satellite altimetry or radar measurements. If advected particles rapidly adjust their own velocity to that of a background flow and do not affect the flow properties, they are called passive and satisfy simple advection equations (1.3). Solutions of those equations can be chaotic in the sense of exponential sensitivity to small variations in initial conditions and/or control parameters as in Eq. (1.1), even if the Eulerian velocity field is supposed to be absolutely deterministic. It means that even a simple time-periodic deterministic velocity field may cause practically unpredictable particle trajectories, the phenomenon known as chaotic advection [1, 8].

In the first two chapters, we introduce and analyze in detail some simple deterministic models of chaotic oceanic flows. We do not assume that the reader is familiar with dynamical systems theory and theory of chaos. The most important mathematical notions, used in the text, are **bolded** and explained in the Glossary in the end of the book.

The other chapters are devoted to modeling large-scale mixing and transport in the Northwestern Pacific Ocean and in some adjacent marginal seas. In this book, we are interested, mainly, in mesoscale processes on the scale of ten kms and more which are simulated by integrating trajectories of artificial tracers advected by altimetry-derived AVISO velocity fields or by velocity fields generated in high-resolution numerical models of circulation. In Chap. 3, we review briefly the present state of operational and satellite oceanography. The fourth chapter is a methodological one, where we describe the Lagrangian tools used to simulate and analyze mixing and transport. They include a number of Lagrangian indicators and specific Lagrangian maps which we compute to plot and visualize a large amount of information. We present a method for computing finite-time Lyapunov exponents and a brief description of the so-called Lagrangian coherent structures.

In the rest of the book, we focus on specific features and phenomena. The fifth chapter is devoted to Lagrangian statistical analysis of near-surface transport of subtropical waters in the Japan Sea based on altimeter data. In the sixth chapter, we apply Lagrangian tools to study mesoscale eddies, rotating coherent features which exist almost everywhere in the ocean. It is shown here how to analyze by Lagrangian methods the formation, structure, evolution, and splitting of large mesoscale eddies over the Kuril–Kamchatka trench in the Northwestern Pacific Ocean based on altimetric velocity field and hydrological in situ observations of those eddies. To study the vertical structure of eddies in the ocean, we need not the altimetric but a numerical velocity field generated in an eddy-resolved multilayered circulation model. We use a regional model to analyze from a Lagrangian perspective the vertical structure of simulated deep-sea anticyclonic eddies in the Japan Sea constrained by the bottom topography.

The last part of the book deals with applications of elaborated methods and tools to some practical problems. In the seventh chapter, we apply the developed Lagrangian approach to simulate propagation of Fukushima-derived radionuclides advected by altimetric velocity field in the Northwestern Pacific Ocean. The results of the simulation are compared with in situ measurements of levels of ^{134}Cs and ^{137}Cs concentrations just after the accident and 15 months later. Different

Lagrangian diagnostics are used to reconstruct the history and origin of synthetic tracers imitating measured seawater samples collected inside the mesoscale eddies with the risk to be contaminated.

In the eighth chapter, we introduce the notion of a Lagrangian front which is defined as a boundary between waters with strongly different values of a Lagrangian indicator. The Lagrangian fronts can be identified in a given velocity field by computing Lagrangian maps. We study the connection of Lagrangian fronts with fishing grounds and catch locations for Pacific saury and neon flying squid in a region in the Northwestern Pacific Ocean with one of the richest fisheries in the world. This diagnostic is shown to be a new indicator for potential fishing grounds. In this chapter, we also review recent studies on the foraging behavior of top marine predators as great frigatebirds, elephant seals, and Mediterranean fin whales and their relationship with specific Lagrangian coherent structures.

The book is intended for graduate and postgraduate students and research scientists and for those oceanographers, physicists, and applied mathematicians who are interested in applications of ideas and methods of dynamical systems theory to geophysical fluid dynamics. It is our great pleasure to thank our co-workers who participated in obtaining some of the results presented in this book: Andrey Andreev, Pavel Fayman, Vladimir Goryachev, Konstantine Koshel, Vyacheslav Lobanov, Denis Makarov, Vladimir Ponomarev, and Eugene Samko.

Finally, we wish to thank the Russian Foundation for Basic Research (project nos. 16–05–00213, 15–35–20105 mol_a_ved, 13–05–00099, 13–01–12404, 12–05–00452 and 11–05–98542), the Russian Science Foundation (project no. 16–17–10025), Far Eastern Branch of the Russian Academy of Sciences (project nos. 15–I–1–003, 15–I–4–041, 15–I–1–047 and 12–I–23–05), the Ministry of Education and Science of the Russian Federation, and the POI FEBRAS Program "Nonlinear dynamical processes in the ocean and atmosphere" (project no. 0201363045) for a multiyear support of our research.

Vladivostok, Russia   Sergey V. Prants
2016   Michael Yu. Uleysky
                                                                            Maxim V. Budyansky

Contents

Acronyms

ACE	Anticyclonic eddy
CE	Cyclonic eddy
CJT	Cross-jet transport
CTD	Conductivity, temperature, density
FTLE	Finite-time Lyapunov exponent
FNPP	Fukushima Nuclear Power Plant
LCS	Lagrangian coherent structure
LF	Lagrangian front
MHIM	Marine Hydrophysical Institute Model
PDF	Probability density function
R/V	Research vessel
SSH	Sea surface height
SST	Sea surface temperature

References

1. Aref, H.: The development of chaotic advection. Phys. Fluids **14**(4), 1315–1325 (2002). 10.1063/1.1458932
2. Bennett, A.: Lagrangian Fluid Dynamics. Cambridge Monographs on Mechanics. Cambridge University Press, Cambridge (2006)
3. Griffa, A., Kirwan Jr., A., Mariano, A., Özgökmen, T., Rossby, H. (eds.): Lagrangian Analysis and Prediction of Coastal and Ocean Dynamics. Cambridge University Press, Cambridge (2007)
4. Haller, G.: Lagrangian coherent structures. Annu. Rev. Fluid Mech. **47**, 137–162 (2015). 10.1146/annurev-fluid-010313-141322
5. Haller, G., Yuan, G.: Lagrangian coherent structures and mixing in two-dimensional turbulence. Physica D **147**(3–4), 352–370 (2000). 10.1016/S0167-2789(00)00142-1
6. Koshel', K.V., Prants, S.V.: Chaotic advection in the ocean. Physics-Uspekhi **49**(11), 1151–1178 (2006). 10.1070/PU2006v049n11ABEH006066

7. Makarov, D., Prants, S., Virovlyansky, A., Zaslavsky, G.: Ray and Wave Chaos in Ocean Acoustics: Chaos in Waveguide. Series on Complexity, Nonlinearity and Chaos, vol. 1. World Scientific, Singapore (2011). 10.1142/9789814273183_fmatter
8. Ottino, J.M.: The Kinematics of Mixing: Stretching, Chaos, and Transport. Cambridge Texts in Applied Mathematics, vol. 3. Cambridge University Press, Cambridge (1989)
9. Pierrehumbert, R.T., Yang, H.: Global chaotic mixing on isentropic surfaces. J. Atmos. Sci. **50**(15), 2462–2480 (1993). 10.1175/1520-0469(1993)050<2462:GCMOIS>2.0.CO;2
10. Polvani, L.M., Wisdom, J., DeJong, E., Ingersoll, A.P.: Simple dynamical models of Neptune's great dark spot. Science **249**(4975), 1393–1398 (1990). 10.1126/science.249.4975.1393
11. Rom-Kedar, V., Leonard, A., Wiggins, S.: An analytical study of transport, mixing and chaos in an unsteady vortical flow. J. Fluid Mech. **214**, 347–394 (1990). 10.1017/s0022112090000167
12. Samelson, R.M., Wiggins, S.: Lagrangian Transport in Geophysical Jets and Waves: The Dynamical Systems Approach. Interdisciplinary Applied Mathematics, vol. 31. Springer, New York (2006). 10.1007/978-0-387-46213-4
13. Virovlyansky, A.L., Makarov, D.V., Prants, S.V.: Ray and wave chaos in underwater acoustic waveguides. Phys. Usp. **55**(1), 18–46 (2012). 10.3367/ufne.0182.201201b.0019
14. Weiss, J.B., Provenzale, A. (eds.): Transport and Mixing in Geophysical Flows. Lecture Notes in Physics, vol. 744. Springer, New York (2008). 10.1007/978-3-540-75215-8
15. Wiggins, S.: The dynamical systems approach to Lagrangian transport in oceanic flows. Annu. Rev. Fluid Mech. **37**(1), 295–328 (2005). 10.1146/annurev.fluid.37.061903.175815

Chapter 1
The Dynamical Systems Theory Approach to Transport and Mixing in Fluids

1.1 Chaotic Advection

One-dimensional pendulum under the influence of a periodic force and in the absence of any noisy perturbation can rotate or oscillate periodically at some initial conditions and/or at some values of the control parameters and is able to rotate and/or oscillate irregularly at another ones. In the irregular regime of motion, a distance between initially close trajectories in the phase space grows exponentially in time

$$\|\delta \mathbf{r}(t)\| = \|\delta \mathbf{r}(0)\| e^{\Lambda t}, \tag{1.1}$$

where Λ is a positive number, known as the maximal **Lyapunov exponent**. This quantity characterizes asymptotically as $t \to \infty$ the average rate of that separation, and $\| \cdot \|$ is a norm of the position vector $\mathbf{r}$. It immediately follows from (1.1) that it is impossible to forecast the pendulum position $\mathbf{r}$ beyond the so-called predictability horizon

$$T_{\text{pred}} \simeq \frac{1}{\Lambda} \ln \frac{\|\Delta_{\mathbf{r}}\|}{\|\Delta_{\mathbf{r}}(0)\|}, \tag{1.2}$$

where $\|\Delta_{\mathbf{r}}\|$ is a confidence interval and $\|\Delta_{\mathbf{r}}(0)\|$ is a practically inevitable inaccuracy in specifying pendulum's initial position. The deterministic dynamical system with at least one positive maximal Lyapunov exponent for almost all initial positions and momenta (in the sense of nonzero measure) is called fully chaotic. Theory of chaos is a branch of dynamical systems theory developed from a geometric approach to differential equations due to Henri Poincaré [23], Andronov [1], and many others.

© Springer International Publishing AG 2017
S.V. Prants et al., *Lagrangian Oceanography*, Physics of Earth and Space Environments, DOI 10.1007/978-3-319-53022-2_1

The phase space of a typical chaotic Hamiltonian system contains **"islands" of stability** embedded in a stochastic sea. The dependence of the predictability horizon T_{pred} on the lack of our knowledge of exact location is logarithmic, i.e., it is much weaker than on the measure of dynamical instability quantified by Λ. With any reasonable degree of accuracy on specifying initial conditions, there is a time interval beyond which the forecast is impossible, and that time may be comparatively small for chaotic systems. It is the ultimate reason why the exact weather forecast is impossible no matter how perfect detectors for measuring initial parameters and how powerful computers we have got.

Methods of dynamical systems theory have been actively used in the last 30 years to describe advection of passive particles in fluid flows on a large range of scales, from microfluidic flows to ocean and atmospheric ones. If advected particles rapidly adjust their own velocity to that of a background flow and do not affect the flow properties, then they are called "passive" and satisfy simple equations of motion

$$\frac{d\mathbf{r}}{dt} = \mathbf{v}(\mathbf{r}, t), \tag{1.3}$$

where $\mathbf{r} = (x, y, z)$ and $\mathbf{v} = (u, v, w)$ are the position and velocity vectors at the point (x, y, z), respectively. This formula just means that the Lagrangian velocity of a passive particle [the left side of Eq. (1.3)] equals to the Eulerian velocity of the flow at the location of that particle [the right side of Eq. (1.3)]. In fluid mechanics by passive particles, one means water (air) small parcels with their properties or small foreign bodies in a flow. The vector equation (1.3) in nontrivial cases is a set of three nonlinear deterministic differential equations whose **phase space** is a physical space for advected particles. Solutions of those equations can be chaotic in the sense of exponential sensitivity to small variations in initial conditions and/or control parameters as in Eq. (1.1).

As to dynamical chaos in ideal fluids, it was Arnold [6] who firstly suggested chaos in streamlines (and, therefore, in trajectories) for a special class of three-dimensional steady flows, the so-called Arnold–Beltrami–Childress flow. It was confirmed numerically in [14] and investigated analytically and numerically later in [13].

The term "chaotic advection" has been coined by Aref [2, 3] who realized that advection equations for two-dimensional flows may have a **Hamiltonian form**. For incompressible planar flows, the velocity components can be expressed in terms of a stream function. The equations of motion (1.3) have now the Hamiltonian form

$$\frac{dx}{dt} = u(x, y, t) = -\frac{\partial \Psi}{\partial y}, \quad \frac{dy}{dt} = v(x, y, t) = \frac{\partial \Psi}{\partial x}, \tag{1.4}$$

with the stream function Ψ playing the role of a Hamiltonian. The coordinates (x, y) of a particle are canonically conjugated variables. All time-independent one degree of freedom Hamiltonian systems are known to be integrable. It means that all fluid particles move along streamlines of a time-independent stream function in a regular way. Equations (1.4) with a time-periodic stream function can be nonintegrable, giving rise to chaotic particle's trajectories. Chaotic advection has been studied analytically and numerically in a number of simple models and in laboratory experiments (for a review see [3, 22]).

Since the phase plane of the dynamical system (1.4) is the physical space for fluid particles, many abstract mathematical objects from dynamical systems theory (stationary points, where the velocity is zero, **Kolmogorov–Arnold–Moser (KAM) tori**, stable and unstable manifolds, periodic and chaotic trajectories, etc.) have their material analogues in fluid flows. It is well known that besides "trivial" elliptic **stationary points**, the motion around which is stable, there are hyperbolic stationary points which organize fluid motion in their neighborhood in a specific way. In a steady flow the hyperbolic points are typically connected by the **separatrices** which are their stable and unstable invariant **manifolds** (see Fig. G2). In a time-periodic flow they are replaced by the corresponding periodic hyperbolic trajectories with two associated invariant manifolds which in general intersect each other transversally (Fig. G3) resulting in a complex manifold structure known as **homo-or heteroclinic tangles**. The fluid motion in these regions is so complicated that it may be strictly called chaotic, the phenomenon known as "chaotic advection." Initially close fluid particles in such tangles rapidly diverge providing very effective mechanism for mixing. The phase space of a typical chaotic open flow consists of different kinds of invariant sets, **KAM tori** with regular trajectories of fluid particles, **cantori**, and a chaotic saddle set embedded into a stochastic sea with chaotic trajectories [8, 9, 21, 28].

Stable and unstable **manifolds** are important organizing structures in the flow because they attract and repel fluid particles not belonging to them and partition the flow into regions with qualitatively distinct regimes of motion. Invariant manifold in a 2D flow is a material line, i.e., it is composed of the same fluid particles in the course of time. By definition, stable, W_s, and unstable, W_u, manifolds of a hyperbolic trajectory $\gamma(t)$ are material surfaces in the extended phase space (x, y, t) consisting of the trajectories asymptotical to $\gamma(t)$ at $t \to \infty$ (W_s) and $t \to -\infty$ (W_u). On a 2D-plane (x, y), they are complicated curves infinite in time and space (in theory) that act as barriers to fluid transport (see, e.g., [26, 30]).

Experiments with chaotic flows provide a visualization of some abstract mathematical notions of dynamical chaos including **KAM tori**, **stationary points**, **stable and unstable manifolds**, etc. Description and illustrations of some experiments with laboratory cavity flows can be found in the nice book by Ottino [22]. In one of them, the setup, consisting of a rectangular tank with two opposing walls which can move in a steady or time-dependent manner, produces a two-dimensional planar velocity field [22]. The stretching and folding pattern of chaotic mixing in that flow is clearly demonstrated in Fig. 1.1 which illustrates time evolution of two blobs of dye. The lower blob, placed originally inside an "island" of stability (a KAM tori), is trapped inside that "island," whereas the upper one, placed originally nearby a hyperbolic point in the chaotic "sea," undergoes significant stretching after a few periods of wall oscillations. Thus, the "regular" blob just rotates without significant deformation, whereas the "chaotic" one stretches and folds many times. Its strong

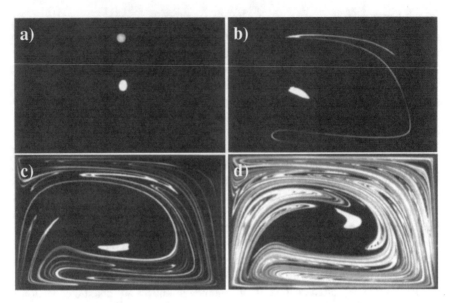

Fig. 1.1 Time evolution of two blobs of dye in the experiment with a time-periodic chaotic flow
[22]. The *lower* blob was placed originally inside an **"island" of stability** and the *upper* one—
nearby a hyperbolic point in the chaotic "sea." The first one is trapped inside the "island." The
upper blob undergoes significant stretching after a few periods. Courtesy of J.M. Ottino

deformation is caused by the presence of an unstable manifold of the corresponding
hyperbolic point. In fact, the pattern in Fig. 1.1b–d gives an approximate image of
that manifold.

Let the stream function takes the form

$$\Psi(t) = \Psi_0 + \xi\Psi_1(t), \qquad (1.5)$$

where Ψ_0 is a stationary component and $\Psi_1(t)$ is a perturbation with the period
T_0, ξ is a small amplitude of perturbation. Onset of dynamical chaos can be
illustrated with a simple periodic flow with the phase portrait shown schematically
in Fig. 1.2. The specific example of such a flow with a fixed point vortex and a
background periodic current will be studied in Sect. 1.2. The main object here is
a loop which passes through a saddle (hyperbolic) point and separates finite and
infinite trajectories of fluid particles. Motion along the separatrix is infinite in time
because the saddle stationary point is a specific trajectory and it cannot merge with
another trajectory, the **separatrix** one. Without a perturbation, the "stable whisker"
of the separatrix, along which the phase point approaches the saddle, coincides
with the unstable one along which it moves away. It is not surprising that such an
exceptional structure breaks down under even a weak perturbation.

As it was found firstly by Poincaré [23], a separatrix splitting may occur in
Hamiltonian systems under an arbitrary small perturbation. Under a perturbation,

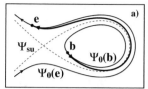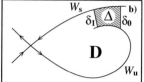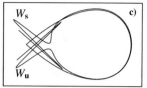

Fig. 1.2 Stochastic layer appears in a Hamiltonian system under a perturbation: (**a**) the unperturbed separatrix loop (*dashed line*) with the stream function Ψ_{su}, unperturbed trajectories of fluid particles inside and outside the loop (*thin solid curves*) with the stream functions $\Psi_0(b)$ and $\Psi_0(e)$, respectively, and the particle's trajectory under a perturbation (*thick curve*), (**b**) splitting of the stable W_s and unstable W_u manifolds of the saddle, (**c**) transversal intersections of those manifolds as they look schematically on the Poincaré section

stable and unstable manifolds of a saddle point may intersect each other transversally, and a stochastic layer appears around the unperturbed separatrix. It is an universal phenomenon in chaotic Hamiltonian systems with this layer to be a "seed" of **Hamiltonian chaos**.

The **instability of motion** near the unperturbed separatrix has the following simple reason: the frequency of oscillations or rotations far away from the unperturbed separatrix depends weakly on stream function (energy), and its small variations result in small variations in phase for the period of oscillations. The period of oscillations near the unperturbed separatrix tends to infinity, and small variations in frequency there may cause large changes in phase. This is the ultimate reason for a local instability of trajectories in a stochastic layer.

Following to Kozlov [17], we present here a visual Poincaré proof of splitting and intersection of stable and unstable manifolds in the **homoclinic** case with a single saddle point shown in Fig. 1.2a. Let W_s and W_u be the lines of intersection of stable and unstable manifolds of the saddle point with the plane $t = 0$. We first suppose that they do not intersect each other under a sufficiently small perturbation, and there is a small gap between those lines (Fig. 1.2b). Let us consider a line segment δ_0 with one end at W_s and the other end at W_u. It shifts over the period T_0 counterclockwise to be the segment δ_1 with both its ends lying at W_s and W_u. The initial area $\mathbf{D} + \Delta$ decreases to $\mathbf{D}$, but this contradicts area conservation in the phase space of Hamiltonian systems. The contradiction is resolved if W_s and W_u intersect each other as shown schematically in Fig. 1.2c.

Separatrix splitting gives rise to important consequences for fluid flows because it breaks down impermeable barriers and allows transport between domains not communicating in the unperturbed flow. The estimates of instant and mean fluxes between those domains and the exchange rates can be given with the help of the so-called Melnikov integral [19] which characterizes the instantaneous distance between the splitting separatrices Δ and the stochastic layer width. In the absence of perturbation, fluid particles move along the streamlines Ψ_0 (thin solid lines in Fig. 1.2a). Under the perturbation with a stream function Ψ_1, particle b, placed at $t = 0$ on the streamline $\Psi_0(b)$ inside the separatrix loop, moves for a time t

to position corresponding to another streamline $\Psi_0(e)$ (Fig. 1.2a). To compute the difference between the two unperturbed streamlines $\Delta\Psi_0 = \Psi_0(e) - \Psi_0(b)$ we need to compute the derivative

$$\frac{d\Psi_0(x(t), y(t))}{dt} = \xi\frac{\partial\Psi_1(t)}{\partial x}\frac{\partial\Psi_0}{\partial y} - \xi\frac{\partial\Psi_1(t)}{\partial y}\frac{\partial\Psi_0}{\partial x} \equiv \xi\{\Psi_1(t), \Psi_0\}, \quad (1.6)$$

which is found with the help of advection equations (1.4) and the definition of the Poisson bracket $\{\Psi_1(t), \Psi_0\}$. The total variation of Ψ_0 is

$$\Delta\Psi_0 = \xi\int\{\Psi_1(t), \Psi_0\}dt, \quad (1.7)$$

where the integral is taken along the particle's trajectory $(x(t), y(t))$. It is an exact formula.

It should be stressed that the curves, illustrating stable and unstable manifolds of the saddle trajectory in Fig. 1.2c, are not trajectories but schematic Poincaré sections of the surfaces of those stable and unstable manifolds. Figure 1.2c illustrates the first stage of oscillations of the curves. They oscillate with increasing amplitude when approaching the saddle and become closer and closer to each other with increasing number of transversal crossings. The transversal crossings of stable and unstable manifolds produce multiple foldings and stretching of material lines in fluids providing effective mixing.

Mixing is a key concept both in hydrodynamics and in dynamical systems theory which can be defined in a strict mathematical sense. Let us consider the basin A with a circulation where there is a domain B with a dye occupying at $t = 0$ the volume $V(B_0)$. Let us consider a domain C in A. The volume of the dye in the domain C at time t is $V(B_t \cap C)$, and its concentration in C is given by the ratio $V(B_t \cap C)/V(C)$. The definition of full mixing is that in the course of time in any domain $C \in A$ we will have the same dye concentration as for the entire domain A, i.e., $V(B_t \cap C)/V(C) - V(B_0)/V(A) \to 0$ as $t \to \infty$. In dynamical systems theory, the full or global mixing is achieved when a small blob of the phase-space fluid is transformed into a long intricate filament occupying all energetically accessible domain in the phase space. The mixing measures are the **Lyapunov exponents**. In real flows mixing due to flow kinematics is accompanied by molecular diffusion.

Chaotic advection in theory is a chaotic mixing in a deterministic velocity field. In real flows there are inevitable random fluctuations of that field. If they are small as compared to mean regular values, it is reasonably to call the corresponding phenomenon as "chaotic advection," because typical patterns are similar to those in purely deterministic flows but become just more fuzzy [18].

1.2 Chaotic Scattering of Fluid Particles at a Point Vortex Embedded in a Time-Periodic Background Flow

1.2.1 Invariant Sets of the Flow

In this paragraph we consider a kinematic model of chaotic advection consisting of a point fixed vortex and a time-periodic current flow. It is a toy model introduced in [10] to simulate transport and mixing of water masses by topographical eddies in the ocean over seamounts. The point vortex is supposed to be embedded in a background flow composed of a steady current and a small periodic component. The corresponding normalized stream function [10]

$$\Psi = \ln \sqrt{x^2 + y^2} + x(\varepsilon + \xi \sin t) \tag{1.8}$$

generates the advection equations

$$\dot{x} = -\frac{y}{x^2 + y^2}, \qquad \dot{y} = \frac{x}{x^2 + y^2} + \varepsilon + \xi \sin t, \tag{1.9}$$

where dot denotes differentiation with respect to dimensionless time t. Without perturbation ($\xi = 0$), the advection equations (1.9) can be solved in quadratures in the polar coordinates $x = \rho \cos \varphi$ and $y = \rho \sin \varphi$ as follows:

$$\varepsilon \, dt = \left[1 - \left(\frac{\Psi_0 - \ln \rho}{\varepsilon \rho} \right)^2 \right]^{-1/2} d\rho, \tag{1.10}$$

where $\Psi_0 = \varepsilon \rho \cos \varphi + \ln \rho$.

The phase portrait of the integrable system is a collection of finite and infinite streamlines separated by the loop passing through a saddle point with the coordinates $(-1/\varepsilon; 0)$. A few streamlines are shown in Fig. 1.3. Depending on initial positions, particles either rotate inside the separatrix loop, moving along closed streamlines, or move along infinite streamlines outside the loop. Period of rotation of particles around the point vortex inside the loop decreases to zero for particles approaching the singular point and increases to infinity for particles approaching the separatrix loop. In the integrable system, the stable manifold of the saddle point, along which particles move towards the saddle, and its unstable manifold, along which they move away from the saddle, coincide.

Under a perturbation, the stationary saddle point becomes a saddle periodic trajectory which is represented on a **Poincaré section** with the dimensionless period 2π by a **fixed point**. On the Poincaré section, there exist stable and unstable manifolds of the saddle point with trajectories approaching the saddle periodic motion when $t \to \pm\infty$. Under a typical perturbation, these manifolds transverse each other on the Poincaré section infinitely many times producing a complicated **homoclinic structure**.

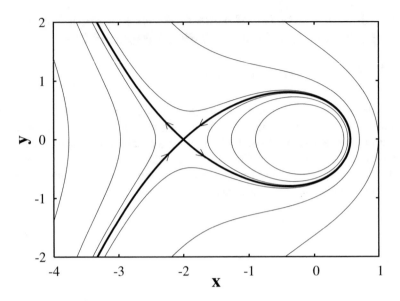

Fig. 1.3 The phase portrait of the unperturbed flow (1.9), $\varepsilon = 0.5$

A large number of tracers are distributed homogeneously in the region ($x_0 \in$ [−0.9, −0.85]; $y_0 \in$ [−0.1, 0.1]). Integrating the advection equations (1.9) for each of them, we fix positions of all the tracers at $t_n = nT_0$ ($n = 1, 2, \ldots$). The corresponding Poincaré section of the flow is shown in Fig. 1.4a. We distinct a free-stream region with incoming and outgoing components, mixing region, and vortex core as the sets of trajectories for which the number of times they wind around the point fixed vortex is zero, finite and infinite, respectively.

Now, let us describe the invariant sets of the flow which make up its building blocks. Particles, belonging to different sets, exhibit qualitatively different behavior. The simplest examples of invariant sets are entire phase space, a stationary point, a periodic orbit (see the article **"Poincaré section"** in the Glossary), and any orbit defined on the time interval $[-\infty, +\infty]$.

The set of invariant curves, shown in Fig. 1.4a, represents sections of **KAM tori**. It is the set of periodic and quasiperiodic tracer trajectories around the vortex center (see the article **"Trajectories"** in the Glossary). On the Poincaré section they make up families of nested closed smooth curves. Most of them lie inside the vortex core. Other families of invariant curves make up "islands" with regular trajectories centered at elliptic points. The "islands" are located both in the vortex core and in the mixing region. The "islands" of stability appear due to **nonlinear resonances** of various orders between particle motions in the vortex and the 2π-periodic perturbation. The largest "island" in the chaotic sea to the west of the vortex core arises from the half-integer (π-periodic) primary resonance and is surrounded by secondary higher order resonances. Figure 1.4b clearly depicts the secondary resonance with period 10π. It is typical for Hamiltonian systems that large "islands"

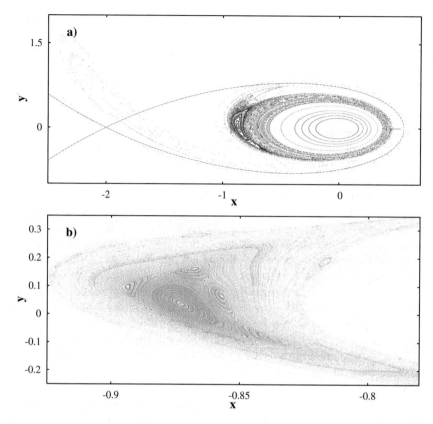

Fig. 1.4 (**a**) Plane of the Poincaré section with the unperturbed separatrix loop. (**b**) A fragment of the Poincaré section illustrating a secondary resonance with the period 10π in a neighborhood of the half-integer primary resonance with the period π. Parameters are $\varepsilon = 0.5$ and $\xi = 0.1$

are surrounded by "islands" of smaller size and those ones, in turn, are surrounded by "islands" of even more smaller size, etc. (see the article **"Hamiltonian chaos"** in the Glossary).

The phase-space topology strongly depends on the values of the control parameters ε and ξ. Their relative values determine the orders of nonlinear resonances in the system and, therefore, the location, number, and size of the "islands" of stability. As the value of ε/ξ increases, the vortex core, occupied by regular trajectories, grows in size, and the orders of surviving resonances increase while the mixing region shrinks correspondingly.

The vortex core also contains "islands" of regular motion and chains of "islands." A zoom shows that these chains are surrounded by narrow stochastic layers [8, 9]. The vortex core is preserved for any combination of ε and ξ, i.e., it is a robust structure. Since the particle rotation frequency in the vortex core is much higher than the perturbation frequency, the perturbation can be treated as adiabatic with respect to most trajectories inside the core. They are regular except for those lying in the neighborhoods of broken resonance separatrices which make up very narrow stochastic layers. The KAM tori are impermeable barriers and obstructions for tracer transport.

It is known [7] that even a small perturbation may cause the appearance of the so-called **cantori** replacing some KAM tori, primarily those with rotation numbers that do not satisfy the Diophantine condition of the **KAM theorem**. The cantori are invariant sets having Cantor-like structure with gaps which are characterized by a topological dimension at least lower than the measure of a curve. Motion on them is quasiperiodic. However, cantori are unstable and, therefore, have stable and unstable manifolds. Unlike KAM tori, cantori are permeable for tracers. The material line in the incoming-flow region, intersecting a stable manifold $W_s(\Sigma)$ of the so-called chaotic invariant set Σ, contains fluid particles that reach the boundary region between the mixing region and the vortex core in the course of time and rotate there as if they would limited by an invariant KAM curve. They percolate in time to the opposite side of the cantori through its gaps and stay there for a while. The process repeats many times until the particles escape that region. It is very difficult to identify cantori in numerical experiments, because they are strongly unstable, but they manifest themselves as domains with a large density of points on Poincaré sections. The existence of an invariant set of cantori and sticky domains around imply that mixing is inhomogeneous, as manifested, in particular, in power-law decay of trapping-time probability distribution functions and in singularities of scattering functions [8, 9].

The chaotic invariant set Σ is a set of all trajectories (except of interiors of KAM tori and KAM tori themselves) that never leave the mixing region. The set consists of an infinite number of periodic **trajectories** and aperiodic chaotic ones. All trajectories in this set are unstable. Passive particles with initial conditions belonging to Σ remain in the mixing region as $t \to \infty$ or $t \to -\infty$. The Poincaré section of Σ is a fractal set of points with zero Lebesgue measure. Most tracers from the incoming flow sooner or later leave the mixing region with the outgoing flow. But the behavior of some of them is dictated by the presence of Σ. These tracers follow trajectories in Σ, wandering for a long time in their neighborhoods and eventually leave the mixing region.

Each trajectory in the chaotic set and, therefore, the entire set has both stable, $W_s(\Sigma)$, and unstable, $W_u(\Sigma)$, **manifolds**. Following trajectories in $W_s(\Sigma)$, particles from the incoming flow enter the mixing region and remain there forever. It was mentioned above that the corresponding initial conditions are a set of measure zero. Particles with initial conditions that are close to trajectories of the stable manifold follow them for a long time and eventually deviate from them, leaving the mixing region along the unstable manifold. Thus, there is a unique opportunity to determine the important properties of Σ by measuring the characteristics of scattering particles and to observe unstable manifolds directly in laboratory experiments and even their approximations in geophysical flows.

Unstable manifolds can be visualized by various methods. An intricate **fractal** curve approaching $W_u(\Sigma)$ in the course of time is formed as a result of the deformation of a blob with many tracers chosen at the intersection of the incoming flow with the stable manifold [8]. A similar picture can be observed in laboratory experiments with dye streaks (see illustrations in the book [16]). Passive particles are advected along the fractal curve of the unstable manifold, which is a kind of attractor in a Hamiltonian system (there are no "classical" attractors in incompressible flows). Direct computation of the so-called trapping maps [11] provides an image of the stable manifold $W_s(\Sigma)$. In the unperturbed flow with $\xi = 0$, the border line between the vortex core and the free flow has a finite length equal to the length of the separatrix loop, but it is infinite in a flow with a periodic perturbation.

1.2.2   Geometry of Chaotic Scattering and Its Fractal Properties

In this paragraph geometry of transport of passive particles in the mixing region is analyzed. We will study evolution of the straight material segment chosen at the line $y = -6$ in the incoming stream with the left endpoint at the intersection of this line with the southern "whisker" of the perturbed separatrix loop at the time $3\pi/2 + 2\pi m$ and the right endpoint at the intersection with that "whisker" at $\pi/2 + 2\pi m$, where $m = 0, 1, \ldots$. Snapshots of that line at the instants $t = 8\pi$, 9π, 10π, and 11π are shown in Fig. 1.5. At $t = 0$, point A was at the intersection of the perturbed separatrix and the line $y = -6$, and point G—at the intersection of this line and the separatrix at the time moments $\pi/2 + 2\pi m$. The particles with initial positions $x_0 < x_0(A)$ and $x_0 > x_0(G)$ are immediately washed away into the free-stream region (dotted segments in Fig. 1.5). Particles A and G move along the stable manifold, come into a neighborhood of the saddle periodic trajectory, and remain in the mixing region for a long time (which is infinite in theory).

Let us compute the total number, n, of turns around the vortex executed by the tracers placed initially at the segment AG before they escape into the outcoming free-stream region (the part of the plane above the line $y = 6$). The graph of $n(x_0)$, shown in Fig. 1.6, is a complicated hierarchy of sequences of fragments of the material line AG generated by infinite intersections of the stable and unstable manifolds with the material line of initial conditions as it rotates around the vortex. The black segments at the level $n = 0$ contain the tracers which escape from the mixing region and reach the line $y = 6$ without a complete turn around the vortex. The black segments at the nth level mark the tracers which escape from the mixing region after n turns around the vortex.

There are sequences of segments for each $n \geqslant 0$, which are called "epistrophes" following reference [20]. The epistrophes make up a hierarchy. The endpoints of each segment at the level n are the limit points of a level-$n + 1$ epistrophe. For example, there is a single epistrophe at the level $n = 0$ converging at the point A. The endpoints of each segment of this epistrophe at the level $n + 1$ generate epistrophes b, c, d, e, g, etc., converging at the corresponding limit points (see Fig. 1.6).

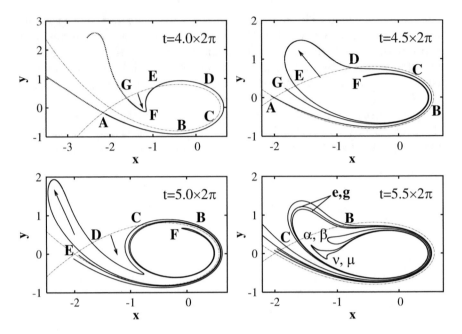

Fig. 1.5 Deformation of a straight material line, chosen in the incoming flow, in the course of time. Formation of the "lobes" from epistrophes and strophes of the fractal shown in Fig. 1.6. The unperturbed separatrix loop is shown for reference

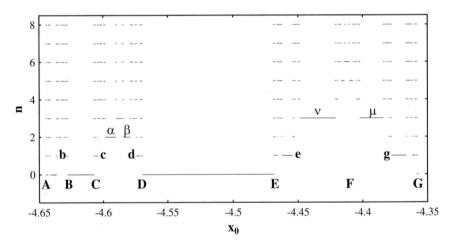

Fig. 1.6 Fractal set of the initial positions, x_0, of incoming-flow tracers which escape from the mixing region after n turns around the point vortex

Analyzing the graph $n(x_0)$, we find the following laws: (1) each epistrophe converges at a limit point in the segment under consideration; (2) the endpoints of each segment in a level-n epistrophe are the limit points of a level-$n + 1$ epistrophe;

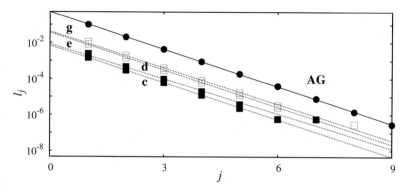

Fig. 1.7 Decrease in the epistrophe-segment length l_j for $n = 0$ (AG) and $n = 1$ (c, d, e, g) with the segment index j in the semilogarithmic scale

(3) the lengths of segments in an epistrophe decrease in geometric progression q, and (4) the common ratio of all progressions is equal to the maximal Lyapunov exponent for the saddle trajectory $\Lambda = -\frac{1}{2\pi} \ln q$. The lengths l_j of an epistrophe segment as a function of its index j for the zeroth-level epistrophe and the first-level epistrophes c, d, e, and g are shown in Fig. 1.7. The slopes of all the graphs are equal to $\ln q$ [8], i.e., the segment lengths in each epistrophe decrease in geometric progression $l_j = l_0 q^j$ with $q \approx 0.2$.

The **fractal** in Fig. 1.6 is not strictly self-similar, because it contains segments, called "strophes," that do not belong to the epistrophes. Some of them are labeled by Greek letters in the graph. Thus, the fractal is characterized by a partial self-similarity: each level contains both self-similar epistrophe sequences and additional elements (strophes) that are preserved in the asymptotic limit and do not fit into a regular structure. The fractal in Fig. 1.6 provides a comprehensive illustration of tracer transport.

The material line AG stretches and bends as it winds around the point vortex, and then a part of it begins to fold as particles, rotating around the vortex, accelerate, while other particles decelerate in a neighborhood of the saddle point. Figure 1.5 illustrates the formation of the first fold at $t = 8\pi$. Segment DE is associated with the first segment of the zeroth-level epistrophe (tracers that have not made a complete turn around the vortex). Segment EFG is represented by an empty space in Fig. 1.6 generating an infinite sequence of strophes and epistrophes at the higher levels. After the period of time $t = 10\pi$, the second "lobe" develops in the material line. After the time interval $t = 11\pi$, two new lobes begin to develop in the stretched portion of the first lobe corresponding to epistrophe segments e and g at the level $n = 1$. The particles in these lobes escape together with the lobe BC before they complete their second turn around the vortex giving rise to the second "finger" in Fig. 1.5. Furthermore, the snapshot taken at $t = 11\pi$ shows the lobes that subsequently develop into the strophes a and b at the level $n = 2$ and into the strophes ν and μ at the level $n = 3$. These strophes give rise to four lobes that combine with zero- and first-level epistrophe segments to form the third "finger."

This process repeats iteratively, i.e., the part, corresponding to an epistrophe segment and an empty segment in Fig. 1.6, unwind off the material segment's "tail" in a neighborhood of the saddle point with each turn around the vortex. This scenario describes the formation of epistrophes and strophes at all nonzero levels, except that each level-n epistrophe segment generates two level-$n + 1$ epistrophes. In experiments with dye tracks, these events can be seen as a periodic formation of lobe pairs. In the course of time, dye streamlines develop into a self-similar pattern (see Fig. 1.5) in the sense that new "fingers" with an increasing number of lobes appear in each subsequent period. This process reveals a kind of order hidden in chaos.

Now let us compute the time moments T, when the particles, initially distributed along the segment AG at the line $y = -6$, reach the line $y = 6$. Figure 1.8 demonstrates the typical scattering function $T(x_0)$ with an uncountable number of singularities which are unresolved in principle. The inset in Fig. 1.8 shows a zoom by the factor of 20 of one of these singularity zones. Successive magnifications confirm a self-similarity of the function with increasing values of the trapping times T. Figure 1.9 gives an example of a trajectory with a very long trapping time equal to $T \simeq 1581$ in units of the perturbation period.

Sticky boundaries of **"islands" of regular motion**, embedded in a stochastic sea outside the vortex core and cantori there, act as a kind of **dynamical traps** providing all the spectrum of values of the trapping times up to infinity. For example, the scattering function for the particles, belonging to the segment DE of the zeroth-level epistrophe in Fig. 1.6, have a U-like form [8]. The endpoints D and E separate particles that fall into the stable and unstable manifolds. A similar role is played by the endpoints of all elements of the strophes and epistrophes. These points make up a set of points that remain in the mixing region forever. The scattering function, corresponding to strophes, has a similar U-like form but with a more pronounced

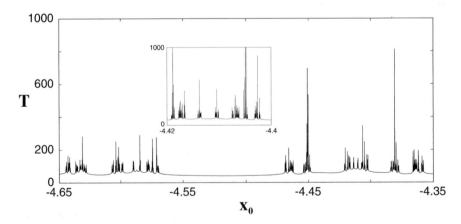

Fig. 1.8 Fractal-like dependence of the trapping time T on initial particle's positions x_0 with the *inset* showing a 20-fold magnification of one of the segments with singularities

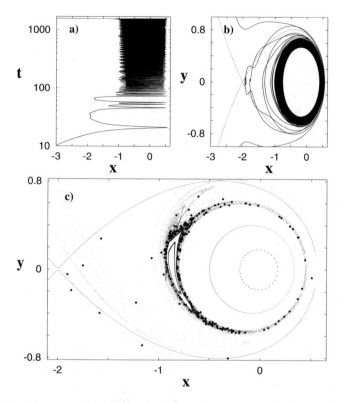

Fig. 1.9 Stickiness of a passive particle initially located at $(x_0 = -4.358034, y_0 = -6)$ around the boundary between the vortex core and the long "island." (**a**) The time dependence of its x coordinate and (**b**) its trajectory. (**c**) The particle's locations on the Poincaré section. The trapping time of that particle, $T \simeq 1581$, is very long

asymmetry. The fractal-like scattering function $T(x_0)$ with an uncountable number of singularities is a typical feature of chaotic scattering in various systems, from billiards [21] and fluids [27, 28] to cold atoms in a laser field [4, 5, 24, 25]. The similar fractal-like picture of chaotic transport and mixing has been found in kinematic models for periodically meandering jet currents [12, 15, 29] (see Chap. 2).

Another typical feature for chaotic scattering in Hamiltonian systems with inhomogeneous **phase space** is the so-called heavy tails for probability distribution functions (PDF). As to our model flow, the respective PDF, $N(T)$, for a large number of tracers demonstrates initially exponential decay followed by a power-law decay at its tail, $N(T) \sim T^{-\gamma}$, with the characteristic exponent $\gamma \simeq 2.8$. While a Poissonian distribution is expected for fully chaotic mixing, a power-law dependence indicates the presence of "islands" of regular motion in the flow with "sticky" boundaries (see the article **"Dynamical traps"** in the Glossary). Particles trapped for longer times give an enhanced contribution to the tail of PDFs.

References

1. Andronov, A., Vitt, A., Khaikin, S.: Theory of Oscillators, International Series of Monographs in Physics, vol. 4. Pergamon Press, Oxford (1966)
2. Aref, H.: Stirring by chaotic advection.   J. Fluid Mech. **143**(-1), 1–21 (1984). 10.1017/S0022112084001233
3. Aref, H.: The development of chaotic advection. Phys. Fluids **14**(4), 1315–1325 (2002). 10.1063/1.1458932
4. Argonov, V.Y., Prants, S.V.: Fractals and chaotic scattering of atoms in the field of a standing light wave. J. Exp. Theor. Phys. **96**(5), 832–845 (2003). 10.1134/1.1581937
5. Argonov, V.Y., Prants, S.V.: Theory of chaotic atomic transport in an optical lattice. Phys. Rev. A **75**(6), 063428 (2007). 10.1103/physreva.75.063428
6. Arnold, V.: Sur la topologie des écoulements stationnaires des fluides parfaits. C. R. Hebd. Seances Acad. Sci. **261**, 17–20 (1965). http://gallica.bnf.fr/ark:/12148/bpt6k4022z/f18.item. r=.zoom [in French]
7. Arnold, V.I., Kozlov, V.V., Neishtadt, A.I.: Mathematical Aspects of Classical and Celestial Mechanics. Encyclopaedia of Mathematical Sciences, vol. 3, 3rd edn. Springer, New York (2006). 10.1007/978-3-540-48926-9
8. Budyansky, M., Uleysky, M., Prants, S.: Chaotic scattering, transport, and fractals in a simple hydrodynamic flow. J. Exp. Theor. Phys. **99**, 1018–1027 (2004). 10.1134/1.1842883
9. Budyansky, M., Uleysky, M., Prants, S.: Hamiltonian fractals and chaotic scattering of passive particles by a topographical vortex and an alternating current. Physica D **195**(3–4), 369–378 (2004). 10.1016/j.physd.2003.11.013
10. Budyansky, M.V., Prants, S.V.: A mechanism of chaotic mixing in an elementary deterministic flow. Tech. Phys. Lett. **27**(6), 508–510 (2001). 10.1134/1.1383840
11. Budyansky, M.V., Uleysky, M.Y., Prants, S.V.: Fractals and dynamic traps in the simplest model of chaotic advection with a topographic vortex. Dokl. Earth Sci. **387**(8), 929–932 (2002)
12. Budyansky, M.V., Uleysky, M.Y., Prants, S.V.: Lagrangian coherent structures, transport and chaotic mixing in simple kinematic ocean models. Commun. Nonlinear Sci. Numer. Simul. **12**(1), 31–44 (2007). 10.1016/j.cnsns.2006.01.008
13. Dombre, T., Frisch, U., Greene, J.M., Hénon, M., Mehr, A., Soward, A.M.: Chaotic streamlines in the ABC flows. J. Fluid Mech. **167**, 353–391 (1986). 10.1017/s0022112086002859
14. Hénon, M.: Sur la topologie des lignes de courant dans un cas particulier. C. R. Hebd. Seances Acad. Sci. Séer. A **262**, 312–314 (1966). http://gallica.bnf.fr/ark:/12148/bpt6k6236863n/f326. item [in French]
15. Koshel', K.V., Prants, S.V.: Chaotic advection in the ocean.   Physics-Uspekhi **49**(11), 1151–1178 (2006). 10.1070/PU2006v049n11ABEH006066
16. Koshel, K.V., Prants, S.V.: Chaotic Advection in the Ocean. Institute for Computer Science, Moscow (2008) [in Russian]
17. Kozlov, V.: Symmetries, topology and resonances in Hamiltonian mechanics. UdGU, Izhevsk (1995). [in Russian]
18. Makarov, D., Uleysky, M., Budyansky, M., Prants, S.: Clustering in randomly driven Hamiltonian systems. Phys. Rev. E **73**(6), 066210 (2006). 10.1103/PhysRevE.73.066210
19. Melnikov, V.: On the stability of the center for time periodic perturbations. Trans. Moscow Math. Soc. **12**, 1–57 (1963)
20. Mitchell, K.A., Handley, J.P., Tighe, B., Delos, J.B., Knudson, S.K.: Geometry and topology of escape. I. Epistrophes. Chaos **13**(3), 880–891 (2003). 10.1063/1.1598311
21. Ott, E.: Chaos in dynamical systems, 2nd edn. Cambridge University Press, Cambridge (2002)
22. Ottino, J.M.: The Kinematics of Mixing: Stretching, Chaos, and Transport. Cambridge Texts in Applied Mathematics, vol. 3. Cambridge University Press, Cambridge (1989)
23. Poincaré, H.: New methods of celestial mechanics. NASA Technical Translation, vol. 450–452. NASA, Springfield (1967)

24. Prants, S.V.: Chaos, fractals, and atomic flights in cavities. JETP Lett. **75**(12), 651–658 (2002). 10.1134/1.1503331
25. Prants, S.V., Uleysky, M.Y.: Atomic fractals in cavity quantum electrodynamics. Phys. Lett. A **309**(5–6), 357–362 (2003). 10.1016/S0375-9601(03)00208-1
26. Samelson, R.M., Wiggins, S.: Lagrangian Transport in Geophysical Jets and Waves: The Dynamical Systems Approach. Interdisciplinary Applied Mathematics, vol. 31. Springer, New York (2006). 10.1007/978-0-387-46213-4
27. Sommerer, J.C., Ku, H.C., Gilreath, H.E.: Experimental evidence for chaotic scattering in a fluid wake. Phys. Rev. Lett. **77**(25), 5055–5058 (1996). 10.1103/physrevlett.77.5055
28. Tél, T., de Moura, A., Grebogi, C., Károlyi, G.: Chemical and biological activity in open flows: a dynamical system approach. Phys. Rep. **413**(2–3), 91–196 (2005). 10.1016/j.physrep.2005.01.005
29. Uleysky, M.Y., Budyansky, M.V., Prants, S.V.: Effect of dynamical traps on chaotic transport in a meandering jet flow. Chaos **17**(4), 043105 (2007). 10.1063/1.2783258
30. Wiggins, S.: Chaotic Transport in Dynamical Systems. Interdisciplinary Applied Mathematics, vol. 2. Springer, New York (1992). 10.1007/978-1-4757-3896-4

Chapter 2
Chaotic Transport and Mixing in Idealized Models of Oceanic Currents

2.1 Chaotic Advection with Analytic Geophysical Models: Introductory Remarks

The study of chaotic advection in the oceanic and atmospheric flows began with simple kinematic and dynamically consistent models of large-scale mixing and transport. The velocity field in a kinematic model is a given function of spatial coordinates and time designed in order to imitate some features of geophysical flows. Kinematic models allow to obtain some analytic results and quantitative estimates. They are attractive also due to their simplicity, generality, and the possibility of revealing the underlying geometric structures responsible for Lagrangian transport and mixing [4–7, 14, 17, 18, 23, 24, 31–35, 42, 43]. But kinematical stream functions are not solutions of dynamical equations of motion, and the potential vorticity, as a rule, is not conserved in kinematic models.

The kinematics of an incompressible flow is described by a stream function. If the stream function is specified disregarding the laws of fluid motion, then we deal with a kinematic model. The stream function for a dynamically consistent model must satisfy relations following from dynamical equations governing fluid motion.

As far as we know, the first nontrivial geophysical two-dimensional and dynamically consistent model with chaotic advection, based on a classical Kida vortex model [16], has been considered and analyzed in [12, 29, 30]. It is the model of an elliptic vortex subjected to a linear deformation consisting of shear and rotational components. In the case of a stationary deformation, the model permits various regimes of motion depending on the values of parameters of the deformation flow and the initial alignment of the ellipse against the exterior strain [16]. Fluid particles near an oscillating elliptic vortex, embedded in a steady deformation flow, can

---

The original version of this chapter was revised. An erratum to this chapter can be found at DOI 10.1007/978-3-319-53022-2_9

© Springer International Publishing AG 2017
S.V. Prants et al., *Lagrangian Oceanography*, Physics of Earth and Space
Environments, DOI 10.1007/978-3-319-53022-2_2

move chaotically [12, 29, 30], because the oscillating elliptic vortex generates a time-periodic perturbation to particle's motion. Similar chaotic dynamics occurs in the case of an ellipsoid vortex model [19, 20, 47], which is a generalization of the elliptic vortex model to a linear vertical stratification of the baroclinic external flow.

A class of dynamically consistent models admitting chaotic advection has been proposed by V.F. Kozlov [21] using the concept of a background flow in geophysical hydrodynamics. This concept allows to obtain a class of dynamically consistent stream functions with specified basin forms, bottom topography and debits at boundaries. Using that concept, chaotic advection has been studied for a variety of dynamically consistent models in the barotropic and baroclinic ocean [48–54]. Simplified analytic models do not pretend to describe quantitatively complicated mixing and transport processes in the real ocean and atmosphere, but they are useful in finding underlying geometric structures and typical mechanisms for onset of chaotic mixing and transport in geophysical flows.

2.2   Chaotic Transport and Mixing in a Kinematic Model of a Meandering Jet Current

2.2.1   The Model Flow and Unstable Periodic Trajectories

Meandering jet is a fundamental structure in laboratory and geophysical fluid flows. Strong oceanic (atmospheric) currents are meandering jets separating water (air) masses with distinct physical properties. For example, the Gulf Stream separates the colder and fresher slope ocean waters from the salty and warmer Sargasso Sea ones. In this section we study chaotic mixing and transport in the simple kinematic model of a meandering jet. In the context of dynamical systems theory, chaotic advection is Hamiltonian chaos in two-dimensional incompressible flows with equations of motion (1.4) for passive particles advected by the flow.

The model of a two-dimensional incompressible flow, we consider in this section, has been introduced in [34] as a toy kinematic model of transport and mixing of passive particles in meandering jet currents in the ocean. Among the variety of models of shear flows, one of the simplest ones is a Bickley jet with the velocity profile $\sim \mathrm{sech}^2 y$ and a running wave imposed. The phase portrait of such a flow in the frame of reference, moving with the phase velocity of the running wave, is shown in Fig. 2.1. The flow consists of three distinct regions, the eastward jet (J), the circulations (C), and the westward peripheral currents (P) to the north and south from the jet, separated from each other by the northern and southern ∞-like separatrices. A simple periodic modulation of the wave's amplitude can break up these separatrices and produce stochastic layers at the place of them. As a result, chaotic mixing and transport of passive particles might occur.

The stream function in the laboratory frame of reference is the following:

$$\Psi'(x', y', t') = -\Psi'_0 \tanh \left(\frac{y' - a \cos k(x' - ct')}{\lambda \sqrt{1 + k^2 a^2 \sin^2 k(x' - ct')}} \right), \qquad (2.1)$$

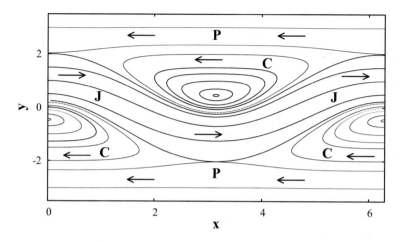

Fig. 2.1 The phase portrait of the kinematic model flow (2.2). The first frame with streamlines in the circulation (C), jet (J), and peripheral currents (P) zones is shown. The parameters are: $A_0 = 0.785$, $C = 0.1168$, and $L = 0.628$

where the hyperbolic tangent produces the Bickley-jet profile, the square root provides a constant width of the jet λ, and a, k, and c are amplitude, wavenumber, and phase velocity of the running wave, respectively. The normalized stream function in the frame of reference moving with the velocity c is

$$\Psi = -\tanh\left(\frac{y - A\cos x}{L\sqrt{1 + A^2\sin^2 x}}\right) + Cy,$$

(2.2)

where $x = k(x' - ct')$ and $y = ky'$ are new scaled coordinates. The normalized jet's width $L = \lambda k$, wave's amplitude $A = ak$, and phase velocity $C = c/\Psi_0'k$ are the control parameters. The advection equations (1.4) with the stream function (2.2) have the following form in the comoving frame of reference:

$$\dot{x} = \frac{1}{L\sqrt{1 + A^2\sin^2 x}\,\cosh^2\theta} - C, \qquad \dot{y} = -\frac{A\sin x(1 + A^2 - Ay\cos x)}{L\left(1 + A^2\sin^2 x\right)^{3/2}\cosh^2\theta},$$

$$\theta = \frac{y - A\cos x}{L\sqrt{1 + A^2\sin^2 x}},$$

(2.3)

where dot denotes differentiation with respect to the scaled time $t = \Psi_0'k^2t'$.

The flow with the stream function (2.1) is steady in the comoving frame of reference, and its phase portrait is shown in Fig. 2.1. There are southern and northern sets of elliptic fixed points: $x_e^{(s)} = 2\pi n$, $y_e^{(s)} = -L\,\mathrm{arcosh}\sqrt{1/LC} + A$ and $x_e^{(n)} = (2n + 1)\pi$, $y_e^{(n)} = L\,\mathrm{arcosh}\sqrt{1/LC} - A$, respectively, and the southern and northern sets of hyperbolic (saddle) fixed points: $x_s^{(s)} = (2n + 1)\pi$, $y_s^{(s)} = -L \operatorname{arcosh} \sqrt{1/LC} - A$ and $x_s^{(n)} = 2\pi n$, $y_s^{(n)} = L \operatorname{arcosh} \sqrt{1/LC} + A$, respectively, where $n = 0, \pm 1, \ldots$.

The perturbation is provided by a periodic modulation of the wave's amplitude

$$A(t) = A_0 + \varepsilon \cos(\omega t + \varphi). \qquad (2.4)$$

The equations of motion (2.3) are symmetric under the following transformations: $t \to t$, $x \to \pi + x$, $y \to -y$ and $t \to -t$, $x \to -x$, $y \to y$. Due to the first symmetry, the motion can be considered in the northern chain of the circulation cells on the cylinder with $0 \leqslant x \leqslant 2\pi$. The part of the phase space with $2\pi n \leqslant x \leqslant 2\pi(n+1)$, $n = 0, \pm 1, \ldots$, is called a frame. The first frame is shown in Fig. 2.1. The values of the following control parameters are fixed in our simulation: $L = 0.628$, $A_0 = 0.785$, $C = 0.1168$, $T_0 = 2\pi/\omega = 24.7752$, $\varphi = \pi/2$. The only varying parameter is the perturbation amplitude ε.

In this paragraph we analyze in detail origin and bifurcations of typical classes of unstable periodic trajectories in the two-dimensional incompressible flow (2.2). Periodic trajectories play an important role in organizing dynamical chaos both in Hamiltonian and dissipative systems. **Stable** periodic trajectories organize a regular motion inside islands of stability in the **phase space**. **Unstable** periodic trajectories form a skeleton around which chaotic dynamics is organized. The motion nearby an unstable periodic trajectory is governed by its stable and unstable **manifolds**. Owing to the density property, the unstable periodic trajectories influence even the asymptotic dynamics. Order and disorder in a chaotic regime are produced eventually by an interplay between sensitivity to initial conditions and regularity of the periodic motion.

To locate periodic trajectories in the phase space we fix values of the control parameters and compute with a large number of particles the Euclidean distance $d = \sqrt{[x(t_0 + mT_0) - x(t_0)]^2 + [y(t_0 + mT_0) - y(t_0)]^2}$ between particle's position at an initial moment of time t_0 and at the moments of time $T = mT_0$, where $m = 1, 2, \ldots$. The data are plotted as a period-m return map that shows by color the values of d for particles with initial positions $[x(t_0), y(t_0)]$. At the first stage, we select a large number of points where the function $d(x(t_0), y(t_0))$ may have local minima. Then we apply the method of a downhill simplex to localize the minima in neighborhoods of those points. There are such minima among them for which $d = 0$ with a given value of m. The procedure allows to detect both unstable periodic trajectories and stable periodic trajectories not only in periodically perturbed Hamiltonian systems but also for any chaotic system.

Return maps with $m = 1, 2, 3, 4$, and 12 have been computed as an example. The main efforts are devoted to analysis of period-4 unstable periodic trajectories because it is not a trivial task to detect the unstable periodic trajectories when the corresponding resonances cannot be identified on **Poincaré sections**. It is the case with $m = 4$. The period-4 return map is shown in Fig. 2.2a with the cross marking location of the period-1 saddle orbit and black dots marking initial positions of twenty six trajectories of the period-4 unstable orbits with a relative accuracy 10^{-13}–10^{-14}. For comparison, we demonstrate in Fig. 2.2b locations of those dots on the

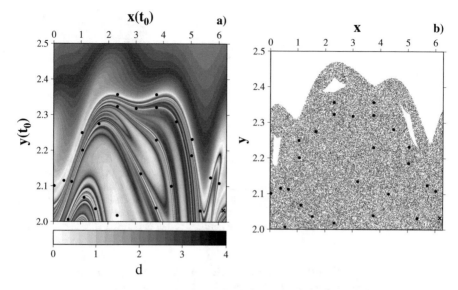

Fig. 2.2 (**a**) The period-4 return map representing the distance d between particle's position $x(t_0)$ and $y(t_0)$ at t_0 and its position at $t_0 + 4T_0$. The *cross* in the *bottom right* corner marks location of the saddle orbit and the *dots* mark initial positions of trajectories of the period-4 unstable orbits. (**b**) Poincaré section surface of the northern separatrix layer with positions of those *dots*

Poincaré section surface. It is evident that they cannot be prescribed to any structures in the phase space and cannot be identified by inspection of the Poincaré section.

The saddle points of the unperturbed equations of motion (2.3), $(x_s^{(n)}, y_s^{(n)})$, and $(x_s^{(s)}, y_s^{(s)})$ become period-1 saddle trajectories under the periodic perturbation (2.4). It is shown in our paper [43] how to compute and analyze them.

2.2.2   Origin and Bifurcations of Period-4 Unstable Orbits

In this paragraph we analyze the origin of the period-4 unstable orbits and their **bifurcations** that occur with changing the perturbation amplitude ε. Using the period-4 return map (Fig. 2.2a), we located all the period-4 unstable orbits and the initial positions of four trajectories for each of them. Orbits differ by the type of motion of passive particles and their length l which is a length of the corresponding trajectories for the time interval $4T_0$. All the unstable periodic trajectories are closed curves on the cylinder $0 \leqslant x \leqslant 2\pi$, and their topology may be very complicated. In the physical space, the unstable periodic trajectories can be classified as rotational ones (particles rotate in the same frame), ballistic ones (particles move ballistically from frame to frame), and rotation-ballistic ones (particles may rotate for a while in a frame, then move ballistically through a few frames and even change their direction of motion).

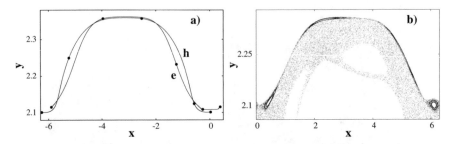

Fig. 2.3 (**a**) Two $C_{WB}^{4:1}$ western ballistic period-4 unstable orbits at $\varepsilon = 0.0785$ are marked by the *bold points*. The *curves* are the $4T_0$-fragments of the corresponding trajectories. Each orbit has four trajectories coinciding on the x–y plane but not in the extended phase space. The e and h orbits were born from the elliptic and hyperbolic points of the northern ballistic resonance $4 : 1$, respectively. (**b**) The Poincaré section surface at $\varepsilon = 0.005$ with the northern ballistic resonance $4 : 1$ consisting of four ballistic islands along the northern border of the separatrix layer

2.2.2.1   The $C_{WB}^{4:1}$ Class: Western Ballistic Unstable Periodic Orbits Associated with a 4 : 1 Western Ballistic Resonance

The shortest ones among all the period-4 unstable orbits are western ballistic orbits. The particles belonging to the $C_{WB}^{4:1}$ class move in a periodic way to the west in the northern separatrix layer which appears between the northern C and P regions in Fig. 2.1 as a result of the perturbation. With the help of the return map in Fig. 2.2a, we located two $C_{WB}^{4:1}$ orbits with initial positions of four periodic trajectories belonging to each of them. In order to track out the origin of the e and h orbits, shown in Fig. 2.3a at $\varepsilon = 0.0785$, we decrease the value of the perturbation amplitude ε, compute the corresponding period-4 return maps, locate the $C_{WB}^{4:1}$ orbits, and measure their length.

The result may be resumed as follows. The western ballistic **nonlinear resonance** 4 : 1 with four elliptic and four hyperbolic points appears under the perturbation with a very small value of ε. It is manifested as four ballistic islands along the northern border of the separatrix layer on the Poincaré section surface in Fig. 2.3b at $\varepsilon = 0.005$. Two of the islands are so thin that they are hardly visible in the figure. With increasing ε, the size of the resonance islands decreases, and they vanish at the critical value $\varepsilon \approx 0.016$ (see Fig. 2.2b where there are no signs of that resonance at $\varepsilon = 0.0785$). The orbit, associated with the elliptic fixed points of the western resonance 4 : 1, loses its stability and bifurcates into the $C_{WB}^{4:1}$ unstable orbit of period-4 which is denoted by the symbol e. Its length practically does not change with increasing ε (see Fig. 2.4). The length l of the h orbit, associated with the hyperbolic points of that resonance, changes dramatically at $\varepsilon \approx 0.0715$ increasing fastly after this point because of appearing a meander and a loop on the h orbit nearby the saddle orbit [43].

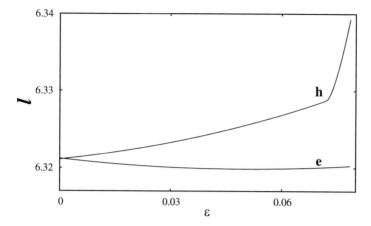

Fig. 2.4 Dependence of the length l of the e and h $C_{WB}^{4:1}$ western ballistic period-4 unstable orbits on the perturbation amplitude ε

2.2.2.2 The $C_{EB}^{4:1}$ Class: Eastern Ballistic Unstable Periodic Orbits Associated with a 4 : 1 Eastern Ballistic Resonance

The eastern ballistic unstable orbits of period-4 lie in the southern separatrix layer between J and C regions (see Fig. 2.1) to the north from the jet. Particles move along corresponding trajectories to the east in a periodic way. The origin and bifurcations of the $C_{EB}^{4:1}$ orbits are similar to the western ballistic ones. They appear from elliptic and hyperbolic points of the eastern ballistic resonance 4 : 1 as ε increases. The elliptic orbit of this resonance loses it stability and bifurcates into an unstable periodic orbit. The trajectories of the hyperbolic orbit changes its topology, transforming from a bell-like curve at $\varepsilon = 0.01$ to the curve with a meander and a loop nearby the saddle orbit at $\varepsilon = 0.0785$ (see Fig. 2.5). The length l of the h orbit decreases with increasing ε up to the bifurcation point $\varepsilon \approx 0.0275$, after which it increases fastly due to a complexification of the trajectory's form (see Fig. 2.5).

2.2.2.3 The $C_{R}^{4:1}$ Class: Orbits Associated with a 4 : 1 Rotational Resonance

Four different orbits in the $C_{R}^{4:1}$ class, located with the help of the period-4 return map, are shown in Fig. 2.6 at $\varepsilon = 0.0785$ along with unstable periodic trajectories for each of them. The corresponding fluid particles rotate in the same frame along closed curves. The genesis of the $C_{R}^{4:1}$ orbits is the following. A 4 : 1 rotational resonance appears under a small perturbation (2.4). As the amplitude ε increases, its elliptic orbit loses stability and bifurcates at $\varepsilon \approx 0.003$ into an unstable periodic orbit which we denote by the symbol e. The hyperbolic orbit of the 4 : 1 resonance h bifurcates at $\varepsilon \approx 0.040945$ into two hyperbolic orbits, h_1, h_2, and an elliptic orbit

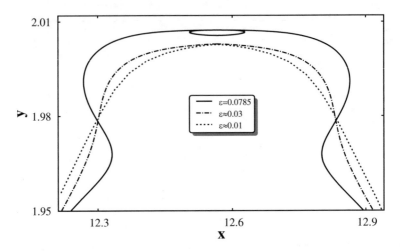

Fig. 2.5 Metamorphoses of the h $C_{EB}^{4:1}$ eastern ballistic unstable periodic trajectory as the perturbation amplitude ε changes: the *solid line* at $\varepsilon = 0.0785$, *dashed line* at $\varepsilon = 0.03$, and *dotted line* at $\varepsilon = 0.01$

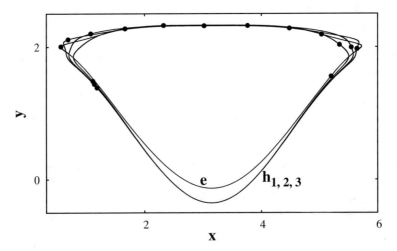

Fig. 2.6 Four 4 : 1 rotational unstable periodic orbits of the $C_R^{4:1}$ class marked by the *bold points*. The *curves* are the corresponding trajectories. The e and $h_{1,2,3}$ orbits were born from the elliptic and hyperbolic points of the 4 : 1 rotational resonance, respectively ($\varepsilon = 0.0785$)

h_3 in the centers of four stability islands. It is a pitchfork bifurcation. In Fig. 2.7a we show by arrows movement of the $C_R^{4:1}$ orbits as ε decreases. The pitchfork bifurcation point is shown as a bold point. As ε increases further, the elliptic orbit h_3 loses its stability and becomes a hyperbolic $C_R^{4:1}$ orbit. The bifurcation diagram $l(\varepsilon)$ is shown in Fig. 2.7b.

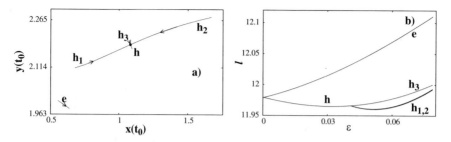

Fig. 2.7 (**a**) Movement of the rotational 4 : 1 unstable periodic orbits with decreasing the perturbation amplitude is shown by the *arrows*. The *bold point* is a pitchfork bifurcation point at $\varepsilon \approx 0.040945$ where the hyperbolic orbit of the resonance, h, bifurcates into the three different orbits h_1, h_2, and h_3. (**b**) The bifurcation diagram $l(\varepsilon)$

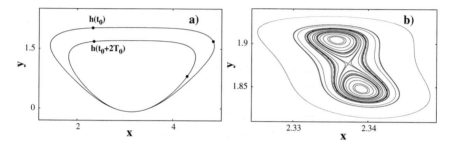

Fig. 2.8 (**a**) The 2 : 1 rotational unstable periodic orbits marked by the *bold points* ($\varepsilon = 0.0785$). The *curves* are the corresponding trajectories. (**b**) Poincaré section surface of the 2 : 1 rotational resonance at $\varepsilon = 0.0668$ (just after the period-doubling bifurcation)

2.2.2.4 The $C_R^{2:1}$ Class: Orbits Associated with a 2 : 1 Rotational Resonance

The class $C_R^{2:1}$ consists of two rotational unstable periodic orbits, associated with a 2 : 1 rotational resonance, with elliptic and hyperbolic orbits and two periodic trajectories on each of them. The resonance appears under a small perturbation. At $\varepsilon \approx 0.0665$, the elliptic orbit of this resonance undergoes a period-doubling bifurcation into a period-4 elliptic orbit with four trajectories and a period-2 hyperbolic orbit with two trajectories. The latter is shown in Fig. 2.8a at $\varepsilon = 0.0785$. The Poincaré section surface of the 2 : 1 rotational resonance at $\varepsilon = 0.0668$ (just after the bifurcation) is shown in Fig. 2.8b. By further increasing ε, we see that the resonance 2 : 1 vanishes, and the period-4 elliptic orbit loses its stability and bifurcates into a period-4 unstable periodic orbit. To clarify this bifurcation we fix a point at the moment of time t_0 on the upper branch of the orbit in Fig. 2.8a and another point on the lower branch at $t = t_0 + 2T_0$ and scan their positions when decreasing the perturbation amplitude ε.

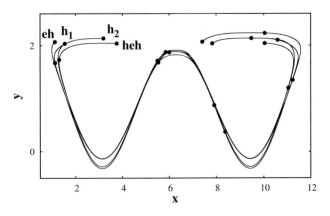

Fig. 2.9 Four rotational-ballistic period-4 unstable orbits (*eh*, h_1, h_2, and *heh*) marked by the *bold points*. The *curves* are $4T_0$-fragments of the corresponding trajectories

2.2.2.5 The $C_{RB}^{4:1}$ Class: Rotation-Ballistic Unstable Periodic Orbits Associated with a Rotational-Ballistic Resonance

The class $C_{RB}^{4:1}$ consists of four period-4 unstable orbits shown in Fig. 2.9a. We call those orbits as rotational-ballistic ones because the corresponding particles begin to move to the west in one frame, then turn to the east and travel in the southern separatrix layer to the next frame, fulfill one turnover in this frame, and repeat their motion to the east. Genesis of the $C_{RB}^{4:1}$ orbits differs from genesis of the other classes of period-4 unstable orbits. Each of the resonances, associated with $C_{RW}^{4:1}$, $C_{RE}^{4:1}$, $C_{R}^{4:1}$ and $C_{R}^{2:1}$, appears under an infinitely small perturbation amplitude ε. The rotation-ballistic motion of period-m and the corresponding orbits cannot in principle appear in the flow below some critical value of ε (which depends on m) because the width of the stochastic layers (which increases with increasing ε) should be large enough in order that particles would have enough time to travel a corresponding distance to the east and west. The genesis and evolution of the $C_{RB}^{4:1}$ unstable periodic orbits are shown schematically on the bifurcation diagrams in Fig. 2.10. In Fig. 2.10a we demonstrate movement of the $C_{RB}^{4:1}$ orbits on the phase plane with decreasing the perturbation amplitude ε. The dependence of the lengths of those orbits on ε is shown in Fig. 2.10b.

The rotational-ballistic 4 : 1 resonance appears at a critical value of the perturbation amplitude $\varepsilon = \varepsilon_1 \approx 0.040715$ due to a saddle-center bifurcation and manifests itself as 4 small islands of stability on the corresponding Poincaré section surface. One of these islands is shown in Fig. 2.11a. If ε increases further, the elliptic orbit *eh* in the centers of the RB 4 : 1 resonance loses its stability and becomes a hyperbolic rotational-ballistic period-4 orbit. The hyperbolic orbit h of the rotational-ballistic resonance at $\varepsilon = \varepsilon_2 \approx 0.0433$ undergoes a pitchfork bifurcation into two hyperbolic rotational-ballistic period-4 orbits h_1 and h_2 and the elliptic orbit *heh* in the centers of a new period-4 rotational-ballistic resonance,

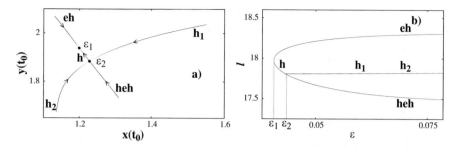

Fig. 2.10 (**a**) At the saddle-center bifurcation point $\varepsilon = \varepsilon_1 \approx 0.040715$, there appears a rotational-ballistic 4 : 1 resonance with the elliptic orbit *eh* which loses its stability with increasing ε. The hyperbolic orbit *h* of the rotational-ballistic resonance bifurcates at $\varepsilon = \varepsilon_2 \approx 0.0433$ into two hyperbolic period-4 unstable orbits h_1 and h_2 and the elliptic orbit *heh* which loses its stability with further increasing ε. (**b**) The bifurcation diagram $l(\varepsilon)$

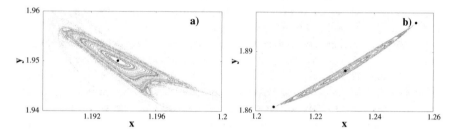

Fig. 2.11 (**a**) One of the stability islands of the rotational-ballistic 4 : 1 resonance appearing after a saddle-center bifurcation at $\varepsilon = 0.04072 \gtrsim \varepsilon_1$ with the elliptic point *eh* in its center. (**b**) One of the stability islands of the rotational-ballistic 4 : 1 resonance appearing after a pitchfork bifurcation at $\varepsilon = 0.0435 \gtrsim \varepsilon_2$ with the elliptic point *heh* in its center

one of whose stability islands is shown in Fig. 2.11b. Under further increasing ε, the elliptic orbit *heh* loses its stability and becomes a hyperbolic rotational-ballistic period-4 orbit.

Rotational and ballistic resonant islands are well-known objects in the phase space of Hamiltonian systems. Rotation-ballistic resonant islands of stability are not so well known [43]. They are expected to appear as well in other jet-flow models (kinematic and dynamic ones) under specific values of control parameters. Like ballistic islands, the rotation-ballistic islands should affect transport and its statistical characteristics that will be demonstrated in the next section. Due to the presence of rotation phase of motion, the mean drift velocity of rotational-ballistic particles is smaller than that for ballistic particles. It is expected that boundaries of the rotational-ballistic islands are specific **"dynamical traps"** that should affect transport and statistical properties of passive particles (see the next section). Properties of rotational-ballistic islands traps may differ from properties of rotational-island's traps and ballistic-islands ones because of different locations of the corresponding hyperbolic points around the islands. They appear to be between islands in chains of rotational and ballistic islands, whereas in the case of rotational-ballistic islands they are situated either at one (a saddle-center bifurcation) or both sides (a pitchfork bifurcation) of a given island.

Period-1 saddle orbits play a crucial role in chaotic transport and mixing of passive particles. Stable and unstable manifolds of unstable periodic orbits are material curves of complicated forms which cannot be crossed by particle's trajectories. Their role in transport and mixing of passive particles in the meandering jet flows is significant because they separate trajectories with different dynamical and topological properties. The unstable periodic orbits found in this section act as dynamical traps where particles and their trajectories may spend a rather long time before escaping. It was checked by computing distributions of the escape times for the period-4 unstable orbits and the saddle orbit. The rotational unstable periodic orbits should contribute to statistics of comparatively short flights (see Fig. 2.16). Unstable periodic orbits with larger values of the period can be classified into three big groups: ballistic, rotational, and rotation-ballistic ones. The origin and bifurcations of the orbits in each class can be studied in a similar way, but it may require larger computational efforts.

The results obtained do not depend critically on exact form of the stream function and chosen values of the control parameters. Unstable periodic orbits in other kinematic and dynamical models of geophysical jets can be detected, located, and classified in a similar way. The unstable periodic orbits form a skeleton for chaotic advective mixing and transport in fluid flows, and the knowledge of them (at least, lower-period ones) allows us to analyze complex, albeit rather regular, stretching and folding structures in the flows.

2.2.3 Chaotic Zonal Transport and Dynamical Traps

In this section we study and discuss a phenomenological connection between dynamical, topological, and statistical properties of chaotic mixing and transport in the kinematic jet-flow model considered in the preceding section and the impact of the so-called **dynamical traps** on transport. Following to G. Zaslavsky [46], the dynamical trap is a domain in the phase space of a Hamiltonian system where a particle (or its trajectory) can spend arbitrary long finite time, performing almost regular motion, despite the fact that the full trajectory is chaotic in any appropriate sense. In fact, it is the definition of a quasitrap. Absolute traps, where particles could spend an infinite time, are possible in Hamiltonian systems only as a zero-measure set. The dynamical traps appear due to a stickiness of trajectories to some singular domains in the phase space, largely, to the boundaries of resonant islands, saddle trajectories, and cantori. Up to now, there are no classification and description of the dynamical traps. G. Zaslavsky described two types of dynamical traps in Hamiltonian systems: hierarchical-island's traps around chains of resonant islands and stochastic-layer traps which are stochastic jets inside a stochastic sea where trajectories can spend a very long time.

It is convenient to characterize chaotic mixing and transport in terms of zonal flights. A zonal flight is a motion of a particle between two successive changes of signs of its zonal velocity, i.e., the motion between two successive events with $\dot{x} = u = 0$. Particles (and corresponding trajectories) in chaotic jet flows can be classified in terms of the lengths of flights x_f as follows. The trajectories with $|x_f| < 2\pi$ correspond to the particles moving in the same frame or in neighbor frames. There are particles moving chaotically in the global stochastic layer in the same frame forever, but they are of a zero measure. Among the particles with inter-frame motion, there are regular and chaotic ballistic ones. Regular ballistic trajectories can be defined as those which cannot have two flights with $|x_f| > 2\pi$ in succession. They correspond to particles moving in regular regions of the phase space persisting under the perturbation (eastward motion in the jet and western motion in the peripheral current) and those moving in the stochastic layer (trajectories belonging to ballistic islands). Chaotic trajectories have complicated distributions over the lengths and durations of flights.

In the laboratory frame of reference (2.1), all the fluid particles move to the east together with the jet flow, and a flight is motion between two successive events when the particle's zonal velocity U is equal to the meander's phase velocity c. If $U < c$, the corresponding particle is left behind the meander (it is a western flight in the comoving frame of reference). If $U > c$, it passes the meander (an eastern flight in the comoving frame of reference). Short flights with $|x_f| < 2\pi$ (motion in the same spatial frame in the comoving frame of reference) correspond to motion in the laboratory frame of reference when two successive events $U = c$ occur on the space interval less than the meander's spatial period $2\pi/k$. Ballistic flights between the spatial frames in the comoving frame of reference with $|x_f| > 2\pi$ correspond to motion in the laboratory frame of reference when particles move through more than one meander's crest between two successive events with $U = c$.

2.2.3.1 Turning Points

As in [31], we will characterize statistical properties of chaotic transport by probability density functions (PDFs) of lengths of flights $P(x_f)$ and durations of flights $P(T_f)$ for a number of very long chaotic trajectories. Both regular and chaotic particles may change many times the sign of their zonal velocity $\dot{x} = u$. From the condition $\dot{x} = 0$ in Eq. (2.3), it is easy to find equations for the curves which are loci of the turning points

$$Y_{\pm}(x, A) = \pm L\sqrt{1 + A^2 \sin^2 x}\ \text{arsech}\ \sqrt{LC\sqrt{1 + A^2 \sin^2 x} + A\cos x}. \qquad (2.5)$$

Let us consider the northern curve, i.e., Eq. (2.5) with the positive sign. Taking into account that the perturbation has the form (2.4), we realize that all the northern turning points are inside a strip confined by the two curves of the form (2.5) with $A = A_0 \pm \varepsilon$. Let us analyze the derivative over the varying parameter A

$$\frac{\partial Y}{\partial A} = \cos x + \frac{ACL^2 \sin^2 x}{2F} \left(2 \operatorname{arsech} \sqrt{F} - \frac{1}{\sqrt{1 - F}} \right), \qquad (2.6)$$

where $F = LC\sqrt{1 + A^2 \sin^2 x}$. If the derivative at a fixed value of x does not change its sign on the interval $A_0 - \varepsilon \leqslant A \leqslant A_0 + \varepsilon$, then Y varies from $Y(x, A_0 - \varepsilon)$ to $Y(x, A_0 + \varepsilon)$, and for each value of y we have a single value of the perturbation parameter A. However, there may exist such values of x for which the equation $\partial Y / \partial A = 0$ has a solution on the interval mentioned above. In this case one may have more than one value of A for a single value of y. Thus, the width of the strip, containing turning points, is defined by the values of Y at the extremum points and at the endpoints of the interval of values of A. In Fig. 2.12b we show the turning points of a single chaotic trajectory on the cylinder $0 \leqslant x \leqslant 2\pi$ confined between two corresponding curves.

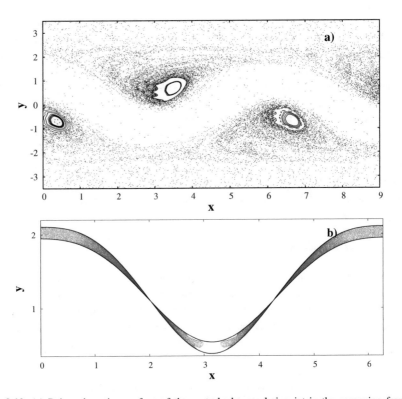

Fig. 2.12 (**a**) Poincaré section surface of the perturbed meandering jet in the comoving frame of reference. The parameters of the steady flow are: the jet's width $L = 0.628$, the meander's amplitude $A_0 = 0.785$, and its phase velocity $C = 0.1168$. The perturbation amplitude and frequency are: $\varepsilon = 0.0785$ and $\omega = 0.2536$. (**b**) Turning points of a single chaotic trajectory on the cylinder $0 \leqslant x \leqslant 2\pi$ are located in a strip confined by the two curves (2.5) with $A = A_0 \pm \varepsilon$

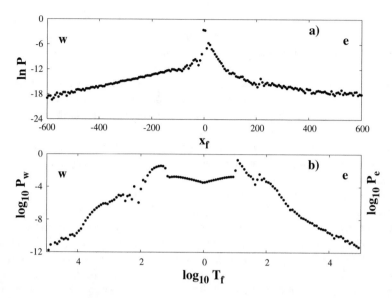

Fig. 2.13 Probability density functions for (**a**) lengths x_f and (**b**) durations T_f of westward (w) and eastward (e) flights. The PDFs $P_w(T_f)$ and $P_e(T_f)$ are normalized to the number of westward ($4.23 \cdot 10^7$) and eastward ($4 \cdot 10^7$) flights, respectively. It is a statistics for five tracers with the computation time $t = 5 \cdot 10^8$ for each one

In the numerical simulation we use a Runge–Kutta integration scheme of the fourth order with the constant time step $\Delta t = 0.0247$. Numerical experiments with tracers initially placed in the stochastic layer have carried out. It was found that statistical properties of chaotic transport practically do not depend on the number of tracers provided that the corresponding trajectories are sufficiently long ($t \simeq 10^8$). The PDFs for the lengths x_f and durations T_f of flights for five tracers with the computation time $t = 5 \cdot 10^8$ for each tracer are shown in Fig. 2.13a, b, respectively, both for eastward (e) and westward (w) motion. Both $P(x_f)$ and $P(T_f)$ are complicated functions with local extrema decaying in a different manner in different ranges of x_f and T_f. The main aim of our study of chaotic transport is to figure out basic peculiarities of the statistics and attribute them to specific zones in the phase space, namely to dynamical traps strongly influencing the transport.

2.2.3.2 Rotational-Island's Traps

It is well known that in nonlinear Hamiltonian systems a complicated structure of the phase space with islands, stochastic layers, and chains of islands, immersed in a stochastic sea, arises under a perturbation due to a variety of **nonlinear resonances** and their overlapping. Motion is quasiperiodic and stable inside those islands. Boundaries of the islands are absolute barriers to transport: particles cannot go through them neither from inside nor from outside. Invariant curves of the unperturbed system (see Fig. 2.1) are destroyed under the perturbation (2.4) (see Fig. 2.12a). As the perturbation strength ε increases, a closed invariant curve with the frequency f is destroyed at some critical value of ε. If f/ω is a rational number, the corresponding curve is replaced by an island chain, while the curves with some irrational frequencies are replaced by **cantori**. There are uncountably many cantori forming a complicated hierarchy. Numerical experiments with Hamiltonian systems with different number of degrees of freedom provide an evidence for the presence of strong partial barriers to transport around the island's boundaries (for review, see [46]) which manifest themselves on Poincaré section surfaces as domains with increased density of points.

With appropriate values of the control parameters, there exist in each frame a vortex core (which is an island of the primary resonance $\omega = f$) immersed into a stochastic sea, where there are six islands of a secondary resonance emerged from three islands of the primary resonance $3f = 2\omega$ (see Fig. 3 in [31]). Chains of smaller-size islands are present around the vortex core and the secondary-resonance islands. Particles belonging to all of these islands (including the vortex core) rotate in the same frame performing short flights with the lengths $|x_f| < 2\pi$. So we will call them "rotational islands" and distinguish from the so-called ballistic islands which will be considered below.

Stickiness of particles to boundaries of rotational islands has been demonstrated in [31]. It means that fluid particles can be trapped for a long time in a singular zone nearby borders of the rotational islands which we will call "rotational-island's traps." To illustrate the effect of the rotational-island's traps, we demonstrate in Figs. 2.14 and 2.15 the Poincaré sections of a chaotic trajectory in the frame $0 \leqslant x \leqslant 2\pi$ sticking to the vortex core and to the secondary-resonance islands, respectively.

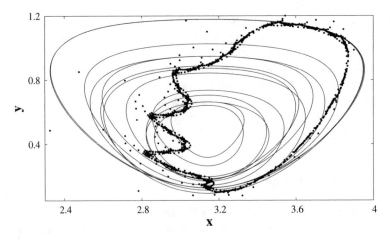

Fig. 2.14 The vortex-core trap. Poincaré section of a chaotic trajectory in the frame $0 \leqslant x \leqslant 2\pi$ with a fragment of a trajectory

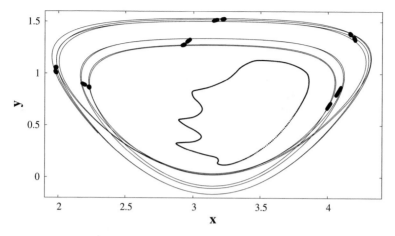

Fig. 2.15 The secondary-resonance island's trap. A fragment of a chaotic trajectory sticking to the islands is shown. The Poincaré section of that trajectory is shown by *points*

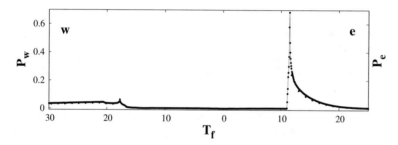

Fig. 2.16 The PDFs for the eastward (e) and westward (w) flights with the length shorter than 2π. The PDFs $P_w(T_f)$ and $P_e(T_f)$ are normalized to the number of westward ($4.19 \cdot 10^7$) and eastward ($3.7 \cdot 10^7$) flights, respectively. It is a statistics for five tracers with the computation time $t = 5 \cdot 10^8$ for each one

The contour of the vortex core is shown in Fig. 2.15 by the thick line. The points are tracks of the particle's position at the moments of time $t_n = 2\pi n / \omega$ (where $n = 1, 2, \dots$) and the thin curves are fragments of the corresponding trajectory on the phase plane. Increased density of points indicates the presence of dynamical traps near the boundaries of the rotational islands. Contribution of the vortex-core rotational-island's traps (Fig. 2.14) to chaotic transport is expected to be much more significant than for the rotational-island's traps of the other islands (Fig. 2.15).

It is reasonable to suppose that rotational-island's traps contribute to the statistics of short flights. By short flights we mean the flights with the length shorter than 2π. Part of the full PDF $P(T_f)$ (Fig. 2.13b) for eastward (e) and westward (w) short flights is shown in Fig. 2.16 separately. There are a comparatively small number of the eastward flights with $T_f < 11$. Let us note the prominent peak of the corresponding PDF at $T_f \simeq 11$ followed by an exponential decay. As to the westward short flights, there are two small local peaks around $T_f \simeq 17$ and 21.

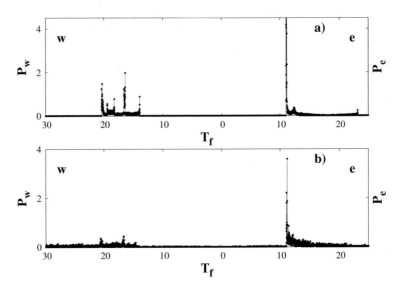

Fig. 2.17 The vortex-core trap PDFs of durations T_f of the eastward (e) and westward (w) flights. (a) A regular quasiperiodic trajectory with the duration $t = 2 \cdot 10^5$ inside the vortex core close to its boundary. Both the PDFs are normalized to the number $8 \cdot 10^3$ of corresponding flights. (b) A chaotic trajectory with the duration $t = 2 \cdot 10^5$ sticking to the boundary of the vortex core from the outside. Both the PDFs are normalized to $4 \cdot 10^3$ flights

To estimate a contribution of the vortex-core rotational-island's trap to the statistics of short flights, we compute and compare statistics of the durations of flights T_f for two trajectories: a regular quasiperiodic one with the initial position close to an inner border of the vortex core (Fig. 2.17a) and a chaotic one with the initial position close to the vortex-core border from the outside (Fig. 2.17b). Each full rotation of a particle in a frame consists of two flights, eastward and westward, with different values of T_f because of the zonal asymmetry of the flow. The statistics for the chaotic trajectory, sticking to the vortex core (Fig. 2.17b), may be considered as a distribution of durations of flights in the vortex-core rotational-island's trap. The minimal flight duration in this rotational-island's trap is $T_f \simeq 11$ (the flights with smaller values of T_f are rare, and they occur outside the trap). Positions of local maxima of the PDF for the sticking trajectory in Fig. 2.17b correlate approximately with the corresponding local maxima of the PDF for the regular trajectory inside the core in Fig. 2.17a. The similar correlations have been found (but not shown here) between the local maxima of the PDFs for the lengths of flights $P(x_f)$ for the interior regular and sticking chaotic trajectories. These correlations and positions of the peaks prove numerically that short flights with $|x_f| < 2\pi$ and $11 \lesssim T_f \lesssim 21$ may be caused by the effect of vortex-core rotational-island's trap. We conclude from Fig. 2.17b that the vortex-core rotational-island's trap contributes to statistics of the short flights in the range $11 \lesssim T_f \lesssim 20$ for the eastward flights with the prominent peak at $T_f \simeq 11$ and in the range $15 \lesssim T_f \lesssim 21$ for the westward flights with small peaks at $T_f \simeq 17$ and 21.

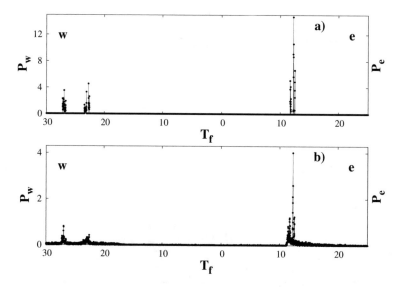

Fig. 2.18 The secondary-resonance island's trap. The PDFs for durations T_f of the eastward (e) and westward (w) flights. (**a**) Regular quasiperiodic trajectory inside the islands with the duration $t = 5 \cdot 10^5$. Both the PDFs are normalized to the number $1.5 \cdot 10^4$ of corresponding flights. (**b**) Chaotic trajectory sticking to the island's boundary from the outside with the duration $t = 5 \cdot 10^5$. $P_w(T_f)$ and $P_e(T_f)$ are normalized to the number of westward ($1.1 \cdot 10^4$) and eastward ($9 \cdot 10^3$) flights, respectively

The effect of the rotational-island's trap of the secondary-resonance islands is illustrated in Fig. 2.15. To find the characteristic times of this rotational-island's trap, we compute two trajectories: a regular quasiperiodic one with the initial position inside one of these islands and a chaotic one with the initial position close to the outer border of the island. The respective PDFs $P(T_f)$, shown in Fig. 2.18a, b, demonstrate strong correlations between the corresponding peaks at $T_f \simeq 12, 23$, and 27. Computed (but not shown here) PDFs $P(x_f)$ for these trajectories confirm the impact of the rotational-island's trap on the statistics of short flights.

2.2.3.3 Saddle Traps

As a result of the periodic perturbation (2.4), the saddle points of the unperturbed system at $x_s^{(n)} = 2\pi n$, $y_s^{(n)} = L \operatorname{arcosh} \sqrt{1/LC} + A$ and at $x_s^{(s)} = (2n + 1)\pi$, $y_s^{(s)} = -L \operatorname{arcosh} \sqrt{1/LC} - A$ $(n = 0, \pm 1, \ldots)$ become periodic unstable saddle trajectories. These hyperbolic trajectories have their own stable and unstable **manifolds** and play a role of specific dynamical traps which we call "saddle traps." In this section we demonstrate that the saddle traps influence strongly on chaotic mixing and transport of passive particles and contribute, mainly, in a short-time statistics of flights.

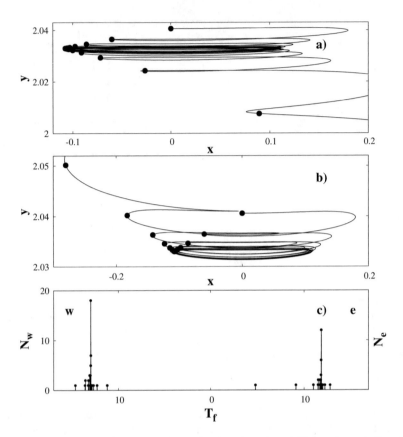

Fig. 2.19 The saddle trap. Fragments of two chaotic trajectories sticking to the periodic saddle trajectory one of which escapes (**a**) to the east and another one (**b**) to the west. (**c**) The number of the eastward (N_e) and westward (N_w) short flights with duration T_f for those two trajectories. Statistics with two trajectories with the duration $t = 10^3$ and the total number of western $N_w = 55$ and eastern $N_e = 51$ flights

Tracers with initial positions close to the stable manifold of a saddle trajectory are trapped for a while performing a large number of revolutions along it. To illustrate the effect of saddle traps, we show in Fig. 2.19a, b fragments of two chaotic trajectories sticking to the saddle trajectory and performing about 20 full revolutions before escaping to the east (Fig. 2.19a) and to the west (Fig. 2.19b). We have managed to detect and locate the corresponding periodic unstable saddle trajectory which is situated in Fig. 2.19a, b in the domain where a few fragments of the chaotic trajectory imposed on each other. Because of the flow asymmetry, duration of eastern flights of a particle along the saddle trajectory $T_e \simeq 11.9$ is shorter than duration of western flights $T_w \simeq 12.9$. The black points are tracks of the particle's positions on the flow plane at the moments of time $t_n = 2\pi n/\omega \simeq 24.8\,n$ (where $n = 1, 2, \ldots$). They belong to smooth curves which are fragments of the stable and unstable manifolds of the saddle trajectory at the chosen initial phase $\phi = \pi/2$.

To estimate the contribution of the saddle traps to statistics of short flights shown in Fig. 2.16, we compute and plot in Fig. 2.19c the number of the eastward (N_e) and westward (N_w) short flights with a given duration T_f for those two chaotic trajectories sticking to the saddle trajectory arising from the saddle point at $x_s = 0$, $y_s \simeq 2.02878$. Each full rotation of the particles consists of an eastward flight with the duration $T_e \simeq 11.9$ and an westward flight with the duration $T_w \simeq 12.9$. The flights with $T_e \simeq 11.9$ contribute to the main peak in Fig. 2.16, and the flights with $T_w \simeq 12.9$—to the "westward" plateau in that figure.

The mechanism of operation of saddle traps can be described as follows. Each saddle trajectory $\gamma(t)$ possesses time-dependent stable $W_s(\gamma(t))$ and unstable $W_u(\gamma(t))$ material manifolds composed of a continuous sets of points through which pass at time t trajectories of fluid particles that are asymptotic to $\gamma(t)$ as $t \to \infty$ and $t \to -\infty$, respectively. Under a periodic perturbation, the stable and unstable manifolds oscillate with the period of perturbation. It was firstly proved by Poincaré that W_s and W_u may intersect each other transversally at an infinite number of homoclinic points through which pass doubly asymptotic trajectories. To give an image of a fragment of the stable manifold of a periodic saddle trajectory, we distribute homogeneously $2.5 \cdot 10^5$ particles in the rectangular $[-0.4 \leqslant x \leqslant 0.45; 2 \leqslant y \leqslant 2.1]$ and compute the time which takes them to escape the rectangular. The color in Fig. 2.20 modulates the time T when particles with given initial positions (x_0, y_0) reach the western line at $x = -1$ or the eastern line at $x = 1$. The particles with initial positions marked by the black and white colors move close to the stable manifold of the saddle trajectory and spend a maximal time near it before escaping. The black and white diagonal curve in Fig. 2.20 is an image of a fragment of the corresponding stable manifold. The particles with initial positions to the north from the curve escape to the west along the unstable manifold of the saddle trajectory whereas those with initial positions to the south from the curve escape to the east along its another unstable manifold.

Fig. 2.20 The saddle-trap map. Color codes the time T which takes for $2.5 \cdot 10^5$ particles with given initial positions (x_0, y_0) to reach the lines at $x = -1$ or $x = 1$ escaping to the west (w) and to the east (e), respectively. The *black* and *white diagonal curve* is an image of a fragment of the stable manifold of the saddle trajectory. The *cross* is the particle's position on that trajectory at the initial time moment. The integration time is $t = 500$

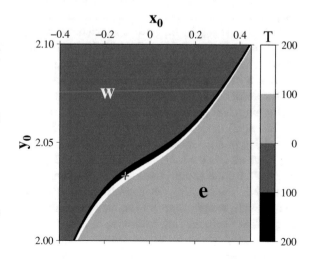

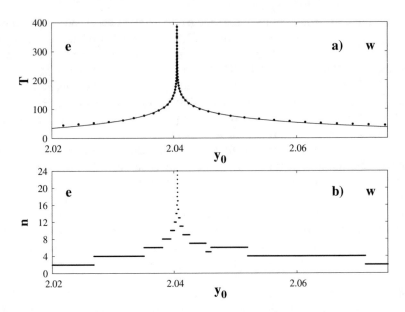

Fig. 2.21 (**a**) The period of time T which takes for the particle with an initial latitude y_0 to quit the saddle trap. (**b**) The number of short flights n that the particle performs before quitting the saddle trap. The ranges of y_0, from which particles quit the trap moving to the west and east, are denoted by w and e, respectively

We have found that particles quit the saddle trap along the unstable manifolds in accordance with specific laws. To find them let us distribute a large number of particles along the segment with $x_0 = 0$ and $2.02 \leqslant y_0 \leqslant 2.06$, crossing the stable manifold W_s, and compute the time T which takes for particles with given initial latitude y_0 to quit the saddle trap. More precisely, $T(y_0)$ is a time moment when a particle with the initial position y_0 reaches the lines with $x = -1$ or $x = 1$. The "experimental" points in Fig. 2.21a fit the law $T_e = (-85.81 \pm 0.04) - (31.216 \pm 0.007) \ln(y_{0s} - y_0)$ for the particles which quit the trap moving to the east and the law $T_w = (-60.61 \pm 0.03) - (28.933 \pm 0.006) \ln(y_0 - y_{0s})$ for those particles which move to the west when quitting the trap, where $y_{0s} = 2.0405755472$ is a crossing point of W_s with the segment of initial positions.

The saddle trap attracts particles and forces them to rotate in its zone of influence performing short flights, the number of which n depends on particle's initial position y_0. The $n(y_0)$ is a step-like function (see Fig. 2.21b) with the lengths of the steps decreasing in a geometric progression in the direction to the singular point, $l_j = l_0 q^{-j}$, where l_j is the length of the jth step and $q \simeq 2.27$ for the western exits and $q \simeq 2.20$ for the eastern ones. It is an analogue to the epistrophe decrease law in Sect. 1.2.2. The seeming deviation from this law in the range $2.045 \leqslant y_0 \leqslant 2.046$ (see a small western segment between two larger ones in Fig. 2.21b) is explained by crossing the initial line $2.02 \leqslant y_0 \leqslant 2.06$ by the curve of zero zonal velocity u.

To have the correct law for the western exits, it is necessary to add two segments of that cut step. The asymmetry of the functions $T(y_0)$ and $n(y_0)$ is caused by the asymmetry of the flow.

2.2.3.4 Ballistic-Island's Traps

Besides the rotational islands with particles moving around the corresponding elliptic points in the same frame, there are ballistic islands situated both in the stochastic layer and in the peripheral currents [31]. Regular ballistic modes correspond to stable quasiperiodic inter-frame motion of particles. Only the ballistic islands in the stochastic layer are important for chaotic transport. Mapping positions of regular ballistic trajectories at the moments of time $t_n = 2\pi n/\omega$ $(n = 1, 2, \ldots)$ onto the first frame, we obtain chains of ballistic islands both in the northern and southern stochastic layers, i.e., between the borders of the northern (southern) peripheral currents and of the corresponding vortex cores. A chain with three large ballistic islands is situated in those stochastic layers. The particles, belonging to these islands, move to the west, and their mean zonal velocity can easily be calculated to be $\langle u_f \rangle = -2\pi/3T_0 = -\omega/3 \simeq -0.0845$. There are also chains of smaller-size ballistic islands along the very border with the peripheral currents.

We have demonstrated in [31] a stickiness of chaotic trajectories to the borders of those three large ballistic islands (see Figs. 6 and 7 in [31]). The Poincaré section with fragments of two chaotic trajectories in the northern stochastic layer is shown in Fig. 2.22a. One particle performs a long flight sticking to the very border with the regular westward current, and another one moves to the west sticking to the very boundaries of three large ballistic islands. A magnification of a fragment of the border and tracks of a sticking trajectory around a smaller-size ballistic island are demonstrated in Fig. 2.22b. Figure 2.22c demonstrates the effective size of the trap of the large ballistic islands with tracks of a sticking trajectory around them.

It is reasonable to suppose that the ballistic-island's traps contribute, largely, to statistics of long flights with $|x_f| \gg 2\pi$. All the ballistic particles, moving both to the west and to east, can finish a flight and make a turn only in the strip shown in Fig. 2.12b. The loci of the corresponding turning points have a complicated fractal-like structure. We consider further only long westward flights, taking place in the northern stochastic layer, because it is much wider than the stochastic layer between the regular central jet and the southern parts of the vortex cores where eastward flights take place.

To distinguish between contributions of the traps of different ballistic islands (and, maybe, other zones in the phase space) to statistics of long flights, we compute (for five long chaotic trajectories up to $t = 5 \cdot 10^8$) the distribution of a number of westward flights with $T_f \geq 10^3$ over the mean zonal velocities $\langle u_f \rangle = x_f/T_f$ of the particles performing such flights. The distribution in Fig. 2.23 has a prominent peak centered at the mean zonal velocity $\langle u_f \rangle \simeq -0.0845$ which corresponds to a large number of long flights of those particles (and their trajectories) which stick

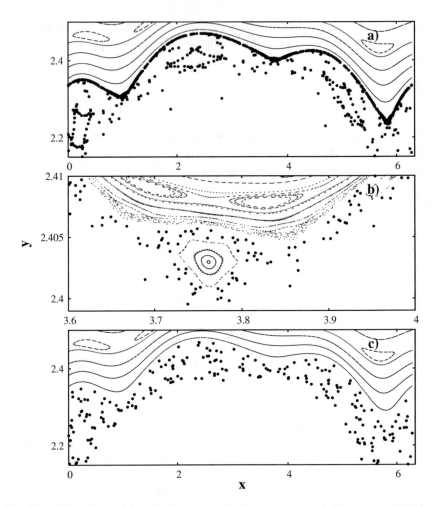

Fig. 2.22 (**a**) Poincaré section of the northern stochastic layer where stickiness to the very border with the regular westward current and to the three large ballistic islands are shown. Increased density of points along the border with the peripheral current is caused by traps of the border ballistic islands one of which is shown in (**b**). (**c**) The trap of the large ballistic islands

to the very boundaries of the large ballistic islands (see Fig. 2.22a) moving with the mean velocity $\langle u_f \rangle \simeq -0.0845$. The flat left wing of the distribution $N(\langle u_f \rangle)$ corresponds to the traps of smaller-size ballistic islands nearby the border with the peripheral current. There are different families of these islands (see one of them in Fig. 2.22b) with their own values of the mean zonal velocity which are in the range $-0.092 \lesssim \langle u_f \rangle \lesssim -0.0845$. Stickiness to the boundaries of the border islands is weaker because they are smaller than the large islands, and their contribution to the statistics of long flights is comparatively small.

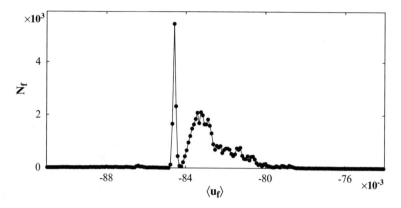

Fig. 2.23 The distribution of a number of long westward flights with $T_f \geq 10^3$ over their mean zonal velocities $\langle u_f \rangle$. The *sharp peak* corresponds to the trap connected with the very boundaries of the large ballistic islands, the *left wing*—to a number of traps of families of the border ballistic islands, and the *right wing*—to the trap situated around the large ballistic islands. Statistics for five tracers with the total number of long westward flights $N_f = 5 \cdot 10^4$ and the computation time $t = 5 \cdot 10^8$ for each tracer

The right wing of the distribution $N(\langle u_f \rangle)$ with $-0.084 \lesssim \langle u_f \rangle \lesssim -0.075$ deserves further investigation. The value $\langle u_f \rangle \simeq -0.075$ is a minimal value of the zonal velocity for long westward flights possible in the northern stochastic layer. Increasing the minimal duration of a flight from $T_f = 10^3$ to $T_f = (2\text{–}5) \cdot 10^3$, we have found splitting of the broad distribution with $-0.084 \lesssim \langle u_f \rangle \lesssim -0.08$ into a number of small distinct peaks. Comparing trajectories with the values of $\langle u_f \rangle$ corresponding to these peaks, we have found that all they move around the large ballistic islands. The particles with smaller values of $\langle u_f \rangle$ used to penetrate further to the south from the islands more frequently than those with larger values of $\langle u_f \rangle$ which prefer to spend more time in the northern part of the dynamical trap connected with those islands. Thus, we attribute the right wing of the distribution $N(\langle u_f \rangle)$ to an effect of the trap situated around the large ballistic islands.

To estimate the contribution of different ballistic island's traps to statistics of long westward flights in Fig. 2.13 we have computed the PDFs $P(x_f)$ and $P(T_f)$ for particles performing westward flights with $x_f \geq 100$ and $T_f \geq 1000$ and with the mean zonal velocity $\langle u_f \rangle$ to be chosen in three different ranges shown in Fig. 2.23: $-0.092 \lesssim \langle u_f \rangle \lesssim -0.085$ (particles sticking to the border islands), $-0.085 < \langle u_f \rangle \lesssim -0.084$ (particles sticking to the very boundary of the three large islands) and $-0.084 < \langle u_f \rangle \lesssim -0.075$ (the trap of the three large islands). All the PDFs $P(x_f)$ decay exponentially but with different values of the exponents equal to $\alpha \simeq -0.005$ and $\alpha \simeq -(0.0014\text{–}0.0018)$ for the traps of border and the large ballistic islands, respectively. The tail of the PDF $P(x_f)$ for westward flights, shown in Fig. 2.13, decays exponentially with $\alpha \simeq -0.0014$. Thus, the contribution of the large ballistic island's trap to statistics of long westward flights is dominant. As to the temporal PDFs $P(T_f)$ for westward long flights, they are neither exponential nor power-law like with strong oscillations at the very tails. The slope for the border ballistic island's traps is again smaller than for the large ballistic island's trap.

We described in this section statistical properties of chaotic mixing and transport of passive particles in the simple kinematic model of a meandering jet flow in terms of dynamical traps in the phase (physical) space. The boundaries of rotational islands (including vortex cores) in circulation zones are dynamical traps (rotational island's traps) contributing, mainly, to statistics of short flights with $|x_f| < 2\pi$. Characteristic times and spatial scales of the rotational island's traps have been shown to correlate with the PDFs for lengths x_f and durations T_f of short flights. The stable manifolds of periodic saddle trajectories play a role of saddle traps with specific values of the lengths and durations of short flights of the particles sticking to the saddle trajectories. The boundaries of ballistic islands in the stochastic layers (including those situated along the border with the peripheral current) are dynamical traps (ballistic island's traps) contributing, mainly, to statistics of very long flights with $|x_f| \gg 2\pi$. Dynamical traps are robust structures in the phase space of dynamical systems in the sense that they present at practically all values of the corresponding control parameters.

2.2.3.5 Fractal Geometry of Mixing

Poincaré sections help to visualize structure of the phase space but not geometry of mixing. In this section we consider the evolution of a material line consisting of a large number of particles distributed initially on the straight line that transverses the stochastic layer at $x = 0$. A typical stochastic layer consists of an infinite number of unstable periodic and chaotic orbits with embedded islands of regular motion. All the unstable invariant sets possess stable and unstable manifolds. When time progresses, particle's trajectories nearby a stable manifold of an invariant set tend to approach the set whereas the trajectories close to an unstable manifold go away from the set. Because of such a very complicated heteroclinic structure, we expect a diversity of particle's trajectories. Some of them are trapped forever in the first eastern frame $0 \leqslant x \leqslant 2\pi$ rotating around the elliptic point. Other ones quit the frame through the lines $x = 0$ or $x = 2\pi$, and then either are trapped there or move to the neighbor frames or return to the first one and so on to infinity.

To get a more deep insight into the geometry of chaotic mixing, we follow the methodology elaborated in Sect. 1.2.2 and compute the time T, particles spend in the neighbor circulation zones $-2\pi \leqslant x \leqslant 2\pi$ before reaching the critical lines $x = 0$, $x = \pm 2\pi$, and the number of times $n/2$ they wind crossing the lines $x = \pm \pi$. The functions $n(y_0)$ and $T(y_0)$ are shown in the upper panel in Fig. 2.24. The upper parts of each function (with $n > 0$ and $T > 0$) represent results for the particles with initial positive zonal velocities which they have simply due to their locations on the material line at $x = 0$. These particles enter the eastern frame and may change the direction of motion many times before leaving the frame through the lines $x = 0$ or $x = 2\pi$. We fix those events for all the particles with $1.9 \leqslant y_0 \leqslant 2.045$. The lower "negative" parts of the functions $n(y_0)$ and $T(y_0)$

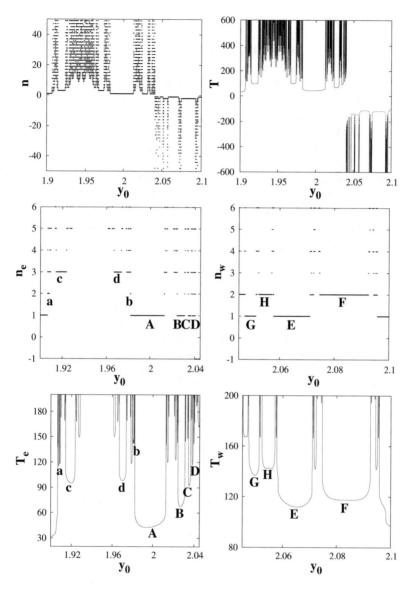

Fig. 2.24 Fractal set of initial positions y_0 of particles that reach the lines $x = 0, \pm 2\pi$ after $n/2$ turns around the elliptic point. The period of time, T, requires to reach the lines $x = 0, \pm 2\pi$. The indices "e" and "w" mean particles moving in the eastward and westward directions, respectively

represent results for the particles with initial negative zonal velocities ($y_0 \geqslant 2.045$) which move initially to the first western frame ($-2\pi \leqslant x \leqslant 0$). In fact, $T_e(y_0)$ and $T_w(y_0)$ are the time moments when a particle with the initial position y_0 quits the eastern or western frame, respectively. Both the functions consist of a number of

smooth U-like segments intermittent with poorly resolved ones. Border points of
each U-like segment separate particles belonging to stable and unstable manifolds
of the heteroclinic structure. The corresponding initial y_0-positions are a set of zero
measure of particles to be trapped forever in the respective frame. A fractal-like
structure of chaotic advection in both the frames is shown in the upper panel in
Fig. 2.24, and its fragments for the first levels are shown in the middle panel for the
eastern and the western fragments separately. Particles with even values of n quit
one of the frames through the border $x = 0$, those with odd n—through the border
$x = 2\pi$ for the eastern frame and $x = -2\pi$ for the western one.

Let us consider in detail the fractal-like structure in the eastern frame keeping in
mind that the results are similar with any other frame. The $n_e(y_0)$-dependence is a
complicated hierarchy of segments of the material line. Following to the preceding
chapter, we call as an "epistrophe" a sequence of segments at the $(n + 1)$-level,
converging to the ends of a segment of a sequence at the nth level, whose lengths
decrease in accordance with a law. At $n_e = 1$, we see in Fig. 2.24 an epistrophe with
segment's length A, B, C, D, and so on decreasing as $l_m = l_0 q^m$ with $q \approx 0.46$.
Letters "a" and "b" in Fig. 2.24 denote the first segments of the epistrophes at the
level $n_e = 2$, whereas "d" and "c"—the first segments of the epistrophes at the level
$n_e = 3$. The respective laws for all those epistrophes are not exponential.

In Fig. 2.25 we demonstrate fragments of evolution of the material line in the
first eastern frame at the moments indicated in the figure. Letters on the line mark
the corresponding segments of the $n_e(y_0)$ and $T_e(y_0)$ functions in Fig. 2.24. As an
example, let us explain formation of the epistrophe ABCD at the level $n_e = 1$. With
the period of perturbation $T_0 = 2\pi/\omega \simeq 8\pi$, a portion from the north end of the
material line leaves the frame through its eastern border. Look at the segments A
and B at $t = 15\pi$ and $t = 23\pi$. They quit the first frame as a fold for the period
$T_0 \simeq 8\pi$. The other segments—C and D (not shown in Fig. 2.25) do the same

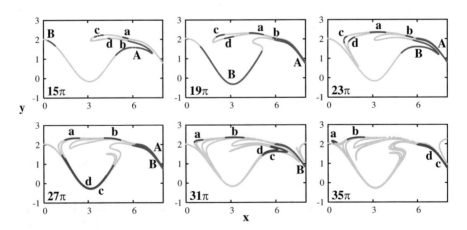

Fig. 2.25 Fragments of the evolution of a material line in the first eastern frame. The fragments
of the fractal in Fig. 2.24 with $n_e = 1, 2, 3$ are marked by the respective letters

job. The epistrophe's segments at the odd levels ($n = 2k - 1 > 1$) quit the frame with the period of perturbation T_0 one by one being folded (c and d segments). The folds of the segments at the $(2k - 1)$-level are exterior with respects to the folds of the segments at the $(2k + 1)$-level. The following empirical law has been found: $T_{2k-1} - T_{2k+1} \simeq 2T_0$, where T_{2k-1} is a time when the first segments of the epistrophes at the level $(2k - 1)$ (A segment with $n_e = 1$) reach the line $x = 2\pi$, and T_{2k+1} is the respective time for the first segments of the epistrophes at the level $2k + 1$ (c and d segments with $n_e = 3$).

Segments of the epistrophes at the even levels ($n = 2k$) leave the frame with the period T_0 as well but through the border $x = 0$ moving to the west. We show the evolution of some of them at the moments $t = 31\pi$ and $t = 35\pi$ in Fig. 2.25. Thus, the material line evolves by stretching and folding, and folds quit the frame in both directions with the period of perturbation.

2.2.4   Chaotic Cross-Jet Transport and Detection of Transport Barriers

Whether or not a zonal jet current provides an effective barrier to meridional or cross-jet transport (CJT), under which conditions the barrier becomes permeable and to which extent, these are crucial problems in physical oceanography and physics of the atmosphere. They must be treated from different points of view. In the straightforward numerical approach based on full-physics nonlinear models, the velocity field is generated as an outcome of a basin circulation model, and a flux across the jet (if any) can be estimated integrating a large number of tracers. The kinematic and linear dynamically consistent models are less realistic, but they allows to identify and analyze different factors which could enhance or suppress that CJT.

As to meandering currents, both the approaches have been applied to study CJT. A simple kinematic model with the basic stream function in the form (2.2) has been shown to reproduce some features of the large-scale Lagrangian dynamics of the Gulf Stream water masses [3]. The phase portrait of advection equations with the meandering Bickley jet (2.1) is plotted in Fig. 2.1 in the frame of reference moving with the meander phase velocity. Time dependence of the meander's amplitude or adding a secondary meander, superimposed on the basic flow, may break the boundaries between distinct regions in Fig. 2.1 producing chaotic mixing and transport between them [14, 26, 31, 34, 42, 43]. Numerical calculations, based on computing the Melnikov function [25], have shown that transport across the jet was much weaker than that between the jet (J), circulation cells (C), and peripheral currents (P) in Fig. 2.1, i.e., a perturbation mixes the water along each side of the jet more efficiently than across the jet core [34].

An attempt to analytically predict the parameter's values for destruction of the transport barrier was made in [26] using the heuristic Chirikov criterion for overlapping resonances [11]. A technique, based on computing the finite-scale Lyapunov exponent, as a function of initial position of tracers, has been found useful in [2] to detect the presence of cross-jet barriers in the kinematic model (2.2). An analysis of CJT, based on lobe dynamics, has been applied in [32] to describe how particles can cross the jet from the north to south and vice versa.

Independently on the work on CJT in the geophysical community, there have been a number of theoretical and numerical investigations of chaotic transport in the so-called area-preserving nontwist maps [8–10, 15, 27, 36, 37, 44, 45]. We mention specially the early study of different reconnection scenario [15, 44] and the first systematic study of CJT in nontwist maps [8–10]. These maps locally violate the twist condition, a map analogue of the non-degeneracy condition for Hamiltonian systems. Nontwist maps are of interest because many important mathematical results, including **KAM** and Aubry—Mather theory, depend on the twist condition. Apart from their mathematical importance, nontwist maps are of a physical interest because they are able to model transition to global chaos. Nontwist maps allow to study different scenarios for this transition: reconnection of separatrices, meandering and breakup of invariant tori, and others.

In this section we develop a method for detecting barriers to CJT (or global chaos in a more general context), apply it to the kinematic model of a meandering jet flow, study changes in its topology under varying the perturbation parameters and scenarios of its destruction.

2.2.4.1   The Amplitude: Frequency Diagram for Cross-Jet Transport

The normalized stream function in the frame of reference moving with the phase velocity of the meander has the form (2.2) with the phase portrait shown in Fig. 2.1. In Fig. 2.26 we plot the frequency map $f(x, y)$ that shows by nuances of the grey color values of the frequency f of particles with initial positions (x, y) in the unperturbed system. Particles moving in the central jet have the maximal value of that frequency, $f_{max} = 1.278$.

Under the perturbation (2.4), the separatrices, connecting saddle points, are destroyed and transformed into stochastic layers. The strength of chaos depends strongly on both the perturbation parameters, the perturbation amplitude ε and

Fig. 2.26 Frequency map represents by color values of the frequency f of particles with initial positions (x_0, y_0) advected by the unperturbed flow

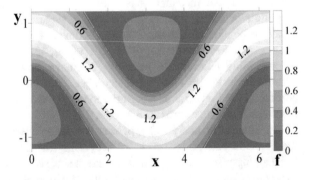

frequency ω. The normalized control parameters are defined just after Eq. (2.2). All these parameters change in a wide range in strong oceanic currents like the Gulf Stream and the Kuroshio [3, 26, 34]: $\lambda \simeq 40$–100 km, $a \simeq 50$–60 km, $2\pi/k \simeq 200$–400 km, $c \simeq 0.1$–0.5 m/s, and the maximal zonal velocity in the jet $u_{max} = \Psi'_0/\lambda \simeq 1$–$1.5$ m/s. So, we get $L \simeq 0.1$–3, $A \simeq 0.7$–2, and $C \simeq 0.02$–0.3. We take the following normalized values of the control parameters that will be used in all the numerical experiments in this section: $A_0 = 0.785$, $C = 0.1168$, and $L = 0.628$.

The equations of motion (2.3) with the stream function (2.2) and the perturbation (2.4) have the symmetry

$$\hat{S}: \begin{cases} x' = \pi + x, \\ y' = -y \end{cases} \tag{2.7}$$

and the time reversal symmetry

$$\hat{I}_0: \begin{cases} x' = -x, \\ y' = y. \end{cases} \tag{2.8}$$

The symmetries (2.7) and (2.8) are involutions, i.e., $\hat{S}^2 = 1$ and $\hat{I}_0^2 = 1$. We recall that the part of the phase space with $2\pi n \leqslant x \leqslant 2\pi(n+1)$, $n = 0, \pm 1, \ldots$, is called a "frame."

The following numerical procedure has been applied to establish the fact of CJT. The advection equations (2.3) for given values of the perturbation amplitude ε and frequency ω and with 20 particles, released nearby the northern saddle point, have been integrated up to the time instant when one of the particles was detected to cross the straight line $y = y_s$ passing through the southern saddle. If after the time $T_{max} = 1000 \cdot 2\pi/\omega$ none of the particles has been able to cross the line $y = y_s$, we state that CJT was absent for given values of the parameters. The ε–ω diagram in Fig. 2.27 shows the values of the parameters for which chaotic CJT exists (white zones). There are a range of frequency values for which transport occurs at surprisingly small values of the perturbation amplitude ε. The absolute minimal value of the perturbation amplitude, at which the chaotic CJT occurs, $\varepsilon_{min} = 0.0218 \approx A_0/36$, corresponds to the frequency $\omega = 1.165$ which is close to the natural frequencies of particles moving in the central jet.

2.2.5 Detecting the Central Invariant Curve

The amplitude–frequency diagram is useful to detect CJT, but its computation is a time-consuming procedure. Moreover, it says nothing about the properties of barrier to transport and mechanism of its destruction. A fractal-like boundary between the colors in Fig. 2.27 reflects an intermittency in appearance and destruction of the

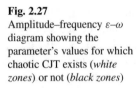

Fig. 2.27
Amplitude–frequency ε–ω
diagram showing the
parameter's values for which
chaotic CJT exists (*white
zones*) or not (*black zones*)

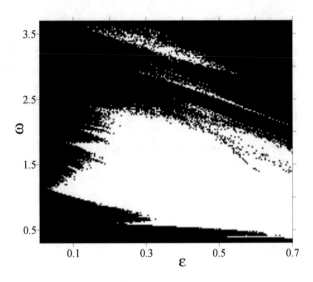

cross-jet barrier when varying ε and ω. Further insight into topology of the barrier
could be obtained if one would be able to find an indicator of CJT, i.e., an object
in the phase space whose form contains an information about permeability of the
barrier.

First of all, we need to give definitions of some basic structures specifying
a cross-jet barrier and its destruction. The central transport barrier is defined
as a strip between the southern and northern unperturbed separatrices confined
by marginal northern and southern ballistic trajectories (excluding trajectories of
ballistic resonances in the stochastic layer). All the trajectories inside the central
transport barrier are ballistic, i.e., the corresponding motion is unbounded. Some of
them are regular, and the other ones are chaotic. The amplitude–frequency diagram
in Fig. 2.27 demonstrates clearly destruction of the central transport barrier at some
values of the perturbation parameters, ε and ω and onset of CJT.

Our Hamiltonian flow with the stream function (2.2) is degenerate, i.e., it violates
the non-degeneracy condition, $\partial f/\partial \Psi \neq 0$, for some values of the natural frequency
of passive particles f and their stream function Ψ in the unperturbed system.
Physically it means that the zonal velocity profile $u_0(y)$ has a maximum. The curve,
for which the twist condition (analogue of the non-degeneracy condition) is violated,
is called a nonmonotonic curve in theory of nontwist maps [45]. In our model
flow (2.2) it is some value of unperturbed stream function along which the frequency
f is maximal (see Fig. 2.26).

Instead of integrating advection equations, we integrate the corresponding
Poincaré map, an orbit of which is defined as a set of points $\{(x_i, y_i)\}_{i=-\infty}^{\infty}$ on the
phase plane such that $\hat{G}_{T_0}(x_i, y_i) = (x_{i+1}, y_{i+1})$, where $\hat{G}_t$ is an evolution operator
over a time interval t, and $T_0 \equiv 2\pi/\omega$ is the period of perturbation. Operator $\hat{G}_{T_0}$
can be factorized as a product of two involutions $\hat{G}_{T_0} = \hat{I}_1 \hat{I}_0$, where $\hat{I}_1 = \hat{G}_{T_0} \hat{I}_0$ is
also a time reversal symmetry [8].

The periodic orbit of period nT_0 ($n = 1, 2, \ldots$) is an orbit such that $(x_{i+n}, y_{i+n}) = (x_i + 2\pi m, y_i)$, $\forall i$, where m is an integer. The invariant curve is a curve invariant under the map. The nonmonotonic curve is not an invariant curve under a perturbation. The winding (or rotation) number w of an orbit is defined as the limit $w = \lim_{i\to\infty}[(x_i - x_0)/(2\pi i)]$, when it exists. The winding number is a ratio between the frequency of perturbation ω and the natural frequency f. Periodic orbits have rational winding numbers $w = m/n$. It simply means that a ballistic passive particle in the flow flies m frames before returning to its initial position x_0 (modulo 2π) after n periods of perturbation. Winding numbers of quasiperiodic orbits are irrational.

Now we are ready to introduce the important notion of a central invariant curve. We define the central invariant curve as a curve which is invariant under the operators $\hat{S}$ and $\hat{G}_{T_0}$. It can be shown that two curves invariant under $\hat{S}$ have at least two common points [36, 37]. The curves, which are invariant under $\hat{G}_{T_0}$, cannot intersect each other. So, the central invariant curve is a unique curve. Following to [36], one can show that the central invariant curve corresponds to a local extremum on the winding number profile with an irrational value of w. Such curves are called "shearless curves" in theory of nontwist maps [45]. The significance of a shearless curve is that it acts as barrier to global transport in the phase space of a nontwist map. The violation of the twist condition leads to existence of more than one orbit with the same winding number arising in pairs on both sides of the shearless curve. Those pairs of orbits can collide and annihilate at certain parameter values. The collision of the orbits involves in phenomenon, which was called as "reconnection" of invariant manifolds of the corresponding hyperbolic orbits [15].

The central invariant curve should not be thought as the last cross-jet barrier curve in the central transport barrier in the sense that it breaks down under increasing the perturbation amplitude in the last turn. Sometimes it is the case, but sometimes it is not. Nevertheless, the central invariant curve serves a good indicator of the strength of the central transport barrier and its topology.

The central invariant curve can be constructed by successive iterations of the so-called indicator points [37]. In our model flow (2.2) with the symmetries (2.7) and (2.8), indicator points are the points $(x_j^{(k)}, y_j^{(k)})$, $k = 1, 2$, which are solutions of the equations

$$\hat{I}_0(x_j^{(1)}, y_j^{(1)}) = \hat{S}(x_j^{(1)}, y_j^{(1)}), \tag{2.9}$$

or

$$\hat{I}_1(x_j^{(2)}, y_j^{(2)}) = \hat{S}(x_j^{(2)}, y_j^{(2)}), \tag{2.10}$$

where index j numerates the points. Equation (2.9) gives a pair of indicator points: $(x_1^{(1)} = \pi/2, y_1^{(1)} = 0)$ and $(x_2^{(1)} = 3\pi/2, y_2^{(1)} = 0)$. Instead of solving Eq. (2.10), we solve the equivalent equation

$$\hat{G}_{T_0}(x, y) = \hat{I}_0\hat{S}(x, y) \equiv (\pi - x, -y). \tag{2.11}$$

If some (x, y) is a solution of (2.11), then $\hat{I}_0(x, y)$ is a solution of (2.10). Equation (2.11) cannot be solved analytically, so we apply the numerical method based on computing a minimum of the function $r(x, y) = ||\hat{G}_{T_0}(x, y) - (\pi - x, -y)||$, where $|| \cdot ||$ is a norm on the cylinder. Since $r(x, y) \geq 0$ for any (x, y), the points with $r(x, y) = 0$ are minima of the function $r(x, y)$. Thus, solution of Eq. (2.11) reduces to searching for a local minimum of the function $r(x, y)$ with the additional condition $r(x, y) = 0$ at the point of the minimum. There are a number of numerical methods for doing that job. We prefer to use the downhill simplex method. In our problem the function $r(x, y)$ always has two minima, i.e., two indicator points transforming to each other under action of the operator $\hat{S}$.

Next, we study iterations, i.e., Poincaré mapping of one of the indicator points (x_0, y_0). If the iterations $(x_i, y_i) = \hat{G}_{T_0}^i(x_0, y_0)$ are confined between invariant curves in a bounded region, the following three cases are possible in dependence on the dimension d of the set (x_i, y_i):

1. The iterations lie on a curve on the phase plane with $d = 1$, which is a central invariant curve.
2. The iterations are an organized set of points with $d = 0$. It means that they constitute either a central periodic orbit or a central almost periodic orbit, that is simply an orbit that could not form a smooth curve on the phase plane for a limited integration time.
3. The iterations form a central stochastic layer with $d = 2$.

If the iterations are not confined by any invariant curves in a bounded region, i.e., they occupy all the accessible phase plane to the south and north from the central jet, then there exists global chaotic transport. Thus, the type of motion of indicator points is an indicator of global chaos and the absence of barriers to CJT.

Possible topologies of the corresponding central transport barrier at the fixed frequency $\omega = 1.2$ and with increasing values of the perturbation amplitude ε are illustrated in Fig. 2.28 plotting iterations of the indicator points computed by the above-mentioned method. The panel (a) illustrates the case when those iterations form a central invariant curve. Another typical situation is shown in Fig. 2.28b, where the iterations fall in small segments filling at $t \to \infty$ a continuous curve which is a central almost periodic orbit. If the iterations fill up not a curve but a bounded region between invariant curves, then there appears a central stochastic layer preventing CJT (Fig. 2.28c). If iterations of the indicator points occupy a region that is not confined by any invariant curves, it means destruction of the central transport barrier and onset of global chaos, i.e., chaos in a large region of the phase space accompanied by CJT.

The indicator points have been found and their iterations have been computed in the following range of the control parameters: $\omega \in [0.95 : 1.5]$ and $\varepsilon \in [0.01 : 1]$. We assume that the iterations are bounded, if their coordinates do not cross the unperturbed separatrices after $5 \cdot 10^4$ iterations. The dimension d_k of the set of those iterations is computed by the box-counting method, where the value of d_k for the box size $e_k = (1/2)^k$ is defined as

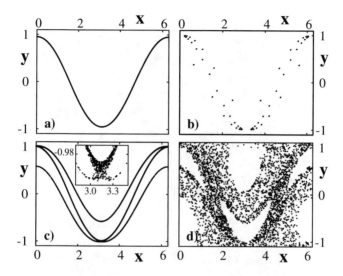

Fig. 2.28 Poincaré mapping of indicator points. In the first three panels the orbit of these points is bounded and there exists a central transport barrier with (**a**) a central invariant curve ($\varepsilon = 0.01$, $\omega = 1.2$), (**b**) a central almost periodic orbit ($\varepsilon = 0.011997277$, $\omega = 1.2$), and (**c**) a central stochastic layer ($\varepsilon = 0.01177$, $\omega = 1.2$) with the *inset* demonstrating a magnification of the small region. (**d**) Destruction of central transport barrier and onset of global chaotic transport as a result of unbounded iterations of indicator points ($\varepsilon = 0.041$, $\omega = 1.2$)

$$d_k = \log_2 \frac{\mathcal{N}_{k+1}}{\mathcal{N}_k}, \tag{2.12}$$

where $\mathcal{N}_k$ is a number of boxes of the size e_k containing set points. The dimension d_k goes to zero with decreasing e_k, and one cannot distinguish in this limit between the central almost periodic orbit and the central stochastic layer at large k. Comparing the values of d_k at different values of k, we have found the empirical value $k = 4$ which is enough to make the difference.

The results of computation of the dimension $d_4(\varepsilon, \omega)$ for a set of iterations of the indicator points are shown in the "bird-wing" diagram in Fig. 2.29. That one and the other "bird-wing" diagrams in the parameter space show the properties of the central transport barrier and the central invariant curve in the range of comparatively small values of the perturbation amplitude ($0.01 \leqslant \varepsilon \leqslant 0.1$) and the frequency ($1.15 \leqslant \omega \leqslant 1.5$) corresponding to particles moving in the central jet (Fig. 2.26). White color corresponds to the regime of global chaotic transport with unbounded motion of iterations of the indicator points. Otherwise, the central transport barrier exists but its topology is different. Grey color means that there exists a central invariant curve with $0.95 \leqslant d_4 \leqslant 1.05$ in the corresponding range of the parameters. White rectangles, which are hardly visible in the main panel (see their magnification on the inset of the figure), means existence of a central almost periodic orbit with $d_4 < 0.95$ and black strips—a central stochastic layer with $d_4 > 1.05$.

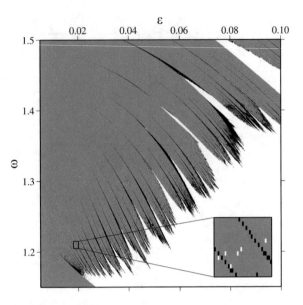

Fig. 2.29 "Bird-wing" diagram of the box-counting dimension $d_4(\varepsilon, \omega)$. *White color* means the regime with global chaotic transport with unbounded motion of indicator points (see Fig. 2.28d). If the motion of indicator points is bounded, then there exists a central transport barrier but its topology may differ. *Grey color* ($0.95 \leqslant d_4 \leqslant 1.05$): regime with a central invariant curve (see Fig. 2.28a). *Small white regions* which are hardly visible inside the *grey* "wing" ($d_4 < 0.95$): regime with a central almost periodic orbit (see Fig. 2.28b). *Black color* ($d_4 > 1.05$): regime with a central stochastic layer (see Fig. 2.28c). *Inset* shows magnification of a small region in the parameter space with visible *white and black regions*

2.2.5.1   Geometry of the Central Invariant Curve and Its Bifurcations

To quantify complexity of the central invariant curve, we define its length l as a sum of the distances between iterations of the indicator points (x_i, y_i) ordered on the phase plane in the following way:

1. The first step. A point B_0, belonging to a set of iterations of the Poincaré map (x_i, y_i), is marked.
2. The $(j + 1)$th step. We find and mark among all the unmarked points that one, B_{j+1}, which minimizes the Euclidean distance $D_j = D(B_j, B_{j+1})$ between B_j and B_{j+1}.
3. The procedure is repeated unless all the points will be marked.

As an output, we have an ordered set of points B_j constituting a central invariant curve. The accuracy is controlled by the quantity $\max D_j$. Large values of this quantity mean that the points are ordered in a wrong way, or a set of points is chaotic. To increase the number of points we use in addition to the original points (x_i, y_i) their "images" $(x_i + \pi, -y_i)$ as well. To minimize the computation time, the points are sorted in accordance with their x coordinates.

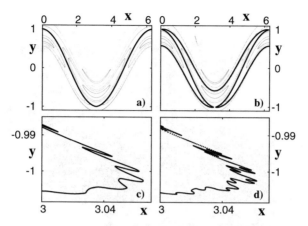

Fig. 2.30 Metamorphosis of the central invariant curve. (**a**) Nonmeandering central invariant curve ($\omega = 1.2$, $\varepsilon = 0.01174929$). (**b**) Meandering central invariant curve of the first order and period T_0 ($\omega = 1.2$, $\varepsilon = 0.01178721$). (**c**) Meandering central invariant curve of the second order and period $79T_0$ ($\omega = 1.2$, $\varepsilon = 0.01179027$). (**d**) Meandering central invariant curve of a higher order ($\omega = 1.2$, $\varepsilon = 0.01179339$)

Figure 2.30 illustrates metamorphosis of the central invariant curve as the perturbation amplitude increases. We start with the central invariant curve, shown in Fig. 2.30a, which we call a "nonmeandering central invariant curve." At the critical value $\varepsilon \approx 0.011758$, invariant manifolds of hyperbolic orbits of two chains of the 1 : 1 resonance islands on both sides of the central invariant curve connect, and after that the central invariant curve becomes a meandering curve of the first order (Fig. 2.30b) and period T_0. The period of central invariant curve's meandering is simply a period of nearby main islands [36, 38]. At the next critical value $\varepsilon = 0.01178721$, reconnection of invariant manifolds of secondary-resonance islands takes place. The corresponding second-order meandering central invariant curve with the period $79T_0$ is shown in Fig. 2.30c. Strongly meandering central invariant curves of higher orders appear with further increasing the perturbation amplitude (Fig. 2.30d).

Some smooth invariant curves inside the central transport barrier break down under perturbation (2.4), and chains of ballistic resonance islands appear at their place. Those islands appear in pairs to the north and south from a central invariant curve due to the flow symmetries (2.7) and (2.8) (see Fig. 2.30a, b). The geometry of the central invariant curve, size and number of the islands, and topology of their invariant manifolds change with variation in the perturbation amplitude ε and frequency ω in a very complicated way.

In Fig. 2.31 we plot in the parameter space the values of the length l of the central invariant curve coding it by nuances of the grey color. White color corresponds to those values of the parameters ε and ω for which CJT exists due to destruction of the central transport barrier. Black color codes the regime with a broken central invariant

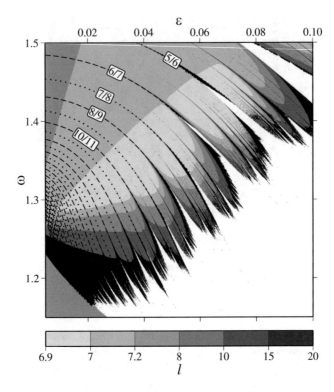

Fig. 2.31 "Bird-wing" diagram showing the length l of the central invariant curve by nuances of the *grey color* in the parameter space (ε, ω). *White zone*: regime with a broken central transport barrier and CJT. *Black color*: regime with a broken central invariant curve but a remaining central transport barrier preventing CJT. Resonant bifurcation curves, along which the central invariant curve winding numbers w are rational, end up in the dips of the "wing." The *dotted and dashed lines* correspond to even and odd resonances, respectively

curve but a remaining central transport barrier that prevents CJT ($d_4 > 1.05$). Dotted and dashed lines on the plot are the resonant bifurcation curves along which the central invariant curve winding number w is rational. The m/n resonant bifurcation curve is the set of values of the control parameters for which a reconnection of invariant manifolds of the $n : m$ resonances takes place. The dotted lines correspond to even resonances with $w = (2k - 1)/2k$, and the dashed lines are odd resonances with $w = 2k/(2k + 1)$, $k = 1, 2, \ldots$. All those curves end up in the dips of the "bird-wing" diagram.

In order to analyze a fractal-like boundary of the "bird-wing" diagram in Fig. 2.31, we cross it horizontally at the frequency $\omega = 1.2$ and consider the plot $l(\varepsilon)$ in the range of interest of ε (Fig. 2.32a). The perturbation frequency $\omega = 1.2$ is close to the maximal frequency $f_{\max}$ of particles in the middle of the jet in the unperturbed flow (see Fig. 2.26). The plot $l(\varepsilon)$ consists of a number of spikes with different heights and widths. In the range of small values of the perturbation

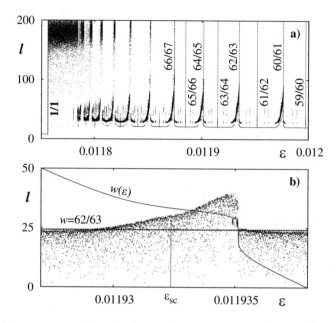

Fig. 2.32 Dependence of the length of the central invariant curve l on the perturbation amplitude at the fixed frequency $\omega = 1.2$. (**a**) General view of the dependence $l(\varepsilon)$ in the range of interest $\varepsilon \in [0.01175 : 0.012]$. The *vertical dotted lines* correspond to rational values of the winding number w of the central invariant curve. The resonance 1/1 appears at $\varepsilon = 0.11756$. The arrangement of the spikes is explained in the text. (**b**) Magnification of one of the wide spikes in panel (**a**). The *solid line* is a winding number profile $w(\varepsilon)$ with the value $w = 62/63$ shown by the *dashed line*

amplitude ($\varepsilon < 0.011756$), the length of the central invariant curve is approximately the same $l \approx 7.35$ (a small fragment of the function $l(\varepsilon)$ is shown in Fig. 2.32a just to the left from the vertical line 1/1). In that range, the central invariant curve is a nonmeandering curve (see Fig. 2.30a) surrounded by smooth invariant curves and 1 : 1 resonance islands with a heteroclinic topology. The size of those islands is comparable with the frame size. The width of the central transport barrier, filled by invariant curves around the central invariant curve, decreases with increasing the perturbation amplitude ε.

The winding number w of the central invariant curve changes under a variation of the perturbation amplitude. At $\varepsilon \simeq 0.011756$, invariant manifolds of the 1 : 1 resonance connect, and a central stochastic layer appears at the place of the central invariant curve. This layer exists up to $\varepsilon \simeq 0.011785$ (see a random set of points in Fig. 2.32a in that range of ε). At $\varepsilon > 0.011785$, the central invariant curve appears again. Now it is a meandering curve of the first order (see Fig. 2.30b) whose length is larger due to reconnection of the 1 : 1 resonance islands. As ε increases further, its length l changes in a wide range. Smooth fragments with approximately the same value of $l \approx 20$ alternate with spikes of different heights and widths. The spikes are condensed, when approaching to the value $w = 1/1$, and overlap in the range $0.011756 \lesssim \varepsilon \lesssim 0.011785$.

The arrangement of the spikes in Fig. 2.32a can be explained using a representation of rational numbers by continued fractions. A continued fraction is the expression

$$c = [a_0; a_1, a_2, a_3, \ldots] = a_0 + \cfrac{1}{a_1 + \cfrac{1}{a_2 + \cfrac{1}{a_3 + \cdots}}}, \qquad (2.13)$$

where a_0 is an integer number and the other a_n are natural numbers. Any rational (irrational) number can be represented by a continued fraction with a finite (infinite) number of elements. The spikes in Fig. 2.32a are arranged in convergent series in such a way that each spike in a series generates a series of spikes of the next order. For example, the series of the integer $n : 1$ resonance has the winding numbers equal to $1/n$ or $[0; n]$ in the continued-fraction representation. Each spike in that series generates a series of resonance spikes of the next order converging to the parent spike. Winding numbers of those resonances are $[0; n, i]$, $i = 2, 3, 4 \ldots$ (at $i = 1$, one gets a spike in the main series because of the identity $[a_0; a_1, \ldots, a_n, 1] \equiv [a_0; a_1, \ldots, a_n + 1]$). The spikes with $[0; 1, i] = i/(i+1)$ converge to the spike of the $1 : 1$ resonance. That is clearly seen in Fig. 2.32a. The direction of convergence of the spikes in a series alternates with the series order: the winding number increases in the series of the first order, decreases in the series of the second order, and increases again in the series of the third order. That is why a chaotic region in Fig. 2.32a is situated to the right from the $1 : 1$ resonance, i.e., in the range of smaller values of w, whereas it is to the left for the series of the second order, i.e., in the range of larger values of w. There also exists an additional hierarchical structure with fractional $1 : n$ resonances, whose frequencies are below the $1 : 1$ resonance frequency and a series with resonances corresponding to the spikes below $[0; 1, i, (1)]$, for example, a clearly visible series of spikes below $[0; 1, 4, (1)]$ converging to the spike $5/6$ in the l-diagram (see Fig. 2.31).

Unfortunately, we could not identify series of the third and higher order because numerical errors in identifying the winding numbers are greater than the distance between the spikes of higher-order series. Unresolved regions on the plot $l(\varepsilon)$ in Fig. 2.32a appear because of a decrease of the distance between the spikes in the same series with increasing series number, a process resembling Chirikov's overlapping of resonances [11].

A magnification of one of the wide spikes is shown in Fig. 2.32b. We plot the winding number profile $w(\varepsilon)$ together with the function $l(\varepsilon)$ for the spike. To illustrate what happens with the central invariant curve and its surrounding with increasing ε, we plot the corresponding Poincaré section surface in Fig. 2.33. In the range $0.0119 \lesssim \varepsilon \lesssim 0.01192$ the lengths of the central invariant curve and surrounding invariant curves increase slowly due to small changes in their geometry (Fig. 2.33a). After a saddle-center bifurcation at $\varepsilon_{sc} \simeq 0.011934$, there appear two chains of homoclinically connected $63 : 62$ islands separated by a meandering central invariant curve (Fig. 2.33b). The amplitude of the meanders of the central

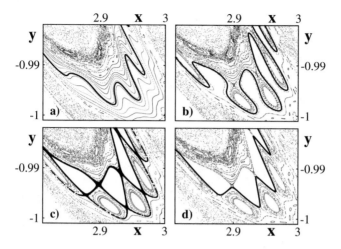

Fig. 2.33 Poincaré sections surfaces at $\omega = 1.2$ and increasing values of ε in the range corresponding to the wide spike with $w = 62/63$ in Fig. 2.32b. (**a**) Meandering central invariant curve surrounded by meandering invariant curves ($\varepsilon = 0.011931$). (**b**) Central invariant curve meandering between the odd 63 : 62 islands born as a result of a saddle-center bifurcation ($\varepsilon_{sc} = 0.011934$). (**c**) Destruction of the central invariant curve due to connection of invariant manifolds of the 63 : 62 islands ($\varepsilon = 0.01193511$). A narrow stochastic layer appears at the place of the central invariant curve. (**d**) The central invariant curve appears again ($\varepsilon = 0.0119352$)

invariant curve increases with further increasing ε in the range $0.011934 \lesssim \varepsilon \lesssim 0.011935$. In that range the central invariant curve disappears and appears again in a random-like manner (see the corresponding fragment on the plot $l(\varepsilon)$ in Fig. 2.32b) due to overlapping of higher-order resonances and reconnection of their invariant manifolds. The example of such a reconnection for the 63 : 62 resonance at $\varepsilon = 0.01193511$ is shown in Fig. 2.33c, where a stochastic layer appears at the place of the central invariant curve. As ε increases further, the central invariant curve appears again but with a smaller number of meanders (Fig. 2.33d).

The other wide spikes on the plot $l(\varepsilon)$ with a similar structure are caused by another odd resonances between the external perturbation and particle's motion along the central invariant curve. Under a resonance of the central invariant curve with the winding number $w = m/n$, we mean reconnection of invariant manifolds of the resonance $n : m$ and onset of a local stochastic layer. The narrow spikes, situated between the wide ones in Fig. 2.32a, correspond to reconnection of even resonances. They are hardly resolved on the plot. Even resonances of higher orders have a smaller effect on central invariant curve geometry then odd resonances. As an example, we illustrate in Fig. 2.34 metamorphosis of the central invariant curve with the winding number $w = 1/2$. The perturbation amplitude is fixed at a rather small value $\varepsilon = 0.015$ and the frequency increases in the range $2.51 < \omega < 2.556 = 2f_{max}$. At $\omega = 2.51$, there are islands of the even 2 : 1 resonance separated by a central invariant curve (Fig. 2.34a). At some critical value

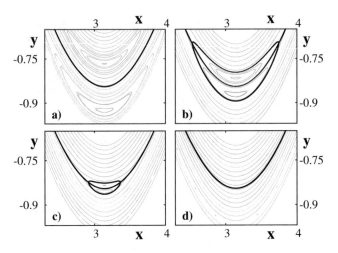

Fig. 2.34 Poincaré section surfaces at $\varepsilon = 0.015$ and increasing values of ω. (**a**) Central invariant curve between islands of the even 2 : 1 resonance ($\omega = 2.51 < 2f_{max} = 2.556$). (**b**) Reconnection of invariant manifolds of that resonance and formation of the vortex pair with a narrow stochastic layer shown by the *bold curves* ($\omega = 2.55$). (**c**) The vortex size decreases with increasing ω ($\omega = 2.555$). (**d**) At some critical value of ω, the vortex pair disappears, and the central invariant curve appears again ($\omega = 2.56$)

of ω, invariant manifolds of the 2 : 1 resonance connect and the islands form a tight vortex-pair structure surrounded by a narrow stochastic layer (see Fig. 2.34b at $\omega = 2.55$). The size of the pair decreases gradually with further increasing ω, and the corresponding hyperbolic orbits approach each other (see Fig. 2.34c at $\omega = 2.555$). At some critical value of ω, hyperbolic and elliptic orbits of the resonance collide and annihilate, and a central invariant curve appears again (see Fig. 2.34d at $\omega = 2.56$). Vortex pairs of the other even resonances are formed in a similar way. The higher is the order of the resonance, the smaller is the vortex size.

We conclude this section by computing winding numbers w of the central invariant curve in the parameter space. The result is shown in the "bird-wing" diagram in Fig. 2.35. The central invariant curve does not exist in the white region where the central transport barrier is broken and CJT takes place. The curves, which end up on the tips of the "feathers of the wing," have winding numbers w with the following continued-fraction representation: $[a_0; a_1, \ldots, a_n, (1)]$. These are the so-called noble numbers which are known to be the numbers that cannot be approximated by continued-fraction sequences to better accuracy than the so-called Diophantine condition (see, for example, [13]). The central invariant curves with noble winding numbers are in a sense the most structurally robust invariant curves, i.e., they may survive under a comparatively large perturbation preventing CJT. The noble curves are arranged in series like the resonant bifurcation curves with rational winding numbers which end up in the dips of the "wing" in the "bird-wing" diagram in Fig. 2.31. For example, the noble series $[0; 1, i, (1)]$ in Fig. 2.35

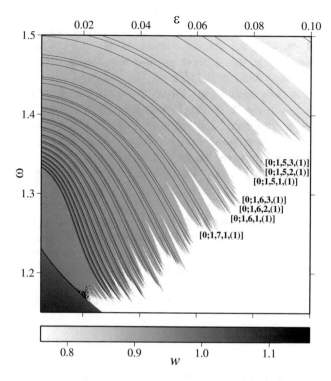

Fig. 2.35 "Bird-wing" diagram $w(\varepsilon, \omega)$ in the parameter space showing values of the winding number w of the central invariant curve by nuances of the *grey color*. *White zone*: regime with broken central transport barrier and CJT. The curves with irrational winding numbers end up on the tips of the "feathers of the wing" (some of them are marked by the corresponding noble numbers), whereas the curves with rational winding numbers (shown in Fig. 2.31) end up in the dips of the "wing"

corresponds to the resonance series $[0; 1, i]$ in Fig. 2.31. In the w diagram we show a few representatives of the noble series $[0; 1, i, (1)]$ and series of the next order $[0; 1, i, j, (1)]$ (see Fig. 2.35 with $j = 2, 3$).

2.2.5.2   Breakdown of Central Transport Barrier

We have studied in the preceding section properties of the central invariant curve which has been shown to be a diagnostic means to characterize a central transport barrier and its destruction. The central transport barrier separates water masses to the south and north from the central jet and prevents their mixing. It is not a homogeneous jet-like layer but consists of chains of ballistic islands, narrow stochastic layers, and meandering invariant curves of different orders and periods (including a central invariant curve) to be confined by invariant curves from the south and north. Those curves break down one after another when increasing the perturbation amplitude ε, producing stochastic layers at their place on both sides of the central jet, until the stochastic layers merge with one another and with stochastic layers around the southern and northern circulation cells producing a global stochastic layer and onset of CJT.

Upon moving along any resonant bifurcation curve with a rational value of the winding number w in the "bird-wing" diagram in Fig. 2.31, we have those values of the perturbation amplitude ε and frequency ω at which the corresponding central invariant curve is broken due to reconnection of invariant manifolds. It does not mean that the central transport barrier is broken as well. That is the case only if we are at the dips of the "wing." Destruction of the central transport barrier for this type of movement in the parameter space is illustrated in Fig. 2.36. We fix a point ($\varepsilon = 0.04889$, $\omega = 1.31625$) on the resonant bifurcation curve with $w = 8/9$ nearby its right edge in Fig. 2.31 and plot the corresponding Poincaré section surface. A narrow stochastic layer, confined between the invariant curves providing a transport barrier, appears on the Poincaré section surface in panel (a) at the place of a broken central invariant curve. The barrier will be broken if one would choose the values of parameters in the white zone in Fig. 2.31. Merging of southern and northern stochastic layers and onset of CJT are shown in panel (b) at $\varepsilon = 0.054$, $\omega = 1.285$.

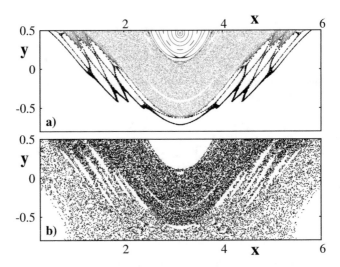

Fig. 2.36 Destruction of the central transport barrier upon moving in the parameter space along a resonant bifurcation curve with the rational winding number $w = 8/9$. (**a**) Narrow stochastic layer on the Poincaré section surface is confined between the invariant curves which provide a transport barrier. The perturbation parameters $\varepsilon = 0.04889$ and $\omega = 1.31625$ are chosen on the curve with $w = 8/9$ nearby its *right edge* (see Fig. 2.31). (**b**) Onset of CJT at the values $\varepsilon = 0.054$ and $\omega = 1.285$ chosen in the *white zone* of that dip

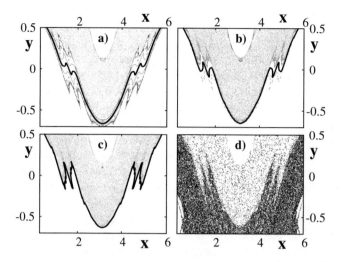

Fig. 2.37 Destruction of the central transport barrier upon moving in the parameter space along a curve with the noble value of the central invariant curve's winding number $w = [0; 1, 5, 1, (1)]$. When approaching a tip of the corresponding "feather of the wing" in Fig. 2.35, one observes on the Poincaré section surface a decrease in the width of the transport barrier with a central invariant curve (*bold curves*) inside. (**a**) $\varepsilon = 0.07017$, $\omega = 1.367875$. (**b**) $\varepsilon = 0.07416$, $\omega = 1.350375$. (**c**) $\varepsilon = 0.0796067$, $\omega = 1.325875$. (**d**) Onset of CJT at the values $\varepsilon = 0.08$, $\omega = 1.3142$ chosen beyond a tip of that "feather" in the white zone in Fig. 2.35

Upon moving along any curve with a noble value of the winding number w in the "bird-wing" diagram in Fig. 2.35, we have those values of the perturbation amplitude ε and frequency ω at which a central invariant curve with the corresponding noble number exists. Destruction of the central transport barrier for the motion in the parameter space along the noble curve with $w = [0; 1, 5, 1, (1)]$ is illustrated in Fig. 2.37. When moving to the tip of the corresponding "feather of the wing" in Fig. 2.35, one observes progressive destruction of invariant curves and decrease of the width of the transport barrier [panels (a) and (b)] unless a single central invariant curve remains as the last barrier to CJT [panel (c)]. Onset of CJT [panel (d)] happens at the values of parameters chosen beyond a tip of that "feather" in the white zone in Fig. 2.35.

Upon moving along any resonant bifurcation curve to the corresponding dip of the "bird-wing" diagram in Fig. 2.31, we find CJT at smaller values of the perturbation amplitude as compared to the case with irrational winding numbers, because in order to provide CJT in the first case it is enough to destruct all the KAM curves, whereas the central invariant curves with irrational and especially noble values of the winding number may deform in a complicated way but still survive under increasing ε up to comparatively large values.

2.2.5.3   Concluding Remarks

Being motivated by the problem of CJT in geophysical flows in the ocean and atmosphere, we have studied in detail topology of a central transport barrier and its destruction in the simple kinematic model of a meandering current with chaotic advection of passive particles (Fig. 2.26) that belongs to the class of non-degeneracy Hamiltonian systems. Direct computation of the amplitude–frequency diagram (Fig. 2.27) demonstrated the onset of CJT at surprisingly small values of the perturbation amplitude ε provided that the perturbation frequency ω was sufficiently large. As an indicator of the strength of the central transport barrier and its topology, we used a central invariant curve which was constructed by iterating indicator points by a numerical procedure borrowed from theory of nontwist maps. The central transport barrier has been shown to exist provided a set of the iterations was bounded. Otherwise, CJT has been observed (Fig. 2.28). The results are presented as a diagram of the box-counting dimension of those sets of iterations in Fig. 2.29.

Geometry of the central invariant curve has been shown to be highly sensitive to small variations in the control parameters near a fractal-like boundary of the diagram (Fig. 2.30). Quantifying complexity of the central invariant curve's form by its length l, we computed the corresponding l-diagram looking like a bird wing with a fractal-like boundary (Fig. 2.31). Resonant bifurcation curves with rational winding numbers m/n end up in the dips of the boundary. Along those curves in the parameter space, invariant manifolds of the corresponding $n : m$ resonances connect providing a destruction of the central invariant curve. Scenarios of the reconnection are different for odd (Fig. 2.33) and even (Fig. 2.34) resonances.

Computing the winding number w of the central invariant curve, we have got an information about those values of the perturbation parameters at which the central transport barrier is strong or weak. Using representation of the values of winding numbers by continued fractions, we were able to order spikes with rational values of w in Fig. 2.32 into hierarchical series of the corresponding resonances of the central invariant curve. The curve, which ends up on the tips of "feathers of the wing" in the winding-number diagram in Fig. 2.35, has noble winding numbers which are so irrational that the corresponding central invariant curves break down in the last turn when varying the perturbation parameters. The noble curves have been found to be arranged in series like the resonant bifurcation curves with rational values of w. Destruction of the central transport barrier is illustrated for two ways in the parameter space: upon moving along the resonant bifurcation curves with rational values of w (Fig. 2.36) and along curves with noble values of w (Fig. 2.37).

In conclusion we address two points that may be important in possible applications of the results obtained. Molecular diffusion in laboratory experiment and turbulent diffusion in geophysical flows are expected to wash out ideal fractal-like structures caused by chaotic advection after a characteristic time scale. The question is what is this scale. As to molecular diffusion in the ocean, the diffusion time-scale L^2/D is very large since the diffusion coefficient is of the order of $D \simeq 10^{-5}$ cm^2/s and L is, at least, of a kilometer scale. The scale of molecular diffusion in laboratory tanks is, of course, much smaller. However, some fractal-like structures have been observed in real laboratory experiments (see, for example, [40] and the book [18] for a recent review of experiments).

The advantage of the kinematic approach is its ability to identify different factors that may enhance or suppress CJT. However, in any kinematic model the velocity field is postulated based on known features of the current while in dynamic models it should obey dynamical equations following from the conservation of potential vorticity [28]. The phenomenon of CJT is considered in the next section with a dynamically consistent model with a zonal Bickley jet and two propagating Rossby waves.

2.3 Chaotic Cross-Jet Transport in a Dynamical Model of a Meandering Jet Current with Propagating Rossby Waves

2.3.1 The Dynamical Model with Rossby Waves

We introduce in this section the dynamic model of a meandering jet current with propagating Rossby waves which obeys equations of motion following from the conservation of potential vorticity. Motion of two-dimensional incompressible fluid in the rotating frame of reference is governed by the equation for conserving potential vorticity

$$(\partial/\partial t + \mathbf{v} \cdot \nabla)\Pi = 0. \tag{2.14}$$

In the quasigeostrophic approximation [28], one gets

$$\Pi = \nabla^2 \Psi + \beta y, \tag{2.15}$$

where β is the Coriolis parameter. The x axis is chosen along the zonal flow, from the west to the east and y—along the gradient from the south to the north. Barotropic perturbations of zonal flows produce Rossby waves which have an essential impact on transport and mixing in the ocean and atmosphere [28]. The stream function is sought in the form

$$\Psi = \Psi_0 + \Psi_{\text{int}} = \Psi_0(y) + \sum_j \Phi_j(y)e^{ik_j(x-c_j t)}, \tag{2.16}$$

where Ψ_0 describes a zonal flow, and Ψ_{int} is its perturbation which is supposed to be a superposition of zonal running Rossby waves. After substituting (2.16) in the equation for the potential vorticity (2.14) and a linearization, we get the Rayleigh–Kuo equation [22]

$$(u_0 - c_j)\left(\frac{d^2\Phi_j}{dy^2} - k_j^2\Phi_j\right) + \left(\beta - \frac{d^2u_0}{dy^2}\right)\Phi_j = 0, \tag{2.17}$$

where the zonal velocity $u_0 = -d\Psi_0/dy$ has a single extremum at $y = 0$. If one takes the zonal velocity profile in the form of a Bickley jet

$$u_0(y) = u_{max}\,\text{sech}^2\,\frac{y}{L}, \tag{2.18}$$

then Eq. (2.17) admits two neutrally stable solutions

$$\Phi_j(y) = A_j u_{max} L\,\text{sech}^2\,\frac{y}{L}, \quad j = 1, 2, \tag{2.19}$$

where u_{max} is the maximal velocity in the flow, L is a measure of its width, and A_j are the wave amplitudes. It is easy to check that (2.18) and (2.19) are compatible with (2.17) if there is the following condition for the phase velocities:

$$c_{1,2} = \frac{u_{max}}{3}(1 \pm \alpha), \quad \alpha \equiv \sqrt{1 - \beta^*}, \quad \beta^* \equiv \frac{3L^2\beta}{2u_{max}}, \tag{2.20}$$

which are connected with the wavenumbers by the dispersion relation $c_{1,2} = u_{max}L^2k_{1,2}^2/6$. Two neutrally stable Rossby waves exist if $\beta L^2/u_{max} < 2/3$.

Thus, the stream function of the zonal flow with two Rossby waves, satisfying the conservation of the potential vorticity, has the form

$$\Psi(x, y, t) = -u_{max}L\left(\tanh\frac{y}{L} - [A_1\cos k_1(x - c_1t) + A_2\cos k_2(x - c_2t)]\,\text{sech}^2\,\frac{y}{L}\right). \tag{2.21}$$

One of the tasks of this section is to present results in the form allowing a comparison with laboratory experiments [39, 41] in which an azimuthal jet at the radius R with Rossby waves with the wavenumbers n_1 and n_2 has been produced:

$$k_{1,2} = \frac{n_{1,2}}{R}, \quad c_{1,2} = \frac{u_{max}L^2}{6R^2}n_{1,2}^2. \tag{2.22}$$

Let it be $n_1 > n_2$, and the wave with n_1 is called the first one. Let the wavenumbers be represented as $n_1 = mN_1$ and $n_2 = mN_2$, where $m \neq 1$ is the greatest common divisor and N_1/N_2 is an irreducible fraction. Introducing new coordinates x', y', and t'

$$x = \frac{(x' + C_2t')R}{m}, \quad y = Ly', \quad t = \frac{R}{mu_{max}}t', \tag{2.23}$$

we rewrite the stream function (2.21) in the frame of reference moving with the phase velocity of the first wave

$$\Psi'(x',y',t') = -\tanh y' + A_1 \operatorname{sech}^2 y' \cos(N_1 x') + A_2 \operatorname{sech}^2 y' \cos(N_2 x' + \omega t') + C_2 y',$$
(2.24)

where

$$\omega \equiv \frac{2N_2(N_1^2 - N_2^2)}{3(N_1^2 + N_2^2)}, \qquad C_2 \equiv \frac{2N_1^2}{3(N_1^2 + N_2^2)}.$$
(2.25)

Thus, we get the stream function (2.24) with the control parameters N_1 and N_2 which are specified by the four experimental parameters: u_{max}, β, L, and R. One can now study CJT with any combination of the wavenumbers n_1 and n_2 that can be realized in a laboratory experiment by adjusting the radius R, the jet width L, the maximal velocity u_{max}, and the Coriolis-like parameter β [39, 41].

The advection equations with the stream function (2.24) have the form

$$\frac{dx}{dt} = -C_2 + \operatorname{sech}^2 y[1 + 2A_1 \tanh y \cos (N_1 x) + 2A_2 \tanh y \cos (N_2 x + \omega t)],$$

$$\frac{dy}{dt} = -\operatorname{sech}^2 y[A_1 N_1 \sin (N_1 x) + A_2 N_2 \sin (N_2 x + \omega t)],$$
(2.26)

where we omitted the primes over x, y, and t. If the amplitude of the second wave is zero, $A_2 = 0$, then the set (2.26) is integrable. We use the values $N_1 = 5$ and $N_2 = 1$ to perform most of the numerical experiments in this section.

Figure 2.38a shows the phase portrait of the stationary flow (if $A_2 = 0$) in a coordinate system moving with the phase velocity of the first wave. By the normalization adopted here, a frame of size 2π spans five wavelengths of the first wave and one wavelength of the second one. There is an eastward jet sandwiched between two chains of five vortices, while the adjacent flows to its south and north are westward in the moving coordinate system. In the steady flow, tracers move along stationary streamlines. If $A_2 > 0$, the set (2.26) has chaotic solutions. The onset of chaos follows a scenario typical for nonlinear Hamiltonian systems: as the separatrices of an integrable system break down, they are superseded by a stochastic layer.

In Fig. 2.38b the central invariant curve is represented by a bold curve with stochastic layers to its north and south where chaotic transport takes place. This Poincaré section surface is obtained by setting $A_2 = 0.09$, in which case the region occupied by the central invariant curve and adjacent invariant curves acts as a barrier to CJT. With increasing A_2, the central invariant curve breaks up and is superseded by a stochastic layer where local chaotic CJT is observed (Fig. 2.38c). The layer is confined between invariant curves, and its average width increases with the wave amplitudes. Under further increasing A_2, the chaotic CJT becomes global (Fig. 2.38d).

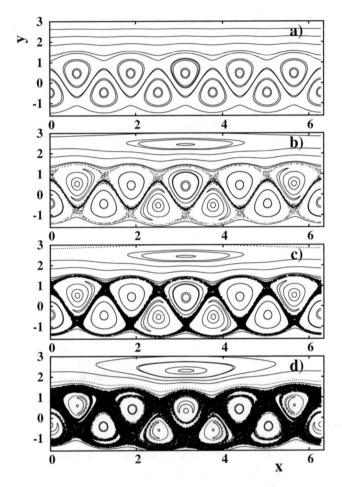

Fig. 2.38 Poincaré section surfaces of a zonal flow modulated by two Rossby waves (2.22) with $N_1 = 5$ and $N_2 = 1$ in a reference frame moving with the phase velocity of the first wave for several values of the perturbation amplitude ($A_1 = 0.2416$). (**a**) $A_2 = 0$ (steady flow): streamline pattern. (**b**) $A_2 = 0.09$: chaotic advection in the zonal direction, but the central invariant curve (the *bold curve*) acts as a barrier to CJT. (**c**) $A_2 = 0.095$: the central invariant curve is destroyed and superseded by a stochastic layer where local chaotic CJT is observed. (**d**) $A_2 = 0.2$: chaotic CJT is global

2.3.2 Mechanisms of Chaotic Cross-Jet Transport for Odd Wavenumbers and Detection of Transport Barriers

The steady flow described by the stream function (2.23) with $A_2 = 0$ is degenerate: $\partial f / \partial \Psi = 0$, where f and Ψ are the frequency of motion and stream function of advected particles in the unperturbed flow. Now, we introduce the quantitative measure of chaotic CJT to be average width $h = S/2\pi$ of the central stochastic layer, where S is the area of the stochastic layer and 2π is the frame length (the jet half-width measured in these units is 1). The chaotic CJT is called to be local if $h \ll 1$. We refer to chaotic CJT as global if $h \gtrsim 1$ (the mixing region and the jet have comparable widths).

Among other consequences, degeneracy implies the nonuniqueness of orbits with a particular winding number, which occur in pairs on both sides of the central invariant curve. As it was mentioned in the preceding section, they may collide and annihilate as control parameters are varied. Collisions of periodic orbits are involved in the reconnection of the invariant manifolds of the corresponding hyperbolic orbits [15]. At odd values of N_1 and N_2, Eq. (2.26) have the same symmetries (2.7) and (2.8) as the advection equations (2.3) for the kinematic model considered in the preceding sections. Recalling that N_1/N_2 is an irreducible fraction, we note that the actual Rossby wavenumbers, $n_1 = mN_1$ and $n_2 = mN_2$, may be even as well. The algorithm for detecting the central invariant curve is the same as to be described in Sect. 2.2.5. The kind of the pattern of iterations of an indicator point signifies the presence or absence of a barrier to CJT.

Figure 2.39 illustrates how the central invariant curve transforms with increasing A_2. If this amplitude is relatively small, the central invariant curve is a nonmeandering curve (see Fig. 2.39a). As the threshold value $A_2 = 0.0605$ is exceeded, the invariant manifolds of the hyperbolic orbits of the 7 : 3 **resonance islands** chains overlap, which results in a breakup of the central invariant curve. As the amplitude increases further, invariant manifolds split and the central invariant curve having

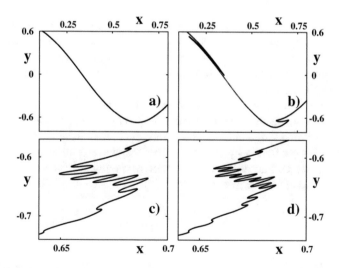

Fig. 2.39 Transformations of the central invariant curve in a zonal flow with two Rossby waves with $N_1 = 5$ and $N_2 = 1$ for $A_1 = 0.243$ and several values of the perturbation amplitude. (**a**) $A_2 = 0.05$: a nonmeandering central invariant curve. (**b**) $A_2 = 0.082$: a central invariant curve with 7 first-order meanders. (**c**) $A_2 = 0.099$: a central invariant curve with 33 second-order meanders. (**d**) $A_2 = 0.1047$: a central invariant curve with higher-order meanders

seven first-order meanders appears (see Fig. 2.39b). The number of meanders of
the central invariant curve is determined by the number of resonance islands in the
reconnecting chains. With even further increase in the amplitude, new meanders
appear in the central invariant curve. Figure 2.39c shows a central invariant curve
with 33 second-order meanders. Figure 2.39d shows a higher-order meandering
central invariant curve with a number of meanders to be difficult to count.

Our numerical simulations revealed two mechanisms of the onset of chaotic CJT.
As the Rossby wave amplitudes increase, the southern and northern stochastic layers
around the central invariant curve become wider. As certain threshold values of the
amplitudes are exceeded, the layers overlap and the central invariant curve breaks
up. It is an amplitude mechanism of the breakup of a central invariant curve and
onset of chaotic CJT which is in general a global transport.

Figure 2.40 illustrates the amplitude mechanism of such a breakup and the onset
of chaotic CJT. Keeping $A_1 = 0.2418$ to be constant, we increase A_2. In Fig. 2.40a,
the iterations of an indicator point in the neighborhood of the 7 : 3 resonance make
up a central invariant curve separating the southern and northern stochastic layers. In
another typical pattern, shown in Fig. 2.40b, the iterations make up small segments
on the phase plane, which merge into an almost periodic continuous central curve
separating the southern and northern stochastic layers as $t \to \infty$. The iterations
in Fig. 2.40c do not fit along any single curve because the central invariant curve
was destroyed by an overlap of the southern and northern stochastic layers. For the
corresponding perturbation amplitude ($A_2 = 0.088$), we have $h \ll 1$, i.e., local
chaotic CJT is observed. Figure 2.40d illustrates a global chaotic CJT with $h \simeq 2$
($A_2 = 0.2$).

Alternatively, the central invariant curve can break up via a resonance mechanism
when the Rossby wave amplitudes are arbitrarily small. Amplitude variations
change the rotation number of the central invariant curve (see Fig. 2.41). If the
rotation number is rational, the central invariant curve breaks up, and a local chaotic
CJT is generally observed.

Since tracers move with the jet, we can reasonably assume that the central
invariant curve breaks up if the maximum frequency of unperturbed motion in the
central-jet region is resonant with the perturbation frequency ω. Using (2.26) to
estimate the former frequency as $f_1 \simeq -C_2 + 1$ and substituting expression (2.25)
for the latter, we find an estimate for the rotation number of the central invariant
curve to be valid at small wave amplitudes [7]:

$$w \approx \frac{f_1}{\omega} \simeq \frac{N_1^2 + 3N_2^2}{2N_2(N_1^2 - N_2^2)}. \qquad (2.27)$$

Setting the right-hand side of (2.27) equal to a rational number, we can find pairs
of N_1 and N_2 such that the central invariant curve is affected by the corresponding
resonance, i.e., the onset of CJT is possible if the wave amplitudes are relatively
small.

With increasing the perturbation amplitude, some smooth invariant curves around
the central invariant curve break up and are superseded by resonance-island chains.

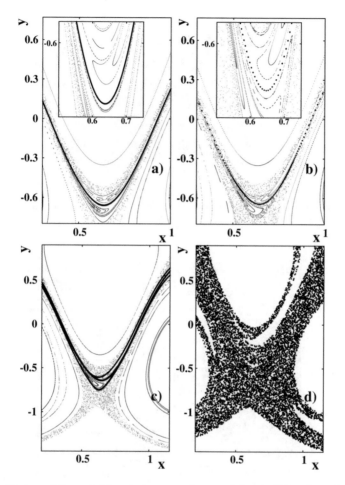

Fig. 2.40 Breakup of the central invariant curve and onset of chaotic CJT at $A_1 = 0.2418$ and several values of the perturbation amplitude. In (**a**) and (**b**) the iterations of an indicator point fit along a central invariant curve ($A_2 = 0.067$) and an almost periodic central orbit ($A_2 = 0.08703$), respectively. In (**c**) and (**d**) the central invariant curve is destroyed, and the iterations lie in a stochastic layer where local and global chaotic CJT are observed ($A_2 = 0.088$ and $A_2 = 0.2$), respectively. The *insets* show magnified images of the phase space near 7 : 3 resonance islands

These chains of island occur in pairs on both sides of the central invariant curve by virtue of the flow symmetry (2.7). The geometry of the central invariant curve, the size and number of resonance islands, and the topology of the corresponding invariant manifolds vary with changing the control parameters in a complicated manner.

As an illustration of the breakup of the central invariant curve and onset of local chaotic CJT via the resonance mechanism, Fig. 2.42 shows Poincaré section surfaces computed for A_2 varying between 0.1016 and 0.1024, while $A_1 = 0.243$ is held constant. As the A_2 value varies within this interval, the invariant manifolds of the

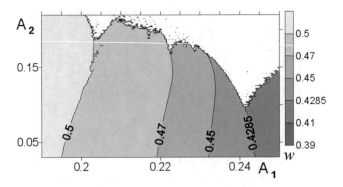

Fig. 2.41 Diagram of the winding number w in the space of the Rossby waves amplitudes A_1 and A_2 with $N_1/N_2 = 5/1$. *White zone* corresponds to the regime with a cross-jet transport

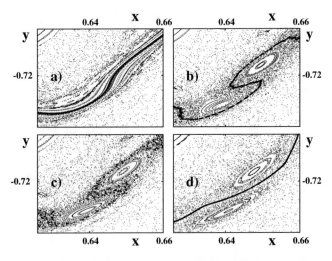

Fig. 2.42 Poincaré section surfaces illustrate a breakup of the central invariant curve and onset of local chaotic CJT via the resonance mechanism at $A_1 = 0.243$ and A_2 varying between 0.1016 and 0.1024. (**a**) $A_2 = 0.102$: a central invariant curve is surrounded by smooth invariant curves. (**b**) $A_2 = 0.1022212 > A_{cr}$: saddle-center bifurcation gives rise to 151 : 64 resonance-island chains on both sides of a central invariant curve with 151 meanders. (**c**) $A_2 = 0.102225$: invariant manifolds of hyperbolic orbits of a high-order resonance overlap, and the central invariant curve is destroyed and superseded by a stochastic layer. (**d**) $A_2 = 0.10236$: a new central invariant curve appears as the manifolds of the hyperbolic orbits of the 151 : 64 resonance split

hyperbolic orbits of the 151 : 64 resonance reconnect and split so that a meandering, disappearing, and reappearing central invariant curve is observed. In Fig. 2.42a the region occupied by the central invariant curve and adjacent invariant curves acts as a transport barrier. As the perturbation amplitude exceeds $A_{cr} = 0.102$, a 151 : 64 resonance-island chain appears on each side of the central invariant curve, which has 151 meanders (Fig. 2.42b). With further increase in A_2, the invariant manifolds of the hyperbolic orbits of the 151 : 64 resonance overlap, the central invariant curve breaks up, and local chaotic CJT is observed. Next, the manifolds of the hyperbolic orbits of the 151 : 64 resonance split, and a new central invariant curve becomes a barrier to CJT.

Two scenarios of the breakup of central invariant curves should be distinguished whether n is even or odd. The corresponding $n : m$ resonances are called even and odd, respectively (m is the number of frames passed by a fluid particle). In the case of an odd resonance discussed above, a saddle-center bifurcation gives rise to a pair of resonance-island chains. With increasing the perturbation amplitude, both the islands and the central invariant curve's meanders grow larger. As the amplitude exceeds a certain threshold value, the invariant manifolds of hyperbolic orbits overlap and the central invariant curve breaks up. With further increase in the amplitude, the manifolds split, and a new central invariant curve appears. In the literature on nontwist maps, this process is called "separatrix reconnection" [15].

In the case of an even resonance, the distance between the resonance-island chains, separated by the central invariant curve, decreases with increasing the perturbation amplitude. As the amplitude exceeds a certain threshold value, the chains overlap transforming into vortex pairs. With further increase in the amplitude, the vortex-pair size gradually decreases until both hyperbolic and elliptic orbits annihilate giving rise to a new central invariant curve.

Figure 2.43 presents an interesting example, where the onset of chaotic CJT occurs as parameters are varied along the bifurcation curve of the winding number $w = 1/2$ (see Fig. 2.41). The m/n-resonance bifurcation curve is a set of points in the (A_1, A_2) parameter space corresponding to collisions of the invariant manifolds of the hyperbolic orbits of the $n : m$ resonance. The point $(0.2038, 0.17)$ on the bifurcation curve corresponds to the formation of a chain of vortex pairs surrounded by invariant curves (Fig. 2.43a). As the amplitudes increase along the bifurcation curve, the invariant curves gradually break up, and the areas of the northern and southern stochastic layers increase. When the amplitudes become sufficiently large, the overlapping layers should be expected to merge into a single layer where unobstructed southward and northward tracer migration is possible. However, this is not always the case.

Figure 2.43b shows how the northern and southern stochastic layers collide at $A_1 = 0.2038$ and $A_2 = 0.205$. Iterations of an indicator point (bold dots) fill only the southern part of the stream rather than the entire layer, contrary to our expectations. We emphasize that no iterations of the indicator point was found on the other side of the boundary between the layers (visible in the figure) during an extremely long time (while as many as five million iterations are performed), even though no invariant curves make this impossible. It can be hypothesized that the breakup of the central invariant curve gives rise to a complicated, dense network of stable and unstable manifolds of hyperbolic orbits that acts as a transport barrier. In effect, Fig. 2.43b illustrates a situation where the central invariant curve breaks up, but no chaotic CJT is actually observed until the Rossby wave amplitudes increase further.

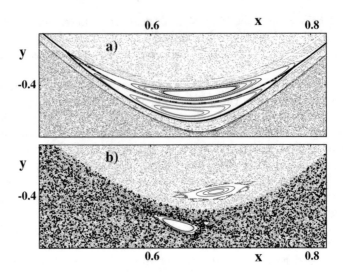

Fig. 2.43 Breakup of the central invariant curve with parameter variation along the bifurcation curve with the winding number $w = 1/2$. (**a**) $A_1 = 0.2038$, $A_2 = 0.17$: the central invariant curve is destroyed by an overlap of the invariant manifolds of the hyperbolic orbits of an odd resonance and is superseded by a stochastic layer sandwiched between smooth invariant curves. (**b**) $A_1 = 0.2038$, $A_2 = 0.205$: the central invariant curve is destroyed, but no chaotic CJT is actually observed because the iterations of an indicator point (*bold dots*) fill only the southern part of the stream. These iterations do not cross the visible border between the layers during an extremely long time even though there are no invariant curves forbidding that

To conclude this section, we present chaotic CJT diagrams computed for Rossby waves with odd wavenumbers and both A_1 and A_2 varying from 0.05 to 0.5. If the iterations of an indicator point fit along a central invariant curve or a central periodic orbit, then CJT is obviously impossible. The corresponding areas in Fig. 2.44 are white. After the wave amplitudes exceed certain values specific to each pair of the wavenumbers, the width of the stochastic layer, that supersedes the destroyed central invariant curve, increases. The layer widths are represented by grayscale values in Fig. 2.44. A local chaotic CJT is observed (gray areas) if $h < 1$. Black areas represent global chaotic CJT with $h > 1$. The shape of the boundary, that separates flow regimes where the central invariant curve persists and chaotic CJT occurs, is of particular importance. The spikes along the boundary, seen in the diagrams, are explained by the existence of resonance-island chains in the neighborhood of the central invariant curve.

2.3.3 Chaotic Cross-Jet Transport for Even-Odd Wavenumbers

Let us consider CJT in a stream parametrized by (2.25) with even-odd pairs of N_1 and N_2. In this case, the set (2.26) does not have the symmetry $\hat{S}$ in (2.7), used in

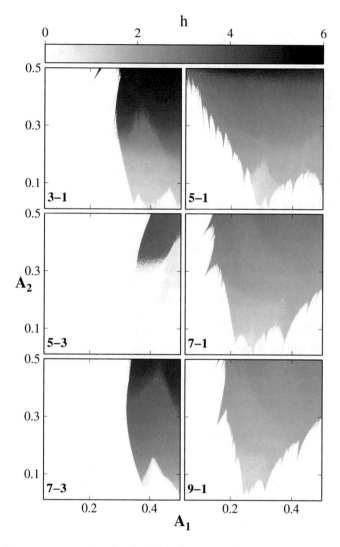

Fig. 2.44 Diagrams representing chaotic CJT in the space of the Rossby wave amplitudes for the odd wavenumber pairs 3–1, 5–1, 5–3, 7–1, 7–3, and 9–1. *Grayscale* values represent the width of the stochastic layer that supersedes the destroyed central invariant curve. *White, gray, and black areas* correspond to the absence of chaotic CJT, local chaotic CJT, and global chaotic CJT, respectively

the previous section to develop a technique for constructing a central invariant curve and to analyze bifurcations and breakup of the curve. We describe here a numerical method for simulating CJT for even-odd pairs of wavenumbers. We determine that a CJT occurs by detecting an overlap of the northern and southern stochastic layers in computations of Poincaré sections starting from initial conditions set on the northern and southern sides of the central jet region, respectively. The overlap of the layers is defined as follows: if each point of the former layer lies in the latter and vice versa, then the layers overlap. When computing time is finite, a complete overlap is obviously impossible, i.e., the layers overlap only partially.

We use the following algorithm to determine that a particular point lies in the stochastic layer. Consider a point $\mathscr{A}$ and a set of points $\mathscr{B}_i$ randomly distributed in a bounded region $\mathscr{D}$. The region $\mathscr{D}$ is not known exactly, whereas the set of $\mathscr{B}_i$ is given. For each $\mathscr{B}_i$, we calculate the distance to its nearest neighbor: $r_i = \min_{j \neq i} \|\mathscr{B}_i, \mathscr{B}_j\|$, where $\|\mathscr{B}_i, \mathscr{B}_j\|$ is the distance between points $\mathscr{B}_i$ and $\mathscr{B}_j$. The point $\mathscr{A}$ is said to lie in $\mathscr{D}$ if the distance from $\mathscr{A}$ to $\mathscr{B}_i$ is smaller than the largest distance between nearest neighbor in the set of $\min_i \|\mathscr{B}_i, \mathscr{A}\| \leq R = \max_i r_i$. Instead of the largest distance R, we actually use the distance R_n such that a fraction n of the r_i values are smaller than R_n. In particular, if $n = 0.95$, then 95% of the points are separated from their nearest neighbors by distances smaller than $R_{0.95}$. This is done to eliminate outliers in the distribution of r_i. On the other hand, when R_n is used as a criterion, the points that do belong to $\mathscr{D}$ are discarded with probability $1 - n$. However, this does not affect the detection of CJT.

To establish that two stochastic layers overlap on Poincaré sections, we calculate and analyze the values of F_1 and F_2, where F_1 is the fraction of particles in the first section that lie in the second stochastic layer and F_2 is the fraction of particles in the second section that lie in the first stochastic layer. The following cases are possible.

1. $F_1 = F_2 = 0$. The stochastic layers are disjoint.
2. $F_1 \simeq n \simeq F_2$. The stochastic layers overlap completely. The value of n is less than 1 as explained above.
3. $F_1 \simeq F_2 \ll 1$. The stochastic layers collide but do not overlap. This happens either if nondegenerate tori break up while the central invariant curve does not or if CJT occurs with a low probability (a breakup of the central invariant curve via even resonance illustrated in Fig. 2.43).
4. Otherwise, the stochastic layers partially overlap when the computing time is too short for a complete overlap to occur.

The diagrams in Fig. 2.45, calculated by the algorithm described above, represent CJT in the space of the Rossby wave amplitudes for the even-odd wavenumber pairs 3–2, 5–4, and 6–5. White areas correspond to wave amplitudes for which the southern and northern stochastic layers are separated by a transport barrier. Grayscale values represent the width of a single stochastic layer that forms when chaotic CJT is observed. For even-odd wavenumber pairs, the Rossby wave amplitudes required for global CJT to occur are on the order of unity, being several times larger than those for odd wavenumbers. We note that, with increasing A_2, a **heteroclinic** flow can transform into a **homoclinic** one with a westward jet. This situation can also be analyzed by using our technique. Rossby waves with amplitudes as large as these ($A_1, A_2 > 1$) can hardly develop in actual jet streams or laboratory experiments.

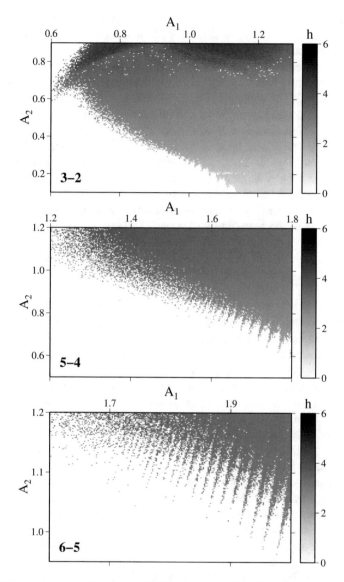

Fig. 2.45 Diagrams representing chaotic CJT in the space of the Rossby wave amplitudes for the even-odd wavenumber pairs 3–2, 5–4, 6–5. *Grayscale* values represent the width of a single stochastic layer. *White areas* correspond to the absence of chaotic CJT

Now, we briefly summarize the results obtained in this paragraph. Representing the Rossby wavenumbers as $n_1 = mN_1$, $n_2 = mN_2$, where $m \neq 1$ is the greatest common divisor of n_1 and n_2 (i.e., N_1/N_2 is an irreducible fraction), we reduce our analysis to two classes of flows: (1) those with odd N_1 and N_2 and (2) those with even-odd pairs of N. The case of an even-even n_1 and n_2 pair being equivalent to one of the above. In the former case, the advection equations (2.26) have a symmetry $\hat{S}$ defined in (2.7), which makes it possible to find the central invariant curve by calculating Poincaré iterations of an indicator point (Fig. 2.39). Depending on the dimension of a minimal cover of the set of iterations, distinction can be made between flows with a central invariant curve and those with a stochastic layer where chaotic CJT is observed (Fig. 2.40). We expose amplitude and resonance mechanisms of a breakup of the central invariant curve and onset of chaotic CJT (Fig. 2.40 and Figs. 2.42 and 2.43, respectively). The latter mechanism leads to different scenarios of the breakup of the central invariant curve for even and odd resonances (Figs. 2.42 and 2.43, respectively). If the wavenumbers are odd, relatively small Rossby wave amplitudes are required for such a breakup and for the onset of a chaotic CJT to occur (Fig. 2.44).

For flows modulated by waves with even-odd wavenumber pairs, where one of the symmetries of the advection equations is lacking, an alternative technique is developed based on detecting an overlap between northern and southern stochastic layers. Simulation shows that the Rossby wave amplitudes, required for global chaotic CJT, are much larger than those corresponding to odd wave-numbers (Fig. 2.40).

Finally, we briefly discuss the feasibility of testing our predictions in laboratory experiments on chaotic advection in rotating flows [1, 39, 41] simulating nonlinear geostrophic atmospheric and oceanic flows. Water was pumped at into a rotating tank through ports in its bottom, made conical to mimic the gradient of the Earth's Coriolis force. As a result, an eastward jet modulated by Rossby waves was produced whose flow field is correctly described by the stream function (2.21). Dye injected on one side of the jet rapidly mixed by chaotic advection in the zonal direction, but did not cross the jet, i.e., a central transport barrier was observed. Measured velocity power spectra suggest that stable jets have been produced in most experiments [1, 39, 41], dominated by a single mode whose wavenumber varies between 3 and 8 depending on the rotation frequency and the pumping rate. One or two weaker modes have also been detected. Experimental and predicted streamlines have been compared for flows with 5–4 and 5–6 wavenumber pairs. It follows from our numerical experiments (Fig. 2.40) that unphysically large amplitudes are required to destroy the transport barrier in flows with even-odd wavenumber pairs. Therefore, the experimental conditions in [1, 39, 41] were such that the transport barrier could not be destroyed.

However, the barrier can be destroyed by creating a jet stream as described in [1, 39, 41] but with an odd-odd Rossby wavenumber pair such as 3–1 or 5–1 (see Fig. 2.44). It is possible in this case to find the flow parameters corresponding to resonances that can lead to a breakup of the corresponding central invariant curve and onset of CJT at relatively small Rossby wave amplitudes. Then varying one of the wave amplitudes, it is possible to destroy the barrier and observe dye filaments crossing the central jet region.

References

1. Behringer, R.P., Meyers, S.D., Swinney, H.L.: Chaos and mixing in a geostrophic flow. Phys. Fluids A **3**(5), 1243 (1991). 10.1063/1.858052
2. Boffetta, G., Lacorata, G., Redaelli, G., Vulpiani, A.: Detecting barriers to transport: a review of different techniques. Physica D **159**(1–2), 58–70 (2001). 10.1016/S0167-2789(01)00330-X
3. Bower, A.S.: A simple kinematic mechanism for mixing fluid parcels across a meandering jet. J. Phys. Oceanogr. **21**(1), 173–180 (1991). http://dx.doi.org/10.1175/1520-0485(1991)021<0173:askmfm>2.0.co;2
4. Budyansky, M., Uleysky, M., Prants, S.: Chaotic scattering, transport, and fractals in a simple hydrodynamic flow. J. Exp. Theor. Phys. **99**, 1018–1027 (2004). 10.1134/1.1842883
5. Budyansky, M., Uleysky, M., Prants, S.: Hamiltonian fractals and chaotic scattering of passive particles by a topographical vortex and an alternating current. Physica D **195**(3–4), 369–378 (2004). 10.1016/j.physd.2003.11.013
6. Budyansky, M.V., Prants, S.V.: A mechanism of chaotic mixing in an elementary deterministic flow. Tech. Phys. Lett. **27**(6), 508–510 (2001). 10.1134/1.1383840
7. Budyansky, M.V., Uleysky, M.Y., Prants, S.V.: Detection of barriers to cross-jet Lagrangian transport and its destruction in a meandering flow. Phys. Rev. E **79**(5), 056215 (2009). 10.1103/physreve.79.056215
8. del Castillo-Negrete, D., Greene, J., Morrison, P.: Area preserving nontwist maps: periodic orbits and transition to chaos. Phys. D Nonlinear Phenomena **91**(1–2), 1–23 (1996). doi:10.1016/0167-2789(95)00257-x
9. del Castillo-Negrete, D., Greene, J., Morrison, P.: Renormalization and transition to chaos in area preserving nontwist maps. Physica D **100**(3–4), 311–329 (1997). 10.1016/s0167-2789(96)00200-x
10. del Castillo-Negrete, D., Morrison, P.J.: Chaotic transport by Rossby waves in shear flow. Phys. Fluids A **5**(4), 948–965 (1993). 10.1063/1.858639
11. Chirikov, B.V.: A universal instability of many-dimensional oscillator systems. Phys. Rep. **52**(5), 263–379 (1979). 10.1016/0370-1573(79)90023-1
12. Dahleh, M.D.: Exterior flow of the Kida ellipse. Phys. Fluids A **4**(9), 1979–1985 (1992). 10.1063/1.858366
13. de Almeida, A.M.O.: Hamiltonian Systems: Chaos and Quantization. Cambridge Monographs on Mathematical Physics. Cambridge University Press, Cambridge (1988)
14. Duan, J., Wiggins, S.: Fluid exchange across a meandering jet quasiperiodic variability. J. Phys. Oceanogr. **26**(7), 1176–1188 (1996). 10.1175/1520-0485(1996)026<1176: feaamj>2.0.co;2
15. Howard, J.E., Hohs, S.M.: Stochasticity and reconnection in Hamiltonian systems. Phys. Rev. A **29**(1), 418–421 (1984). 10.1103/physreva.29.418
16. Kida, S.: Motion of an elliptic vortex in a uniform shear flow. J. Phys. Soc. Jpn. **50**(10), 3517–3520 (1981). 10.1143/jpsj.50.3517
17. Koshel', K.V., Prants, S.V.: Chaotic advection in the ocean. Physics-Uspekhi **49**(11), 1151–1178 (2006). 10.1070/PU2006v049n11ABEH006066
18. Koshel, K.V., Prants, S.V.: Chaotic Advection in the Ocean. Institute for Computer Science, Moscow (2008) [in Russian]
19. Koshel, K.V., Ryzhov, E.A., Zhmur, V.V.: Diffusion-affected passive scalar transport in an ellipsoidal vortex in a shear flow. Nonlinear Process. Geophys. **20**(4), 437–444 (2013). 10.5194/npg-20-437-2013
20. Koshel, K.V., Ryzhov, E.A., Zhmur, V.V.: Effect of the vertical component of diffusion on passive scalar transport in an isolated vortex model. Phys. Rev. E **92**(5), 053021 (2015). 10.1103/physreve.92.053021
21. Kozlov, V.F.: Background currents in geophysical hydrodynamics. Izv. Atmos. Oceanic Phys. **31**(2), 229–234 (1995)

22. Kuo, H.L.: Dynamic instability of two-dimensional nondivergent flow in a barotropic atmosphere. J. Meteorol. **6**(2), 105–122 (1949). 10.1175/1520-0469(1949)006 <0105:diotdn>2.0.co;2

23. Makarov, D., Uleysky, M., Budyansky, M., Prants, S.: Clustering in randomly driven Hamiltonian systems. Phys. Rev. E **73**(6), 066210 (2006). 10.1103/PhysRevE.73.066210

24. Mancho, A.M., Small, D., Wiggins, S.: A tutorial on dynamical systems concepts applied to Lagrangian transport in oceanic flows defined as finite time data sets: theoretical and computational issues. Phys. Rep. **437**(3–4), 55–124 (2006). 10.1016/j.physrep.2006.09.005

25. Melnikov, V.: On the stability of the center for time periodic perturbations. Trans. Moscow Math. Soc. **12**, 1–57 (1963)

26. Meyers, S.D.: Cross-frontal mixing in a meandering jet. J. Phys. Oceanogr. **24**(7), 1641–1646 (1994). 10.1175/1520-0485(1994)024<1641:cfmiam>2.0.co;2

27. Morozov, A.D.: Degenerate resonances in Hamiltonian systems with 3/2 degrees of freedom. Chaos **12**(3), 539–548 (2002). 10.1063/1.1484275

28. Pedlosky, J.: Geophysical Fluid Dynamics, 2nd edn. Springer, New York (1987). 10.1007/ 978-1-4612-4650-3

29. Polvani, L.M., Wisdom, J.: Chaotic Lagrangian trajectories around an elliptical vortex patch embedded in a constant and uniform background shear flow. Phys. Fluids A **2**(2), 123–126 (1990). 10.1063/1.857814

30. Polvani, L.M., Wisdom, J., DeJong, E., Ingersoll, A.P.: Simple dynamical models of Neptune's great dark spot. Science **249**(4975), 1393–1398 (1990). 10.1126/science.249.4975.1393

31. Prants, S.V., Budyansky, M.V., Uleysky, M.Y., Zaslavsky, G.M.: Chaotic mixing and transport in a meandering jet flow. Chaos **16**(3), 033117 (2006). 10.1063/1.2229263

32. Raynal, F., Wiggins, S.: Lobe dynamics in a kinematic model of a meandering jet. I. Geometry and statistics of transport and lobe dynamics with accelerated convergence. Physica D **223**(1), 7–25 (2006). 10.1016/j.physd.2006.07.021

33. Rom-Kedar, V., Leonard, A., Wiggins, S.: An analytical study of transport, mixing and chaos in an unsteady vortical flow. J. Fluid Mech. **214**, 347–394 (1990). 10.1017/s0022112090000167

34. Samelson, R.M.: Fluid exchange across a meandering jet. J. Phys. Oceanogr. **22**(4), 431–444 (1992). 10.1175/1520-0485(1992)022<0431:FEAAMJ>2.0.CO;2

35. Samelson, R.M., Wiggins, S.: Lagrangian Transport in Geophysical Jets and Waves: The Dynamical Systems Approach. Interdisciplinary Applied Mathematics, vol. 31. Springer, New York (2006). 10.1007/978-0-387-46213-4

36. Shinohara, S., Aizawa, Y.: The breakup condition of shearless KAM curves in the quadratic map. Prog. Theor. Phys. **97**(3), 379–385 (1997). 10.1143/ptp.97.379

37. Shinohara, S., Aizawa, Y.: Indicators of reconnection processes and transition to global chaos in nontwist maps. Prog. Theor. Phys. **100**(2), 219–233 (1998). 10.1143/ptp.100.219

38. Simó, C.: Invariant curves of analytic perturbed nontwist area preserving maps. Regular Chaotic Dyn. **3**(3), 180–195 (1998). 10.1070/rd1998v003n03abeh000088

39. Solomon, T.H., Holloway, W.J., Swinney, H.L.: Shear flow instabilities and Rossby waves in barotropic flow in a rotating annulus. Phys. Fluids A **5**(8), 1971 (1993). 10.1063/1.858824

40. Sommerer, J.C., Ku, H.C., Gilreath, H.E.: Experimental evidence for chaotic scattering in a fluid wake. Phys. Rev. Lett. **77**(25), 5055–5058 (1996). 10.1103/physrevlett.77.5055

41. Sommeria, J., Meyers, S.D., Swinney, H.L.: Laboratory model of a planetary eastward jet. Nature **337**(6202), 58–61 (1989). 10.1038/337058a0

42. Uleysky, M.Y., Budyansky, M.V., Prants, S.V.: Effect of dynamical traps on chaotic transport in a meandering jet flow. Chaos **17**(4), 043105 (2007). 10.1063/1.2783258

43. Uleysky, M.Y., Budyansky, M.V., Prants, S.V.: Genesis and bifurcations of unstable periodic orbits in a jet flow. J. Phys. A Math. Theor. **41**(21), 215102 (2008). 10.1088/1751-8113/41/21/215102

44. van der Weele, J., Valkering, T., Capel, H., Post, T.: The birth of twin Poincaré-Birkhoff chains near 1:3 resonance. Physica A **153**(2), 283–294 (1988). 10.1016/0378-4371(88)90007-6

45. Wurm, A., Apte, A., Fuchss, K., Morrison, P.J.: Meanders and reconnection – collision sequences in the standard nontwist map. Chaos **15**(2), 023108 (2005). 10.1063/1.1915960

46. Zaslavsky, G.: Dynamical traps. Physica D **168–169**, 292–304 (2002). 10.1016/s0167-2789 (02)00516-x
47. Zhmur, V.V., Ryzhov, E.A., Koshel, K.V.: Ellipsoidal vortex in a nonuniform flow: Dynamics and chaotic advections. J. Mar. Res. **69**(2), 435–461 (2011). 10.1357/002224011798765204
48. Kozlov, V.F., Koshel, K.V.: Barotropic model of chaotic advection in background ows. Izv. Atmos. Ocean. Phys. **35**(1), 123–130 (1999)
49. Kozlov, V.F., Koshel, K.V.: Some features of chaos development in an oscillatory barotropic ow over an axisymmetric submerged obstacle. Izv. Atmos. Ocean. Phys. **37**(1), 351–361 (2001)
50. Izrailsky, Y.G., Kozlov, V.F., Koshel, K.V.: Some specific features of chaotization of the pulsating barotropic ow over elliptic and axisymmetric seamounts. Phys. Fluids **16**(8), 3173–3190 (2004). 10.1063/1.1767095
51. Ryzhov, E., Koshel, K., Stepanov, D.: Background current concept and chaotic advection in an oceanic vortex ow. Theor. Comput. Fluid Dyn. **24**(1–4), 59–64 (2010). 10.1007/s00162-009-0170-1
52. Ryzhov, E.A., Koshel, K.V.: Estimating the size of the regular region of a topographically trapped vortex. Geophys. Astrophys. Fluid Dyn. **105**(4–5), 536–551 (2010). 10.1080/03091929.2010.511205
53. Ryzhov, E.A., Koshel, K.V.: Interaction of a monopole vortex with an isolated topographic feature in a three-layer geophysical ow. Nonlinear Process. Geophys. **20**(1), 107–119 (2013). 10.5194/npg-20-107-2013
54. Ryzhov, E.A., Sokolovskiy, M.A.: Interaction of a two-layer vortex pair with a submerged cylindrical obstacle in a two layer rotating fluid. Phys. Fluids 28(5), 056,602 (2016). 10.1063/1.4947248

Chapter 3
Oceans from the Space and Operational Oceanography

3.1 Monitoring Oceans with Satellite Sensors

Operational oceanography is defined as the activity of systematic long-term routine measurements of the oceans and atmosphere and their rapid interpretation and dissemination (see, e.g., [13]). Operational oceanography critically depends on availability of near real time and high-quality in situ and satellite data with sufficiently dense space sampling. That requires an adequate global ocean observing system and the rapid transmission of observational data to assimilation centers where powerful computers use modern numerical circulation models to process the data. The outputs from the models are used to generate final data products in the form of electronic maps for many geophysical parameters. Operational oceanography provides accurate description of the present state of the ocean and continuous forecasts of its future condition. The impressive progress in satellite monitoring and development of high-resolution numerical models of ocean circulation have opened up new opportunities in physical oceanography. Satellites provide continuous, near real-time and global data at high space resolution for many oceanic and atmospheric parameters: sea surface temperature (SST) and salinity, sea surface height (SSH) and its anomalies, ocean color and concentration of chlorophyll, winds at the ocean surface, sea ice, wave heights, surface roughness, and others.

There are passive and active (or radar) techniques to observe the ocean. The first one measures natural radiation from the sea, whereas in the second case the satellite sends a signal which is measured after its reflection at the sea surface. Satellite sensors operate at different frequencies depending on the parameter to be measured. Many satellite-derived data products can be found free at the websites [AVISO, OC, NODC, NSIDC, GHRC] and others.

Sea surface temperature is a key parameter which is related to air–sea interaction and different processes in the ocean. The global SST products are provided by a few agencies that use different satellites [NODC, GHRSST, REMSS, COPERNICUS, PODAAC, CERSAT]. To get SST data passive sensors (radiometers) measure the

© Springer International Publishing AG 2017
S.V. Prants et al., *Lagrangian Oceanography*, Physics of Earth and Space Environments, DOI 10.1007/978-3-319-53022-2_3

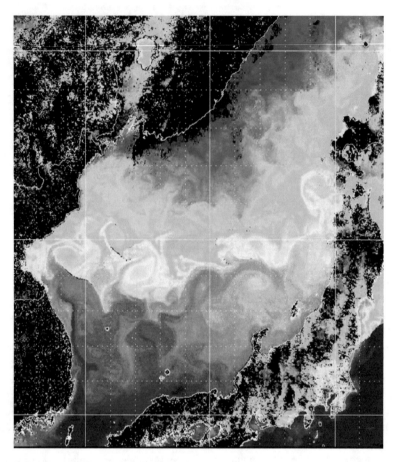

Fig. 3.1 Infrared SST image shows the Subpolar Front separating subtropical (*red*) and subarctic (*blue*) waters in the Japan Sea in winter (Credits NOAA)

natural radiation from the sea surface. Infrared radiometers operate at the waveband (∼3.7–12 μm) where the atmosphere is almost transparent. They provide an almost global coverage and good horizontal resolution up to 1 km and even less. The infrared SST image in Fig. 3.1 shows the Subpolar Front separating subtropical and subarctic waters in the Japan Sea in winter. However, infrared radiometers cannot measure SST below clouds. Passive microwave instruments, working at the wavelengths 1–30 cm, can be used in cloudy situation, but their horizontal resolution is currently only 25 km, and it is difficult to get good results with them close to the coast and sea ice edge. Passive satellite microwave remote sensing of the ocean, land, and atmosphere begun in the former Soviet Union in the 1960s using microwave radiometers on board of the satellites (sputniks) Kosmos-243 (1968) and Kosmos-384 (1970) [3, 14]. The possibilities of the microwave radiometry in measuring SST, wind speed, and sea ice coverage have been shown in these first experiments. The great progress in microwave remote sensing has been achieved with satellites and sensors of new generations [11, 14].

Ocean salinity is a key parameter that relates the Earth global water cycle to ocean circulation. Soil Moisture and Ocean Salinity (SMOS) program is the first satellite mission addressing the challenge of measuring soil moisture and sea surface salinity from space. The SMOS satellite was launched in November 2009. Moisture and salinity decrease the emissivity of soil and seawater, respectively, and thereby affect microwave radiation. An L band (21 cm, 1.4 GHz) microwave interferometric radiometer with aperture synthesis (MIRAS) generates brightness temperature images from which soil moisture and sea surface salinity are computed with the 35 km spatial resolution. Global maps for those geophysical parameters can be found at the websites [CERSAT, COPERNICUS, NODC, SMOS].

Ocean color measurements provide a monitoring of some water constituents and of the chlorophyll concentration that is related to the phytoplankton biomass and primary production. Ocean color is also a tracer of mesoscale and submesoscale processes in the ocean. Satellite-derived ocean color data are used now in fishery and coastal management, in monitoring harmful algal blooms and with ecological aims [OC, NODC, IOCCG, GLOBCOLOUR, COPERNICUS, CERSAT]. The sunlight enters the ocean, is absorbed selectively, is scattered and reflected by phytoplankton and dissolved organic material and suspended inorganic particles and then backscattered through the sea surface. Therefore, ocean color sensors can measure concentration of chlorophyll and those water constituents that interact with the sunlight. There are a few ocean color satellites: MODIS, MERIS, and Sea-viewing Wide Field-of-view Sensor (SeaWiFS). The SeaWiFS was launched on 1st August 1997 and designed for viewing ocean color in visible wavelengths from 400 to 900 nm with the spatial resolution of 0.1–5 km. The SeaWiFS image in Fig. 3.2 shows the chlorophyll a concentration in the global ocean averaged over the period April 2002–April 2012. Ocean color signals in this range can come from depths of 50 m. The main difficulty is to distinguish independently different water constituents. The MODIS (Aqua) image in Fig. 3.3 shows the chlorophyll a concentration near the Falkland Islands on November 22, 2004. The strongly inhomogeneous concentration field demonstrates a picturesque horizontal mixing at the surface with mesoscale and submesoscale eddies and filaments.

A wide variety of sea ice products are derived from passive microwave sensors, visible and infrared sensors. Satellites have provided a continuous record of sea ice since 1979. A spatial resolution of $\sim$50–100 m is now accessible with spaceborne synthetic aperture radars. Information on daily sea ice extent and sea ice edge boundary in Arctic, Antarctic, and high latitudes can be found at the websites [COPERNICUS, PODAAC, SEAICE].

Active or radar satellite instruments operate in microwave bands (1–30 cm) at the frequencies in the range 1–15 GHz. They provide measurements of SST and salinity, SSH and its anomalies, wind speed and direction, sea ice cover, wave height, and surface roughness. Radars can monitor the ocean day and night and in the presence of clouds.

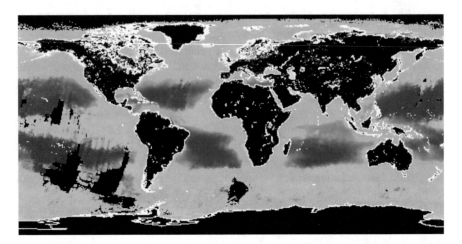

Fig. 3.2 SeaWiFS image shows the chlorophyll *a* concentration in the oceans averaged over the period April 2002–April 2012 (Credits NASA) with the *red color coding* the highest chlorophyll *a* concentration

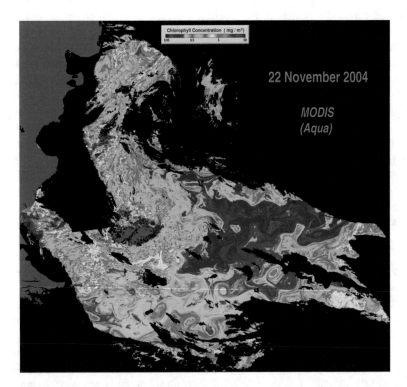

Fig. 3.3 MODIS (Aqua) image shows the chlorophyll *a* concentration near the Falkland Islands on November 22, 2004 (Credits NASA). "*Red*" and "*yellow*" waters are rich in chlorophyll *a*, whereas "*blue*" and "*violet*" waters are poor. *Black color* is cloudy areas, and the *grey color* means land

Speed and direction of ocean near-surface winds can be measured from space using scatterometry [COPERNICUS, WINDS, STAR]. A scatterometer (e.g., SeaWinds on the QuikSCAT satellite working at the frequency 13.4 GHz) is a microwave radar sensor which measures the resonant Bragg scattering produced while scanning the surface of the ocean from an aircraft or a satellite. A smooth surface does not produce a backscattering of a radar pulse. Wind increases a surface roughness and, correspondingly, the reflected signal towards the sensor. The near-surface wind direction is derived by measuring the azimuthal dependence of the scattered signal with respect to the wind direction. The SeaWinds [QUICKSCAT] provides every day measurements of wind speed from 3 to 20 m/s with the current accuracy of 2 m/s and of wind direction with an accuracy of 20° and with current spatial resolution of 25 km. It provides daily and weekly wind stress fields.

3.2 Satellite Altimetry and AVISO Velocity Field

Satellite altimetry has become the most essential observing system for ocean study and operational oceanography (see, e.g., [5]). It provides global and real-time data for measurements of the sea surface height (SSH) above the ocean rest state with high precision (currently around 1 cm) and high space and time resolution. Altimeters are not capable of measuring oceanic current directly, but sea level is directly related to ocean circulation in the geostrophic approximation. The near surface geostrophic current can be calculated from the deviation of sea level from the equigeopotential at the surface (marine geoid).

Altimeter is an active radar emitting very short microwave pulses with known power towards the sea surface at satellite nadir, the point directly beneath the satellite. It measures the round-trip travel time of a radar pulse part of which is reflected back to the satellite. Significant wave heights and near-surface wind speed also can be determined from the backscatter power related to surface roughness. The two main altimeter missions provide along-track measurements every 7 km along repetitive tracks every 9.9 days for the TOPEX/Poseidon, Jason 1 and 2 missions (the distance between tracks is 315 km at the equator) and every 35 days for the ERS, ENVISAT, and Saral/AltiKa missions (the distance between tracks is around 90 km at the equator).

The sea surface height measured by altimetry is given by

$$SSH(x, t) = GEOID(x) + h(x, t) + \varepsilon(x, t), \tag{3.1}$$

where $\varepsilon(x, t)$ represents measurement errors. The geoid is the equipotential surface that would coincide with the mean ocean surface if the oceans and atmosphere were at rest relative to the rotating Earth. It is the shape that the surface of the oceans would take under the influence of Earth's gravitation and rotation only, i.e., in the absence of winds, tides, and other influences. The absolute dynamic topography $h(x, t)$ is the distance between the height of the ocean surface from the geoid. It is caused by ocean waves, tides, currents, and variations in atmospheric pressure.

Dynamic topography refers to the topography of the sea surface related to the dynamics of its own flow. In hydrostatic equilibrium, the surface of the ocean would have no topography, but currents can raise SSH by up to a meter higher over the surrounding area. A clockwise rotation (anticyclone) is found around elevations on the ocean surface in the northern hemisphere and depressions (lows) in the southern hemisphere. Conversely, a counterclockwise rotation (cyclone) is found around depressions in the northern hemisphere and elevations in the southern hemisphere.

The geoid shape is not known accurately enough to estimate globally the absolute dynamic topography. However, one can get the variable part of dynamic topography $h(x, t) - \langle h(x, t) \rangle_T$, where $\langle h(x, t) \rangle_T$ is the mean dynamic topography obtained by averaging over a time T. The sea level anomaly (SLA) is calculated now as

$$\text{SLA}(x, t) = \text{SSH}(x, t) - \langle \text{SSH}(x, t) \rangle_T = h(x, t) - \langle h(x, t) \rangle_T + \varepsilon'(x, t). \qquad (3.2)$$

The global image of the mean dynamic topography, obtained by the TOPEX/ Poseidon altimeter for the period from October 3 to October 12, 1992, is shown in Fig. 3.4.

To get the absolute signal $h(x, t)$ one has to use a climatology and satellite gravimetric data providing accurate geoid. It is estimated currently with the accuracy of 1–2 cm and the spatial resolution of about 150 km [GRACE, GOCE]. Each component of SSH and SLA fields is contaminated by errors of different nature. Operational systems and product data are continually updated with the modern technologies to minimize the errors and to provide more and more accurate data. The SSH and SLA products can be found at the websites [PODAAC, COPERNICUS,

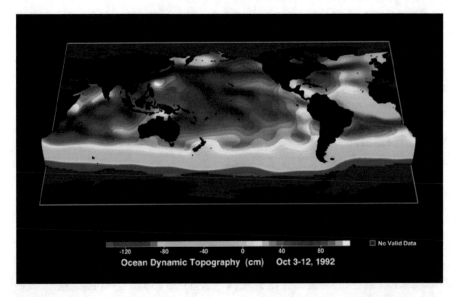

Fig. 3.4 Global image of the mean dynamic topography obtained by the TOPEX/Poseidon altimeter for the period from October 3 to October 12, 1992 (Credits CNES and NASA)

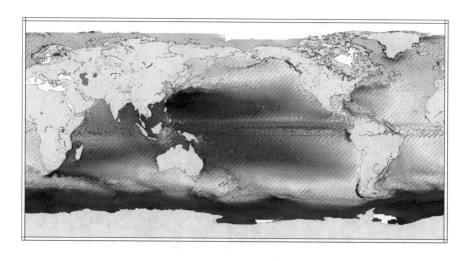

$$-1.4\ -1.2\ -1.0\ -0.8\ -0.6\ -0.4\ -0.2\ \ 0.0\ \ 0.2\ \ 0.4\ \ 0.6\ \ 0.8\ \ 1.0\ \ 1.2\ \ 1.4$$

Fig. 3.5 Global image of the sea level anomalies over the period January 1993–March 2010. *Red areas* are the ones where the SSHs change the most (Credits CNES/CLS)

NODC, AVISO, OSTM]. As an example, the global image of the sea level anomalies over the period January 1993–March 2010 is shown in Fig. 3.5.

Horizontal motions in the ocean on scales of tens km and more are much larger than vertical ones. The ocean is approximately hydrostatic and pressure is determined by the height and density of the water column. At these scales, away from boundaries and outside the surface and bottom layers, horizontal pressure gradients almost exactly balance the Coriolis force. The resulting flow follows lines of constant pressure and is known as geostrophic. The major currents, such as the Gulf Stream, the Kuroshio, and the Antarctic Circumpolar Current are geostrophic currents with a good approximation. Because pressure is related to SSH h, the zonal u_{gs} and meridional v_{gs} velocity components of geostrophic currents can be calculated by the following formula:

$$u_{gs} = -\frac{g}{f}\frac{\partial h}{\partial y}, \quad v_{gs} = \frac{g}{f}\frac{\partial h}{\partial x}, \tag{3.3}$$

where g is gravity, $f = 2\Omega \sin\phi$ is the Coriolis parameter, Ω is the angular speed of the Earth, and ϕ is the latitude.

Daily geostrophic velocities for the world's oceans, provided by the AVISO team [AVISO], approximate geostrophic ocean currents for horizontal distances exceeding a few tens of kilometers, and for times greater than a few days. The AVISO velocity data now covers the period from 1992 to the present time with daily data onto a $1/4° \times 1/4°$ grid. The description of altimeter products AVISO can be

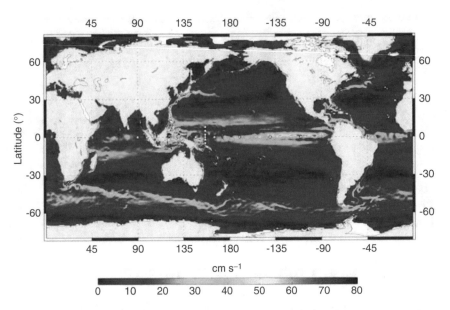

Fig. 3.6 Map of high-resolution currents averaged over the period 2000–2008 and computed from altimetry and scatterometry data (Credits LEGOS/CNRS)

found in the user handbook [2]. The global altimetric velocity field, averaged over the period 2000–2008, is shown in Fig. 3.6.

We extensively use the AVISO velocity field in our Lagrangian simulations. The important issue is its hydrodynamic interpretation which involves answers to the following questions. At which horizontal scales and for which areas in the ocean the AVISO field may be considered to be an adequate approximation? For which vertical layer it is valid?

The geostrophic velocities (3.3) are calculated from the SSH field h which is the sum of the SLA and the mean dynamic topography $\langle h(x, t) \rangle_T$. So, the altimetry errors in measuring all this quantities are the errors of the AVISO velocity field. Altimetric SLAs are measured along satellite tracks. These data are interpolated by the AVISO team onto the regular $1/4° \times 1/4°$ grid in space and daily in time. Those daily SLA maps are used to calculate daily AVISO velocity fields by solving Eq. (3.3). So the AVISO velocity field nominally has the $1/4°$ horizontal spatial resolution.

Only the geostrophic component of the horizontal velocity field is captured by altimetry. Because of comparatively large spacing between satellite tracks, submesoscale processes are not currently resolved by gridded SLAs. Submesoscale is characterized by scales of motion smaller than the Rossby radius of deformation, but large enough to be influenced by Earth's rotation and density stratification. Typical submesoscale structures are fronts, eddies, and filaments with spatial scales of $\simeq 1$–10 km and time scales of $\simeq 1$ day. Thus, altimeter-based results should be considered with a caution when dealing with submesoscale features. Sometimes

Lagrangian structures computed from the FTLE field have been found to be able to reproduce not only mesoscale slowly varying features, but also submesoscale filaments and fronts present in the tracer patterns [1, 4, 7–10]. The agreement between those structures and tracer filaments and fronts means that in some cases submesoscale phenomena appear mostly due to stirring and advection of mesoscale quasi-2D fields. However, local submesoscale phenomena like frontogenesis and ageostrophic instabilities cannot be reproduced correctly in altimeter-based velocity fields. In the near future, development of wide-swath altimetry by both the NASA SWOT and ESA Wavemill programs would take the spatial resolution of the SSH field to a few km, which should significantly improve our ability to identify submesoscale features.

As to relevant areas in seas and oceans which can be adequately represented by altimetry, there are two points. There is a singularity near the equator because the Coriolis parameter f vanishes. So, calculations of the geostrophic velocities inside the equatorial band (5° S–5° N) require special attention [6, 12]. Another errors in altimeter measurements arise near coastal lines, especially for inland seas, due to an inaccuracy in filtering tides. There are some correction methods to subtract tides and improve the SLA products using in situ observations and assimilation of tidal data. However, it is common to consider SLA data in the coastal band of 20–30 km to be unreliable.

The pressure gradient, caused by a sea surface slope, is barotropic, i.e., it does not depend on depth. In principle, it could determine the velocity of currents in a near-surface layer only because the baroclinic pressure gradient may be significant in the pycnocline due to horizontal density gradients. The altimeters cannot measure pure drift currents like Ekman currents which can be calculated from the wind stress forcing fields. Winds exert a stress on the ocean surface, transfer momentum, and drive surface currents. There are algorithms to convert the surface wind forcing to Ekman currents at different depths (see, e.g., [12]). The surface currents can be calculated from a combination of geostrophic currents derived from altimetry, a mean SSH field derived from climatology, and Ekman currents from scatterometry. The resolution is still limited by the wind forcing products and by the time scales of ocean response to winds. For most of the global ocean the geostrophic component dominates whereas the contribution of the Ekman current is less than 5 cm/s [12]. Some global surface current products, directly calculated from satellite altimetry and near-surface winds, can be found at the sites [OSCAR, MERCATOR, CTOH].

3.3   Satellite-Tracked Buoys in the Ocean

Recent advance in satellite technology has also revolutionized measurements taken by buoys drifting in the ocean. Buoys provide real-time information about ocean circulation measuring water conductivity, density, temperature, and other parameters in a high-frequency manner (six hourly or less). They collect this information and transmit it to passing satellites. Satellite sensors detect instantaneous position of a buoy and can measure its velocity fixing changes in its coordinates. To validate

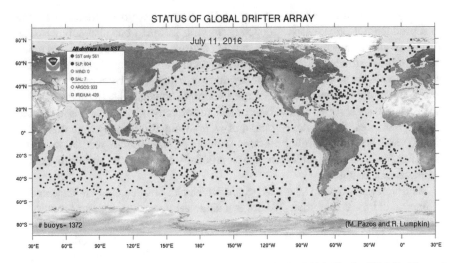

Fig. 3.7 The drifter's positions in the global ocean on July 11, 2016 (Credits NOAA). The *red* drifters measure SST only, the *blue* ones—sea level atmospheric pressure, *green* ones—salinity

our simulations we will use in this book data from two types of buoys, near-surface drifting buoys (drifters) and "diving" Argo floats.

Satellite-tracked drifter is a device floating at 15 m depth that consists of a waterproof container for instruments, transmitters to transmit the collected data, and an underwater sail to catch water currents. They are Lagrangian devices since they follow the flow in a Lagrangian manner. Lagrangian drifter data are available from the international Global Drifter Program [GDP]. The drifter's positions in the global ocean are shown in Fig. 3.7 on July 11, 2016. Drifters do not perfectly follow the water column averaged over the drogue depth. For example, water can downwell, while the drifter is forced to stay at the sea surface. Drifters suffer from some slip in high winds. The resulting speed of the drifter is thus a combination of a large-scale current at 15 m depth, the surface wind-driven flow, and the slip.

Argo floats are designed to return to the surface and telemeter their data to a satellite at the end of each mission [ARGO]. In difference from drifters Argo floats are able to rise and descend in the ocean on a programmed schedule. They provide high-quality temperature and salinity profiles from the upper 2000 m of the ice-free global ocean and currents from intermediate depths. They spend most of their life drifting at depth where they are stabilized by being neutrally buoyant at "the parking depth," typically at a depth of 1000 m. Every 10 days, by changing their buoyancy, they dive to a depth of 2000 m and then move to the sea surface, measuring conductivity and temperature profiles as well as pressure. From these, salinity and density can be calculated. Drifter and buoy observations cover currently most areas in the global ocean at sufficient density to map mean currents at, approximately, one degree resolution.

References

1. Abraham, E.R., Bowen, M.M.: Chaotic stirring by a mesoscale surface-ocean flow. Chaos **12**(2), 373–381 (2002). 10.1063/1.1481615
2. AVISO: SSALTO/DUACS User Handbook: (M)SLA and (M)ADT Near-Real Time and Delayed Time Products, 4.4 edn. (2015). http://www.aviso.altimetry.fr/fileadmin/documents/data/tools/hdbk_duacs.pdf
3. Basharinov, A., Gurvich, A., Egorov, S.T.: Microwave Radiation of the Planet Earth. Nauka, Moscow (1974) [in Russian]
4. d'Ovidio, F., Isern-Fontanet, J., López, C., Hernández-García, E., García-Ladona, E.: Comparison between Eulerian diagnostics and finite-size Lyapunov exponents computed from altimetry in the Algerian basin. Deep-Sea Res. I Oceanogr. Res. Pap. **56**(1), 15–31 (2009). 10.1016/j.dsr.2008.07.014
5. Fu, L.L., Cazenave, A. (eds.): Satellite Altimetry and Earth Sciences: A Handbook of Techniques and Applications. Academic Press, San Diego (2001)
6. Lagerloef, G.S.E., Mitchum, G.T., Lukas, R.B., Niiler, P.P.: Tropical Pacific near-surface currents estimated from altimeter, wind, and drifter data. J. Geophys. Res. Oceans **104**(C10), 23313–23326 (1999). 10.1029/1999jc900197
7. Lehahn, Y., d'Ovidio, F., Lévy, M., Heifetz, E.: Stirring of the northeast Atlantic spring bloom: a Lagrangian analysis based on multisatellite data. J. Geophys. Res. Oceans **112**(C8), C08005 (2007). 10.1029/2006JC003927
8. Prants, S.V., Budyansky, M.V., Uleysky, M.Y.: Identifying Lagrangian fronts with favourable fishery conditions. Deep-Sea Res. I Oceanogr. Res. Pap. **90**, 27–35 (2014). 10.1016/j.dsr.2014.04.012
9. Prants, S.V., Budyansky, M.V., Uleysky, M.Y.: Lagrangian fronts in the ocean. Izv. Atmos. Oceanic Phys. **50**(3), 284–291 (2014). 10.1134/s0001433814030116
10. Prants, S.V., Budyansky, M.V., Uleysky, M.Y.: Lagrangian study of surface transport in the Kuroshio extension area based on simulation of propagation of Fukushima-derived radionuclides. Nonlinear Process. Geophys. **21**(1), 279–289 (2014). 10.5194/npg-21-279-2014
11. Robinson, I.S.: Discovering the Ocean from Space: The Unique Applications of Satellite Oceanography. Springer, Berlin (2010). 10.1007/978-3-540-68322-3
12. Sudre, J., Morrow, R.A.: Global surface currents: a high-resolution product for investigating ocean dynamics. Ocean Dyn. **58**(2), 101–118 (2008). 10.1007/s10236-008-0134-9
13. Traon, P.Y.L.: Satellites and operational oceanography. In: Operational Oceanography in the 21st Century, pp. 29–54. Springer, Dordrecht (2011). 10.1007/978-94-007-0332-2_2
14. Wilson, W., Fellous, J.L., Kawamura, H., Mitnik, L.: A history of oceanography from space. Manual of Remote Sensing. Remote Sensing of the Marine Environment, vol. 6, 3rd edn., pp. 1–31. American Society of Photogrammetry and Remote Sensing, Washington (2006)

List of Internet Resources

[ARGO] International Argo Program Homepage.
    http://www.argo.net/
[AVISO] Archiving, Validation and Interpretation of Satellite Oceanographic (AVISO).
    http://www.aviso.altimetry.fr
[CERSAT] CERSAT — Centre ERS d'Archivage et de Traitement.
    http://cersat.ifremer.fr/
[COPERNICUS] Copernicus — Marine Environment Monitoring Service.
    http://marine.copernicus.eu/

[CTOH] CTOH — Centre of Topography of the Oceans and the Hydrosphere.
http://ctoh.legos.obs-mip.fr/
[GDP] The Global Drifter Program.
http://www.aoml.noaa.gov/phod/dac
[GHRC] Global Hydrology Resource Center.
https://ghrc.nsstc.nasa.gov/home/
[GHRSST] Group for High Resolution Sea Surface Temperature.
https://www.ghrsst.org/
[GLOBCOLOUR] GlobColour Project — The European Service for Ocean Color.
http://www.globcolour.info/
[GOCE] GOCE — Gravity field and Ocean Circulation Explorer.
http://www.esa.int/esaLP/LPgoce.html
[GRACE] GRACE — Gravity Recovery and Climate Experiment.
http://www.csr.utexas.edu/grace/
[IOCCG] International Ocean Colour Coordinating Group.
http://www.ioccg.org/
[MERCATOR] Mercator Océan — Ocean Forecasters.
http://www.mercator-ocean.fr/
[NODC] National Oceanographic Data Center.
https://www.nodc.noaa.gov/
[NSIDC] NASA Distributed Active Archive Center at NSIDC.
https://nsidc.org/daac
[OC] NASA Ocean Color.
http://oceancolor.gsfc.nasa.gov/cms/
[OSCAR] OSCAR — Ocean Surface Current Analyses Real-time.
http://www.oscar.noaa.gov
[OSTM] Ocean Surface Topography from Space.
https://sealevel.jpl.nasa.gov/
[PODAAC] Physical Oceanography Distributed Active Archive Center.
http://podaac-www.jpl.nasa.gov/
[QUICKSCAT] QuickSCAT Mission.
https://winds.jpl.nasa.gov/missions/quikscat/
[REMSS] Remote Sensing Systems.
http://www.remss.com/
[SEAICE] National Snow and Ice Data Center.
https://nsidc.org/data/seaice
[SMOS] SMOS (Soil Moisture Ocean Salinity) — Earth Online.
https://earth.esa.int/smos
[STAR] STAR Ocean Surface Winds Projects.
http://manati.star.nesdis.noaa.gov/
[WINDS] Measuring Ocean Winds from Space.
https://winds.jpl.nasa.gov/

Chapter 4
Lagrangian Tools to Study Transport and Mixing in the Ocean

4.1 Lagrangian Indicators and Lagrangian Maps

The satellite-derived and numerically generated velocity fields are given as discrete data sets, rather than analytical functions. Moreover, the velocity field in the ocean is only known for finite times. Some numerical algorithms are needed to solve advection equations with such data sets. Lagrangian simulation is based on solving advection equations for a large number of synthetic tracers on the Earth sphere in a given velocity field

$$\dot{\lambda} = u(\lambda, \phi, t), \qquad \dot{\phi} = v(\lambda, \phi, t), \tag{4.1}$$

where u and v are angular zonal and meridional velocities, ϕ and λ are latitude and longitude, respectively.

The velocity field is given on a grid. In order to integrate the advection equations (4.1), we need to know velocities between the grid points interpolating a given data in space and time. Thus, the numerical algorithm for solving advection equations is as follows:

1. A bicubical interpolation in space and an interpolation by third order Lagrangian polynomials in time are used. The velocity components are interpolated independently on each other.
2. The velocities obtained are substituted in Eq. (4.1) which are integrated using a fourth-order Runge–Kutta scheme with a fixed time step.
3. The outputs are processed, analyzed, and represented as plots and Lagrangian maps in the geographical coordinates.

Each elementary volume of water can be attributed to physicochemical properties (temperature, salinity, density, radioactivity, etc.) which characterize this volume as it moves. In addition, each water parcel can be attributed to more specific characteristics as trajectory's functions that carry key data but are not physicochemical

© Springer International Publishing AG 2017
S.V. Prants et al., *Lagrangian Oceanography*, Physics of Earth and Space Environments, DOI 10.1007/978-3-319-53022-2_4

properties. We call them "Lagrangian indicators." Some of them, as distance passed
by a fluid particle for some period of time; full, zonal, and meridional displacements
of particles from their original positions; the number of cyclonic and anticyclonic
rotations; time of residence of fluid particles inside a given area and the exit
time; how many times particles have visited different places in a region, will be
extensively used in the rest of the book. The Lagrangian indicator (or function M
in terminology of authors [61]), based on computation of arc length of particle
trajectories and introduced in [61], has been shown to be useful for revealing phase-
space structures in fluid flows [23, 60, 62, 63]. A family of Lagrangian indicators
for characterizing mixing in fluid flows, based on computation of the extreme extent
of a trajectory and some of its variants, has been proposed recently in [65].

The Lagrangian indicators allow to identify water masses moving coherently,
either propagating together or rotating together. The Lagrangian indicators are also
important because they contain information about the origin, history, and fate of
water masses. Even if adjacent waters are indistinguishable, say, by temperature, and
the corresponding satellite SST images indicate no thermal front, the corresponding
water masses can be distinguishable, for example, by their origin, travelling history,
and other factors.

The Lagrangian indicators are computed by solving advection equations forward
and backward in time in order to know the fate and origin of water masses, respectively. When integrating advection equations (4.1) forward in time we compute
particle trajectories to know their fate and when integrating them backward in time
we know origin of particles and their history. In oceanography, many of important
problems concern where tracers have been or came from, rather than where they are
headed for. For example, it is crucial sometimes to find out the source of those oil
spills that wash ashore from sunken ships. Plane crashes and shipwrecks produce
"debris spills" which drift in the sea. The forward-in-time strategy is not suitable
for that. One needs to seed the whole region under study with a huge number of
synthetic particles and track only a small part of them in order to find those ones
which came to a given place in a given time.

Backward-in-time integration of advection equations seems to be much more
effective. The idea is very simple. We populate the area under study with a large
number of particles and advect them in a known velocity field backward in time
for an appropriate period. In the end of this period of time we get positions of all
the particles on a fixed day in the past. In other words, we can identify particle's
positions from known destinations at earlier times. In addition, it is useful to
compute simultaneously some Lagrangian indicators of those tracers which are
functions of their trajectories.

The backward-in-time tracking method is especially efficient to identify sources,
history, and transport pathways of specific water parcels in the ocean. The Tohoku
earthquake on 11 March 2011, followed by the catastrophic tsunami, inflicted
heavy damage on the Fukushima Nuclear Power Plant (FNPP). Large amount
of water contaminated with radionuclides leaked directly into the ocean at the
FNPP location with a subsequent atmospheric fallout on a large area of the ocean
surface. Fukushima-derived radioactive isotopes ^{137}Cs and ^{134}Cs have been detected

over a broad area in the North Western Pacific in 2011, 2012, and even in 2013
[3, 12, 13, 37, 40, 41, 43–46, 52, 66]. The source of that contamination was, of
course, known, but it has not been known where higher concentration of cesium
could be found in the ocean far away from the FNPP and after a long time
after the accident. Unfortunately, the backward-in-time methods are not popular
in oceanography and meteorology. Perhaps, it may be explained by a long-time
tradition, especially in meteorology, to focus on forecast and lack of detailed data
on velocity fields in the ocean and atmosphere. The situation was cardinally changed
in the era of satellites and advanced numerical models.

In order to display the enormous amount of information, we compute Lagrangian
maps introduced in [72, 77, 83–85] which are plots of a specific Lagrangian
indicator versus particle's initial position. They help to visualize different aspects
of large-scale transport and mixing in the ocean. The region under study is seeded
with a large number of synthetic particles whose trajectories are computed forward
or backward in time for a given period of time. The results obtained are processed
to get a data file with the field of a specific Lagrangian indicator in this area. Finally,
its values are coded by color and represented as a map in geographic coordinates.
Sometimes it is informative to impose two or more different Lagrangian indicators
on a single map.

The finite-time displacement of tracers, D, is a distance between the final,
(λ_f, ϕ_f), and initial, (λ_0, ϕ_0), positions of advected particles on the Earth sphere

$$D \equiv R_E \arccos[\sin \phi_0 \sin \phi_f + \cos \phi_0 \cos \phi_f \cos(\lambda_f - \lambda_0)], \qquad (4.2)$$

where R_E is the Earth radius in km. This quantity and its zonal and meridional
components have been shown to be useful in studying large-scale transport and
mixing in various basins, from bays [83] and regional seas [71, 74, 76, 77] to
the ocean scale [1, 16, 72, 74, 75], in quantifying propagation of radionuclides in
the North Western Pacific after the accident at the FNPP [12, 73, 80, 84] and in
finding potential fishing grounds [78, 79, 85].

Color contrast on the drift maps demarcates boundaries between waters which
passed rather different distances before converging. In Fig. 4.1a we show the AVISO
velocity field in the North Pacific Ocean area on the fixed day, November 14, 2010.
The two powerful currents with increased velocity values are clearly visible on the
map: the Kuroshio and its Extension to the east of the Japan coast and the Alaskan
Stream along the Aleutian Islands between the Kamchatka and Alaska peninsulas.
The map in Fig. 4.1a demonstrates the ocean fronts on the planetary and synoptic
scales, including the subarctic frontal zone in the Japan Sea (situated between the
Asia continent and Japan) and low-energetic regions, as for example, the Okhotsk
Sea (to the north of the Japan Sea) excepting for its southern part between the
Sakhalin and the Kuril Islands.

The Lagrangian drift map on 15 May 2011 in Fig. 4.1b shows absolute displacements, D, for 2.25 millions of particles in the North Pacific computed backward in
time for 2 weeks in the altimetric AVISO velocity field. The black color means
that the corresponding water parcels on the map passed much more distance as

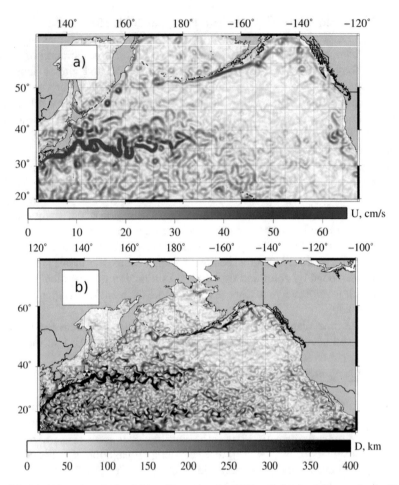

Fig. 4.1 (a) Altimetric velocity field on November 14, 2010 and (b) the drift map for the North Pacific Ocean on May 15, 2011 computed backward in time for 2 weeks

compared to the white-colored particles. Practically all the region is covered by mesoscale eddies of different size including dipole and mushroom-like structures. Some currents, the Kamchatka, the Oyashio, and the Californian currents, look like vortex streets with moving mesoscale eddies each of which is surrounded by a black collar which demarcates the boundary separating the eddy's core from surrounding waters.

Sometimes in order to get a more detailed information about origin of water masses, it is useful to compute latitudinal and longitudinal Lagrangian maps [1, 16, 76, 80]. The domain under study is seeded with synthetic particles which are advected backward in time by a velocity field starting from a given day to a day in the past. The latitude ϕ or longitude λ from which each particle came to its final position during that period of time is coded by color.

The material line technique is another tool to trace origin, history, and fate of water masses [12, 71, 75, 80]. A large number of tracers are placed on the material line, crossing an oceanographic feature under study (an eddy, for example) or a strait, and are evolved backward in time. Different kind of outputs can be obtained with the help of that technique. It is informative to get as an output tracking maps showing by density plots where the corresponding tracers were walking for a given period of time. This method is especially informative fixing material lines along the transects with in situ measurements.

4.2 Hyperbolicity in the Ocean

Hyperbolicity is an important concept in dynamical systems theory (see, e.g., [27, 57, 59, 98]). It is characterized by the presence of expanding and contracting directions in the phase space of a dynamical system. This is a situation with phase trajectories converging in some directions and diverging in other ones. In unperturbed Hamiltonian systems, the separatrices (if they exist) connect either two hyperbolic stationary points or belong to a single hyperbolic **stationary point**. In the case of a heteroclinic connection, the stable branch of the separatrice of one stationary point coincides with the unstable branch of the separatrice of the other point and vice versa. In the case of a homoclinic connection, stable and unstable branches of the separatrice of a single hyperbolic stationary point coincide. In Fig. 4.2 we plot the schematic phase portrait of a flow nearby a hyperbolic stationary point with phase trajectories converging in two directions and diverging in other two.

Under a periodic perturbation, hyperbolic stationary points may transform into periodic hyperbolic trajectories with time-dependent separatrices called stable, $W_s(\gamma)$, and unstable, $W_u(\gamma)$, invariant **manifolds**. If a trajectory has nonzero

Fig. 4.2 The schematic phase portrait of a flow nearby a hyperbolic stationary point with phase trajectories converging in two directions and diverging in other two

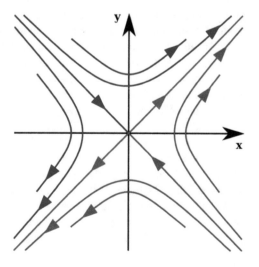

Lyapunov exponents it is said to be hyperbolic [47]. In the extended phase space with time as the third coordinate, $W_s(\gamma)$ and $W_u(\gamma)$ are surfaces filled in trajectories approaching asymptotically to $\gamma(t)$ at $t \to \infty$ (W_s) and $t \to -\infty$ (W_u). Those surfaces intersect each other on the section plane $t = 0$ in an infinite number of homoclinic points producing a heteroclinic or a homoclinic tangle which is a seed of **dynamical chaos** (see, e.g., [27, 57, 98]). Hyperbolic stationary points, hyperbolic trajectories, and stable and unstable invariant manifolds are hyperbolic objects with the dimensions of $n = 0$, $n = 1$, and $n \leq m - 1$, respectively, where m is the dimension of the extended phase space. A blob in the phase space, chosen nearby a hyperbolic object, experiences stretching and folding that typically give rise to chaotic motion even in purely deterministic dynamical systems, the remarkable phenomenon known as deterministic or dynamical chaos [27, 57, 98]. The ultimate reason of that chaos is a local instability of trajectories.

By an instantaneous **stationary point**, one means a point in space where the velocity is zero at a fixed instant of time. As is well known, local stability properties of the stationary points can be characterized by eigenvalues of the Jacobian matrix of a velocity field evaluated at that instant of time. For 2D flows, if the two eigenvalues are real and of opposite sign, then the stationary point is a hyperbolic stationary point. If they are pure imagine and complex conjugated, then one gets an elliptic stationary point. Two zero eigenvalues of the Jacobian matrix mean the existence of a parabolic stationary point. Hyperbolic stationary points are a form of Eulerian hyperbolicity, whereas hyperbolic trajectories are a form of Lagrangian hyperbolicity.

Theory of chaotic advection in fluids [4] is well developed in periodic 2D flows where periodic hyperbolic trajectories are fixed points on the **Poincaré sections**. The stable and unstable manifolds of hyperbolic trajectories can be computed using different techniques and are known to form complex homoclinic or heteroclinic tangles which are "seeds" of chaos. Chaotic advection typically is due to the action of both the deformation part of the velocity field (the strain field), which can permanently stretches and expands a patch of tracer, and the vorticity part which tends to fold it. In elliptic regions, for example, inside eddies, where the relative vorticity dominates over the strain, tracer patterns deform weakly. Deformation regions, where the strain dominates, are called hyperbolic ones. In these regions a tracer pattern stretches horizontally into elongated and thin filaments. Chaotic advection in analytically given periodic 2D flows have been discussed in detail in Chaps. 1 and 2 (see also [50, 51, 89]).

Typical geophysical flows in the ocean and atmosphere and other natural flows are aperiodic. There is no analog of the **KAM theorem** in aperiodic vector fields. Nevertheless, the notions of hyperbolicity and of stable and unstable manifolds are still valid for those fields because they are not connected with nature of the considered time dependence (see, e.g., [47, 56, 97] and other references mentioned in [59]). If hyperbolicity is determined by any means, then the nature of the time dependence plays no role. Once a hyperbolic trajectory is located, the stable and unstable manifold theorem for hyperbolic trajectories immediately applies [59].

On the other hand, the notions of hyperbolicity, stationary points, hyperbolic trajectories and their stable and unstable manifolds, and the very phenomenon of dynamical chaos are strictly defined in the infinite-time limit. In real and numerically generated velocity fields, the hyperbolic objects are of transient nature. Hyperbolic stationary points in steady flows and periodic hyperbolic trajectories in idealistic periodic flows are persistent features. In real-life flows trajectories can gain or lose hyperbolicity over time, i.e., they may be hyperbolic for some time intervals and not to be hyperbolic for other ones. The very definition of stable and unstable manifolds requires an infinite-time limit which is irrelevant for geophysical flows. It is a challenge even to define hyperbolic trajectories and their stable and unstable manifolds in aperiodic flows. The generalization of the dynamical system theory to aperiodic velocity fields, defined on a space-time grid, has been developed recently (see, e.g., [28, 33, 59, 64]).

The AVISO velocity field is provided with a day interval. "Instantaneous" stationary points in such a field are those ones where the velocity is found to be zero on a fixed day. They are not fluid particle trajectories. "Instantaneous" elliptic and hyperbolic stationary points, to be present in an area studied, are indicated on Lagrangian maps in this book by triangles and crosses, respectively. Red (blue) triangles mark elliptic stationary points with anticyclonic (cyclonic) rotations of water around them. The elliptic stationary points, situated mainly in the centers of eddies, are those points around which the motion is stable and circular. The hyperbolic stationary points, situated mainly between and around eddies, are unstable ones with the directions along which waters converge to such a point and other directions along which they diverge (Fig. 4.2).

The stationary points are moving Eulerian features which may undergo **bifurcations** in the course of time. Bifurcation theory, among other things, is interested in behavior of fixed points of vector fields as a parameter is varied. In our case time plays the role of the parameter. One may monitor positions of hyperbolic and elliptic stationary points day by day and look for their movement around in the oceanic flow. Nothing interesting, besides a rearrangement of the flow, occurs if they do not change their stability type. When they do that, there are a few possibilities [59]. In the saddle-node bifurcation, hyperbolic and elliptic stationary points collide and annihilate each other in the course of time. The opposite process could occur as well: two stationary points, one hyperbolic and one elliptic, are born suddenly. After the collision, the stationary points may move apart without changing in number (transcritical bifurcation) or split into three ones (pitchfork bifurcation).

Stable and unstable manifolds of the hyperbolic objects can be simulated by local and global methods. In the local approach, firstly, one locates positions of hyperbolic stationary points. Then it is necessary to identify the hyperbolic trajectories which are situated, as a rule, nearby the hyperbolic stationary points. It can be done by different ways. We prefer to use such a hyperbolic point as the first guess, placing on a fixed day a few material segments oriented at different angles and computing the FTLE for the particles belonging to those segments. Coordinates of the particles with the maximal FTLE give us approximate position of the hyperbolic trajectory nearest to that hyperbolic point on that day. Then we place the patch with a large number of synthetic particles, centered at the hyperbolic trajectory's position, and evolve it forward in time.

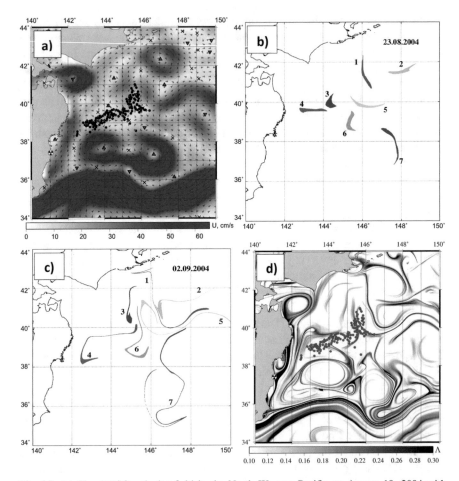

Fig. 4.3 (**a**) The AVISO velocity field in the North Western Pacific on August 19, 2004 with overlaid tuna fishing locations (*dots*) and elliptic (*triangles*), and hyperbolic (*crosses*) "instantaneous" stationary points. (**b**) and (**c**) Evolution of the tracer patches placed on that day at seven hyperbolic stationary points in the region. The patches for 2 weeks delineate the corresponding unstable manifolds, which are approximated by the *black ridges* on the backward-time FTLE map on September 2 (**d**) with the FTLE values in days^{-1}

It is shown in Fig. 4.3b, c how the method works in the Oyashio–Kuroshio frontal zone in the northwestern part of the Pacific Ocean to the east off Japan, where the subarctic waters of the cold Oyashio Current encounter the subtropical waters of the warm Kuroshio Current. This region is known to be one of the richest fisheries in the world. The jet of the Kuroshio Extension in the south, the two mesoscale anticyclonic eddies (ACEs) to the north of the jet, the Kuroshio ring near the Hokkaido Island with the center at 42.4° N, 147° E, and the mesoscale ACE at the traverse of the Tsugaru straight (41.4° N, 142.3° E) are clearly seen in the AVISO velocity field (Fig. 4.3a). Figure 4.3b, c shows how the corresponding unstable manifolds evolve from seven tracer patches placed near chosen seven hyperbolic trajectories. After 1 week only, the patches elongate and display the strongest unstable manifolds in the region. Other examples of manifestation of hyperbolicity in the real ocean can be found in [58, 81, 82, 94].

In the global approach, one seeds the whole area with a large number of synthetic particles and computes the FTLE field which is a commonly used measure of hyperbolicity in oceanic and atmospheric flows (see the next section). It has been shown by Haller [6, 30] that the curves of local maxima of the FTLE field, attributed to initial tracer's positions, approximate stable manifolds when computing advection equations forward in time and unstable ones when computing them backward in time. To compare the results obtained in the local and global approaches, we compute the FTLE field in the same region. Comparing Fig. 4.3c with the backward-time FTLE map on September 2 in Fig. 4.3d, it is clear that the patches delineate the corresponding "ridges" on the backward-time FTLE map. The rectangular patches nos. 1–5 were chosen in the area with nutrient rich waters with comparatively high chlorophyll a concentration, but in the course of time they deformed into narrow filaments penetrated into more oligotrophic waters, poor with nutrients. The passive marine organisms in those fluid patches have been advected along with them into oligotrophic waters attracting fish and marine animals for feeding.

4.3 Finite-Time Lyapunov Exponents

4.3.1 Finite-Time Lyapunov Exponents for an n-Dimensional Vector Field

We describe in this section a general method to compute Lyapunov exponents, introduced in [77], which is valid for n-dimensional vector fields. Let us start with an n-dimensional set of nonlinear ordinary differential equations in the vector form

$$\dot{\mathbf{x}} = \mathbf{f}(\mathbf{x}, t), \quad \mathbf{x} = (x_1, \ldots, x_n),$$
$$\mathbf{f}(\mathbf{x}, t) = (f_1(x_1, \ldots, x_n, t), \ldots, f_n(x_1, \ldots, x_n, t)). \tag{4.3}$$

The Lyapunov exponent at an arbitrary point $\mathbf{x_0}$ is given by

$$\Lambda(\mathbf{x_0}) = \lim_{t \to \infty} \lim_{\|\delta \mathbf{x}(0)\| \to 0} \frac{\ln(\|\delta \mathbf{x}(t)\| / \|\delta \mathbf{x}(0)\|)}{t}, \tag{4.4}$$

where $\delta \mathbf{x}(t) = \mathbf{x_1}(t) - \mathbf{x_0}(t)$, $\mathbf{x_0}(t)$, and $\mathbf{x_1}(t)$ are solutions of the set (4.3), $\mathbf{x_0}(0) = \mathbf{x_0}$. The limit (4.4) exists, is the same for almost all the choices of $\delta \mathbf{x}(0)$, and has a clear geometrical sense: trajectories of two nearby particles diverge (converge) in time exponentially (in average) with the coefficients given by the Lyapunov exponents.

Due to smallness of $\delta \mathbf{x}$, one can linearize the set (4.3) in a vicinity of some trajectory $\mathbf{x_0}(t)$ and obtain a set of time-dependent linear equations [26]

$$\begin{pmatrix} \delta \dot{x}_1 \\ . \\ \delta \dot{x}_n \end{pmatrix} = J(t) \begin{pmatrix} \delta x_1 \\ . \\ \delta x_n \end{pmatrix}, \qquad (4.5)$$

where $J(t)$ is the Jacobian matrix of the set (4.3) along the trajectory $\mathbf{x_0}(t)$

$$J(t) = \begin{pmatrix} \dfrac{\partial f_1(\mathbf{x_0}(t), t)}{\partial x_1} & \cdots & \dfrac{\partial f_1(\mathbf{x_0}(t), t)}{\partial x_n} \\ \cdots\cdots\cdots\cdots \\ \dfrac{\partial f_n(\mathbf{x_0}(t), t)}{\partial x_1} & \cdots & \dfrac{\partial f_n(\mathbf{x_0}(t), t)}{\partial x_n} \end{pmatrix}. \qquad (4.6)$$

Solution of the linear set (4.5) can be found with the help of the evolution matrix $G(t, t_0)$

$$\begin{pmatrix} \delta x_1(t) \\ . . \\ \delta x_n(t) \end{pmatrix} = G(t, t_0) \begin{pmatrix} \delta x_1(t_0) \\ \cdots \\ \delta x_n(t_0) \end{pmatrix}. \qquad (4.7)$$

The evolution matrix obeys the differential equation which can be obtained after substituting (4.7) into (4.5)

$$\dot{G} = JG, \qquad (4.8)$$

with the initial condition $G(t_0, t_0) = I$, where I is the unit matrix. Any evolution matrix has the important multiplicative property

$$G(t, t_0) = G(t, t_1)G(t_1, t_0). \qquad (4.9)$$

One can decompose the evolution matrix as follows:

$$G(t, t_0) = U(t, t_0)\Sigma(t, t_0)V^T(t, t_0), \qquad (4.10)$$

which is known as "a singular-value decomposition." Here U, V are orthogonal, and $\Sigma = \text{diag}(\sigma_1, \ldots, \sigma_n)$ is diagonal matrices. The quantities $\sigma_1, \ldots, \sigma_n$ are called singular values of the matrix G.

The maximal value $\lim_{\|\delta \mathbf{x}(0)\| \to 0} \dfrac{\|\delta \mathbf{x}(t)\|}{\|\delta \mathbf{x}(0)\|}$ for the set (4.5) equals to $\sigma_1(G(t))$. It is the maximal singular value of the matrix $G(t)$. If $\lim_{t \to \infty} \dfrac{\sigma_2(G(t))}{\sigma_1(G(t))} = 0$, where $\sigma_2(G(t))$ is the next (smaller) singular value of the matrix $G(t)$ in magnitude, then (4.4) can be redefined as follows:

$$\Lambda_{max} = \lim_{t \to \infty} \frac{\ln \sigma_1(G(t))}{t - t_0}. \qquad (4.11)$$

The quantity

$$\Lambda = \frac{\ln \sigma_1(G(t))}{t - t_0} \qquad (4.12)$$

is called *the finite-time Lyapunov exponent (FTLE)*. It is the ratio of the logarithm of a maximal possible stretching of a vector to a time interval $t - t_0$. The *instantaneous Lyapunov exponent* Λ_0 is a Lyapunov exponent of the set of linear equations

$$\begin{pmatrix} \delta \dot{x}_1 \\ . \\ \delta \dot{x}_n \end{pmatrix} = J(0) \begin{pmatrix} \delta x_1 \\ . \\ \delta x_n \end{pmatrix}. \qquad (4.13)$$

It is the rate of exponential diverging of trajectories at a given point at a given instant of time.

Equation (4.8) cannot be numerically integrated over a large time because the elements of the corresponding evolution matrix grow exponentially, if one of the Lyapunov exponents is positive. However, we can divide a large time interval on subintervals with the duration which is less or order of the Lyapunov time, $t_\Lambda = 1/\Lambda$, and represent the whole evolution matrix as a product of evolution matrices computed on these subintervals using the property (4.9). We compute this product and the corresponding singular values using the GNU Multiple Precision Arithmetic Library [GMP] in order to preserve the absolute precision of our representation of the evolution matrix.

4.3.2   Singular-Value Decomposition and Evolution Matrix for Two-Dimensional Case

The singular-value decomposition is a representation of any $m \times n$-matrix in the form

$$M = U\Sigma V, \qquad (4.14)$$

where U and V are $m \times m$ and $n \times n$ unitary matrices, respectively, Σ is a diagonal $m \times n$-matrix. The diagonal elements of Σ are singular values of the matrix M. The eigenvectors u and v, such that $Mv = \sigma u$ and $M^*u = \sigma v$ (σ is a singular value of M), are, respectively, left and right singular vectors of the matrix M. If M is real-valued, then its singular values are real as well. U and V are orthogonal matrices. The matrix Σ and its singular-value decomposition are defined to an accuracy of the permutation of singular values. Therefore, one may require to order the singular values of Σ as a nonincreasing sequence, and such a decomposition is unique.

If the matrix M is squared then its singular-value decomposition has a simple geometric meaning. Action of any matrix to a vector can be represented as the following three successive transformations: the first rotation/reflection by the matrix V, a stretching/contraction along the coordinate axis by the matrix Σ, and the second rotation/reflection by the matrix U. Thus, the matrix M transforms a sphere of the unit radius in an ellipsoid with the semiaxis to be equal to singular values directed along the left singular vectors. The right singular vectors are correspondingly pre-images of the ellipsoid's semiaxis.

Let us consider now a 2D flow with 2×2 evolution matrix with the singular-value decomposition

$$G = UDV \Rightarrow \begin{pmatrix} a & b \\ c & d \end{pmatrix} = \begin{pmatrix} \cos\phi_2 & -\sin\phi_2 \\ \sin\phi_2 & \cos\phi_2 \end{pmatrix} \begin{pmatrix} \sigma_1 & 0 \\ 0 & \sigma_2 \end{pmatrix} \begin{pmatrix} \cos\phi_1 & -\sin\phi_1 \\ \sin\phi_1 & \cos\phi_1 \end{pmatrix}. \quad (4.15)$$

Transformations of a circle with the unit radius by those matrices and its singular vectors are shown in Fig. 4.4. Reflection matrices are not used in this decomposition, therefore, singular values can be negative. However, it is clear from general consideration that the evolution matrix of a continuous flow cannot contain reflections.

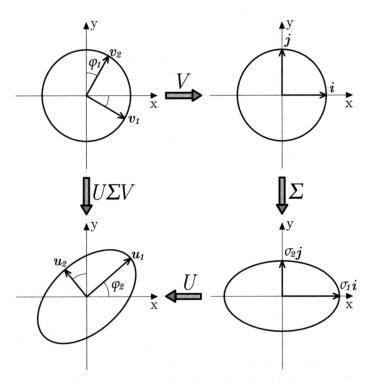

Fig. 4.4 Geometric meaning of the singular-value decomposition of a 2×2 matrix

Multiplying the matrices, one gets the set with four equations and four variables

$$\begin{pmatrix} a & b \\ c & d \end{pmatrix} = \begin{pmatrix} \sigma_1 \cos \phi_1 \cos \phi_2 - \sigma_2 \sin \phi_1 \sin \phi_2 & -\sigma_1 \sin \phi_1 \cos \phi_2 - \sigma_2 \cos \phi_1 \sin \phi_2 \\ \sigma_1 \cos \phi_1 \sin \phi_2 + \sigma_2 \sin \phi_1 \cos \phi_2 & -\sigma_1 \sin \phi_1 \sin \phi_2 + \sigma_2 \cos \phi_1 \cos \phi_2 \end{pmatrix}.$$

(4.16)

Let us introduce the following notations:

$$\alpha = a + d, \quad \beta = a - d, \quad \gamma = c + b, \quad \delta = c - b,$$

$$\xi = \sigma_1 + \sigma_2, \quad \eta = \sigma_1 - \sigma_2, \quad \Phi = \phi_1 + \phi_2, \quad \Psi = \phi_2 - \phi_1.$$

(4.17)

Adding and deducting Eq. (4.16) and using the notations (4.17), we get

$$\alpha = \xi \cos \Phi, \quad \beta = \eta \cos \Psi, \quad \gamma = \eta \sin \Psi, \quad \delta = \xi \sin \Phi.$$

(4.18)

Solution of the set (4.18) is

$$\xi = \sqrt{\alpha^2 + \delta^2}, \quad \eta = \sqrt{\beta^2 + \gamma^2}, \quad \Phi = \text{arctan2} \, (\delta, \alpha), \quad \Psi = \text{arctan2} \, (\gamma, \beta),$$

(4.19)

where arctan2 (y, x) is an angle between the vector (x, y) and the axis x which can be defined as

$$\text{arctan2} \, (y, x) = \begin{cases} \arctan (y/x), & x \geq 0, \\ \arctan (y/x) + \pi, & x < 0. \end{cases}$$

(4.20)

The final solution is

$$\sigma_1 = \frac{\sqrt{(a+d)^2 + (c-b)^2} + \sqrt{(a-d)^2 + (b+c)^2}}{2},$$

$$\sigma_2 = \frac{\sqrt{(a+d)^2 + (c-b)^2} - \sqrt{(a-d)^2 + (b+c)^2}}{2},$$

$$\phi_1 = \frac{\text{arctan2} \, (c - b, a + d) - \text{arctan2} \, (c + b, a - d)}{2},$$

$$\phi_2 = \frac{\text{arctan2} \, (c - b, a + d) + \text{arctan2} \, (c + b, a - d)}{2}.$$

(4.21)

It is evident from the solution that the singular values are ordered in a nonincreasing way, i.e., $\sigma_1 \geq \sigma_2$. The product $\sigma_1 \sigma_2$ defines the ratio of the final area to the initial and equals to Det M. It follows from the definition of a singular-value decomposition that

$$\sigma_1 \geq \frac{\|M\mathbf{x}\|}{\|\mathbf{x}\|} \geq \sigma_2,$$

(4.22)

where $\| \cdot \|$ is the Euclidean norm. In other words, the length of any vector $\mathbf{x}$ is changed under the action of the matrix M in σ_2 times as minimum and in σ_1 times as maximum.

4.4 Lagrangian Coherent Structures

The existence of large-scale coherent structures in quasi-random (turbulent) flows has long been recognized (see, e.g., [39]). Before the coherent structures were found, it was a common opinion that turbulent flows are determined only by irregular vortical fluid motion. Although up to now there is no consensus on a strict definition of coherent structures, they can be considered as connected turbulent fluid masses with phase-correlated (i.e., coherent) vorticity over the spatial extent of the shear layer. Thus, turbulence consists of coherent and phase-random (incoherent) motions with the latter ones superimposed on the former ones. Lagrangian motion may be strongly influenced by those coherent structures that support distinct regimes in a given turbulent flow. The discovery that turbulent flows are not fully random but embody orderly organizing structures was a breakthrough in fluid mechanics.

Haller [29, 30, 32] proposed a concept of distinguished material lines and surfaces for extracting distinguished Lagrangian coherent structures (LCSs) in the flow field. Material line is a curve consisting of fluid particles which is advected by the velocity field. Stable and unstable manifolds are prominent material lines. They can be identified by different methods in steady, periodic, and quasiperiodic flows (see Chaps. 1 and 2) with the velocity field known for all time. Stable and unstable manifolds are a kind of a dynamical skeleton for material patterns in flows. However, they are ill defined in aperiodic flows where the velocity field is known only for finite-time intervals. Haller defines the LCSs in 2D flows during a finite-time interval as the distinguished material lines that repel or attract nearby fluid trajectories at the highest local rate relative to other material lines nearby. In this sense, they are the most influential repelling and attracting material lines. Any fluid flow is composed of a continuum of material surfaces. The idea is to identify among them those with locally the strongest stability which would dominate in advection. The LCS approach provides a means to identify a dynamical skeleton for aperiodic flows over a selected period of time. It should be stressed that LCSs in this book are understood as coherent material lines or surfaces but not coherent regions in the ocean , e.g., vortex cores.

The LCSs are Lagrangian because they are invariant material curves consisting of the same fluid particles. They are coherent because they are comparatively long lived and more robust than the other adjacent structures. They are connected with stable and unstable invariant manifolds of hyperbolic trajectories. A tracer patch, chosen nearby any stable manifold, moves in the course of time to the corresponding hyperbolic stationary point squeezing along that manifold. After approaching that point, the patch begins to stretch along the corresponding unstable manifold. The form of the deformed patch is independent on coordinate frame, whereas the form of the unsteady velocity field may depend strongly on the reference frame. In this sense, the LCSs are frame invariant.

To extract the LCSs, Haller proposed to compute FTLEs. Pierrehumbert and Yang [70] were the first who proposed to compute FTLEs in velocity-field data as a measure of the sensitivity of a current position of a fluid particle to small variations in its initial position. The LCSs are operationally defined as local extrema of the scalar FTLE field, $\Lambda(x, y)$, which characterizes the rate of the fluid particle dispersion over a finite-time interval.

The spatial distribution of the FTLE values can be computed backward and forward in time. A region under study is seeded with a large number of tracers on a grid. The FTLE values are computed by one of the known methods (using the gradient of the flow map [29, 91] or a singular-value decomposition of the evolution matrix [77] described in Sect. 4.3) for all neighboring grid points for a given period of time which depends on the assigned task. Then one plots the spatial distribution of the FTLE coding its values by color. If there were hyperbolic regions in the velocity field for a chosen period of time, then we should get a spatially inhomogeneous FTLE map with "ridges" and "valleys." A ridge is defined as a curve on which the FTLE is locally maximized in the transverse direction [91]. Both the repelling and attracting LCSs can be computed by this way. Integrating the advection equations forward in time and computing the FTLE ridges, we extract repelling LCSs which approximate influential stable manifolds in the area. Expansion in backward time implies contraction in forward time. Therefore, attracting LCSs can be computed analogously but in reverse time. They approximate influential unstable manifolds in the area. A detailed description for computing FTLEs and extracting LCSs from the background FTLE field can be found in [90].

The important issue is how robust the computation of LCS's locations to errors in the velocity data. How velocity-field resolution and random errors affect LCSs? The velocity data contains inherent modelling and measurement errors. Moreover, it is discrete in space and time, and any interpolation introduces additional errors which accumulate during integration. It is natural to ask what we compute, really existing structures or computational artifacts. A quantitative analysis of errors in Lagrangian calculations is a complex problem, because we often do not know the exact nature of those errors.

It has been shown theoretically in [30] that the LCSs are robust to errors in observational or model velocity fields if they are strongly attracting or repelling and exist for a sufficient long time. This is due to the fact that though the particle's trajectories, in general, diverge exponentially from the true trajectories near a repelling LCS, the very LCSs are not expected to be perturbed to the same degree, because errors in the particle trajectories spread along the LCS. So, even large velocity errors in the velocity field lead to reliable predictions on LCSs, as long as the errors remain small in a special time-weighted norm [30]. This norm permits relatively large errors as they are local in time and do not accumulate to large errors when integrated over time.

Several studies have been carried to test the reliability of the Lyapunov technique for extracting the LCSs (see, e.g., [14, 35, 36, 48]), in particular, to test the sensitivity of FTLE ridges to errors in altimetric velocity fields. It has been studied numerically how an additional small noise in advection equations, modelling unknown corrections to the AVISO velocity field, might change the simulation results. The identified LCSs were found to be relatively insensitive to both sparse spatial and temporal resolution and to the velocity-field interpolation method. In general, LCSs have been found to be surprisingly robust against random errors in the velocity data. However, a caution is necessary in extracting LCSs from FTLE ridges, because they can yield false LCSs [31]. In theory, the LCSs are truly Lagrangian entities with the absence of material flux across them. However, the ridges sometimes are not Lagrangian with a flux of material across them.

The comparison of mesoscale LCSs computed with altimetric velocity field in [1, 5, 12, 18, 38, 68, 71, 77, 78, 82] with independent satellite and in situ measurements in different seas and oceans has been shown a good correspondence. The FTLE method was concluded to be reliable for locating boundaries of mesoscale eddies and strong jets. However, submesoscale LCS features were not well resolved from altimetry and should be considered with some caution. A hard work has been done by many people to enlarge the notion of the LCS and invariant manifolds to finite-time realistic flows. The LCSs have been shown in numerous papers to be very useful means to study mixing and transport in the ocean (see, e.g., the focus issues in the journal "Chaos" V. 20 in 2010 and V. 25 in 2015 and papers [1, 2, 5, 6, 6–11, 17, 18, 24, 25, 28–31, 33, 34, 38, 42, 49, 53–55, 67–69, 72, 77, 83–88, 90–93, 95, 96]).

A novel mathematical approach to uncover coherent regions in the ocean has been introduced in [21] and developed in [15, 19, 20, 22]. It is based upon numerically constructing a transfer operator that controls the surface transport of particles. The eigenfunctions of this operator, corresponding to large positive eigenvalues, reveal dominant almost-invariant structures in the surface flow which retain their shape over the time period considered. The finite-time coherent sets provide transport of mass with minimal leakage over a finite-time duration. This technique has been applied to study the spatial and temporal evolution of Agulhas rings, large anticyclonic mesoscale eddies in the Atlantic Ocean [19].

References

1. Abraham, E.R., Bowen, M.M.: Chaotic stirring by a mesoscale surface-ocean flow. Chaos **12**(2), 373–381 (2002). 10.1063/1.1481615
2. Andrade-Canto, F., Sheinbaum, J., Zavala Sansón, L.: A Lagrangian approach to the loop current eddy separation. Nonlinear Process. Geophys. **20**(1), 85–96 (2013). 10.5194/npg-20-85-2013
3. Aoyama, M., Uematsu, M., Tsumune, D., Hamajima, Y.: Surface pathway of radioactive plume of TEPCO Fukushima NPP1 released [134]Cs and [137]Cs. Biogeosciences **10**(5), 3067–3078 (2013). 10.5194/bg-10-3067-2013
4. Aref, H.: Stirring by chaotic advection. J. Fluid Mech. **143**(-1), 1–21 (1984). 10.1017/S0022112084001233
5. Beron-Vera, F.J., Olascoaga, M.J., Goni, G.J.: Oceanic mesoscale eddies as revealed by Lagrangian coherent structures. Geophys. Res. Lett. **35**(12), L12603 (2008). 10.1029/2008GL033957

6. Beron-Vera, F.J., Wang, Y., Olascoaga, M.J., Goni, G.J., Haller, G.: Objective detection of oceanic eddies and the Agulhas leakage. J. Phys. Oceanogr. **43**(7), 1426–1438 (2013). 10.1175/JPO-D-12-0171.1

7. Bettencourt, J.H., Lopez, C., Hernandez-Garcia, E.: Oceanic three-dimensional Lagrangian coherent structures: a study of a mesoscale eddy in the Benguela upwelling region. Ocean Model. **51**, 73–83 (2012). 10.1016/j.ocemod.2012.04.004

8. Bettencourt, J.H., López, C., Hernández-García, E.: Characterization of coherent structures in three-dimensional turbulent flows using the finite-size Lyapunov exponent. J. Phys. A Math. Theor. **46**(25), 254022 (2013). 10.1088/1751-8113/46/25/254022

9. Bettencourt, J.H., López, C., Hernández-García, E., Montes, I., Sudre, J., Dewitte, B., Paulmier, A., Garçon, V.: Boundaries of the Peruvian oxygen minimum zone shaped by coherent mesoscale dynamics. Nat. Geosci. **8**(12), 937–940 (2015). 10.1038/ngeo2570

10. Boffetta, G., Lacorata, G., Redaelli, G., Vulpiani, A.: Detecting barriers to transport: a review of different techniques. Physica D **159**(1–2), 58–70 (2001). 10.1016/S0167-2789(01)00330-X

11. Budyansky, M., Uleysky, M., Prants, S.: Chaotic scattering, transport, and fractals in a simple hydrodynamic flow. J. Exp. Theor. Phys. **99**, 1018–1027 (2004). 10.1134/1.1842883

12. Budyansky, M.V., Goryachev, V.A., Kaplunenko, D.D., Lobanov, V.B., Prants, S.V., Sergeev, A.F., Shlyk, N.V., Uleysky, M.Y.: Role of mesoscale eddies in transport of Fukushima-derived cesium isotopes in the ocean. Deep-Sea Res. I Oceanogr. Res. Pap. **96**, 15–27 (2015). 10.1016/j.dsr.2014.09.007

13. Buesseler, K.O., Jayne, S.R., Fisher, N.S., Rypina, I.I., Baumann, H., Baumann, Z., Breier, C.F., Douglass, E.M., George, J., Macdonald, A.M., Miyamoto, H., Nishikawa, J., Pike, S.M., Yoshida, S.: Fukushima-derived radionuclides in the ocean and biota off Japan. Proc. Natl. Acad. Sci. **109**(16), 5984–5988 (2012). 10.1073/pnas.1120794109

14. Cotte, C., d'Ovidio, F., Chaigneau, A., Levy, M., Taupier-Letage, I., Mate, B., Guinet, C.: Scale-dependent interactions of Mediterranean whales with marine dynamics. Limnol. Oceanogr. **56**(1), 219–232 (2010). 10.4319/lo.2011.56.1.0219

15. Dellnitz, M., Froyland, G., Horenkamp, C., Padberg-Gehle, K., Gupta, A.S.: Seasonal variability of the subpolar gyres in the Southern Ocean: a numerical investigation based on transfer operators. Nonlinear Process. Geophys. **16**(6), 655–663 (2009). 10.5194/npg-16-655-2009

16. d'Ovidio, F., De Monte, S., Alvain, S., Dandonneau, Y., Levy, M.: Fluid dynamical niches of phytoplankton types. Proc. Natl. Acad. Sci. **107**(43), 18366–18370 (2010). 10.1073/pnas.1004620107

17. d'Ovidio, F., Fernández, V., Hernández-García, E., López, C.: Mixing structures in the Mediterranean Sea from finite-size Lyapunov exponents. Geophys. Res. Lett. **31**(17), L17203 (2004). 10.1029/2004GL020328

18. d'Ovidio, F., Isern-Fontanet, J., López, C., Hernández-García, E., García-Ladona, E.: Comparison between Eulerian diagnostics and finite-size Lyapunov exponents computed from altimetry in the Algerian basin. Deep-Sea Res. I Oceanogr. Res. Pap. **56**(1), 15–31 (2009). 10.1016/j.dsr.2008.07.014

19. Froyland, G., Horenkamp, C., Rossi, V., van Sebille, E.: Studying an Agulhas ring's long-term pathway and decay with finite-time coherent sets. Chaos **25**(8), 083119 (2015). 10.1063/1.4927830

20. Froyland, G., Padberg, K.: Almost-invariant sets and invariant manifolds — Connecting probabilistic and geometric descriptions of coherent structures in flows. Physica D **238**(16), 1507–1523 (2009). 10.1016/j.physd.2009.03.002

21. Froyland, G., Padberg, K., England, M.H., Treguier, A.M.: Detection of coherent oceanic structures via transfer operators. Phys. Rev. Lett. **98**(22), 224503 (2007). 10.1103/physrevlett.98.224503

22. Froyland, G., Stuart, R.M., van Sebille, E.: How well-connected is the surface of the global ocean? Chaos **24**(3), 033126 (2014). 10.1063/1.4892530

23. García-Garrido, V.J., Mancho, A.M., Wiggins, S., Mendoza, C.: A dynamical systems approach to the surface search for debris associated with the disappearance of flight MH370. Nonlinear Process. Geophys. **22**(6), 701–712 (2015). 10.5194/npg-22-701-2015

24. García-Olivares, A., Isern-Fontanet, J., García-Ladona, E.: Dispersion of passive tracers and finite-scale Lyapunov exponents in the Western Mediterranean Sea. Deep-Sea Res. I Oceanogr. Res. Pap. **54**(2), 253–268 (2007). 10.1016/j.dsr.2006.10.009

25. Gildor, H., Fredj, E., Steinbuck, J., Monismith, S.: Evidence for submesoscale barriers to horizontal mixing in the ocean from current measurements and aerial photographs. J. Phys. Oceanogr. **39**(8), 1975–1983 (2009). 10.1175/2009JPO4116.1

26. Greene, J.M., Kim, J.S.: The calculation of Lyapunov spectra. Physica D **24**(1–3), 213–225 (1987). 10.1016/0167-2789(87)90076-5

27. Guckenheimer, J., Holmes, P.: Nonlinear Oscillations, Dynamical Systems, and Bifurcations of Vector Fields. Applied Mathematical Sciences, vol. 42. Springer, New York (1983). 10.1007/978-1-4612-1140-2

28. Haller, G.: Finding finite-time invariant manifolds in two-dimensional velocity fields. Chaos **10**(1), 99–108 (2000). 10.1063/1.166479

29. Haller, G.: Distinguished material surfaces and coherent structures in three-dimensional fluid flows. Physica D **149**(4), 248–277 (2001). 10.1016/s0167-2789(00)00199-8

30. Haller, G.: Lagrangian coherent structures from approximate velocity data. Phys. Fluids **14**(6), 1851–1861 (2002). 10.1063/1.1477449

31. Haller, G.: A variational theory of hyperbolic Lagrangian Coherent Structures. Physica D **240**(7), 574–598 (2011). 10.1016/j.physd.2010.11.010

32. Haller, G.: Lagrangian coherent structures. Annu. Rev. Fluid Mech. **47**, 137–162 (2015). 10.1146/annurev-fluid-010313-141322

33. Haller, G., Poje, A.C.: Finite time transport in aperiodic flows. Physica D **119**(3–4), 352–380 (1998). 10.1016/S0167-2789(98)00091-8

34. Haller, G., Yuan, G.: Lagrangian coherent structures and mixing in two-dimensional turbulence. Physica D **147**(3–4), 352–370 (2000). 10.1016/S0167-2789(00)00142-1

35. Harrison, C.S., Glatzmaier, G.A.: Lagrangian coherent structures in the California Current System — sensitivities and limitations. Geophys. Astrophys. Fluid Dyn. **106**(1), 22–44 (2012). 10.1080/03091929.2010.532793

36. Hernández-Carrasco, I., López, C., Hernández-García, E., Turiel, A.: How reliable are finite-size Lyapunov exponents for the assessment of ocean dynamics? Ocean Model. **36**(3–4), 208–218 (2011). 10.1016/j.ocemod.2010.12.006

37. Honda, M.C., Aono, T., Aoyama, M., Hamajima, Y., Kawakami, H., Kitamura, M., Masumoto, Y., Miyazawa, Y., Takigawa, M., Saino, T.: Dispersion of artificial caesium-134 and -137 in the western North Pacific one month after the Fukushima accident. Geochem. J. **46**(1), e1–e9 (2012)

38. Huhn, F., von Kameke, A., Pérez-Muñuzuri, V., Olascoaga, M.J., Beron-Vera, F.J.: The impact of advective transport by the South Indian Ocean Countercurrent on the Madagascar plankton bloom. Geophys. Res. Lett. **39**(6), L06602 (2012). 10.1029/2012GL051246

39. Hussain, A.K.M.F.: Coherent structures — reality and myth. Phys. Fluids **26**(10), 2816–2850 (1983). 10.1063/1.864048

40. Inoue, M., Kofuji, H., Hamajima, Y., Nagao, S., Yoshida, K., Yamamoto, M.: ^{134}Cs and ^{137}Cs activities in coastal seawater along Northern Sanriku and Tsugaru Strait, northeastern Japan, after Fukushima Dai-ichi nuclear power plant accident. J. Environ. Radioact. **111**, 116–119 (2012). 10.1016/j.jenvrad.2011.09.012

41. Inoue, M., Kofuji, H., Nagao, S., Yamamoto, M., Hamajima, Y., Yoshida, K., Fujimoto, K., Takada, T., Isoda, Y.: Lateral variation of ^{134}Cs and ^{137}Cs concentrations in surface seawater in and around the Japan Sea after the Fukushima Dai-ichi nuclear power plant accident. J. Environ. Radioact. **109**, 45–51 (2012). 10.1016/j.jenvrad.2012.01.004

42. Jones, C., Winkler, S.: Chapter 2. Invariant manifolds and Lagrangian dynamics in the ocean and atmosphere. In: Fiedler, B. (ed.) Handbook of Dynamical Systems, vol. 2, pp. 55–92. Elsevier, Amsterdam (2002). 10.1016/S1874-575X(02)80023-6

43. Kaeriyama, H., Ambe, D., Shimizu, Y., Fujimoto, K., Ono, T., Yonezaki, S., Kato, Y., Matsunaga, H., Minami, H., Nakatsuka, S., Watanabe, T.: Direct observation of ^{134}Cs and ^{137}Cs in surface seawater in the western and central North Pacific after the Fukushima Dai-ichi nuclear power plant accident. Biogeosciences 10(2), 4287–4295 (2013). 10.5194/bg-10-4287-2013

44. Kaeriyama, H., Shimizu, Y., Ambe, D., Masujima, M., Shigenobu, Y., Fujimoto, K., Ono, T., Nishiuchi, K., Taneda, T., Kurogi, H., Setou, T., Sugisaki, H., Ichikawa, T., Hidaka, K., Hiroe, Y., Kusaka, A., Kodama, T., Kuriyama, M., Morita, H., Nakata, K., Morinaga, K., Morita, T., Watanabe, T.: Southwest intrusion of ^{134}Cs and ^{137}Cs derived from the Fukushima Dai-ichi nuclear power plant accident in the western North Pacific. Environ. Sci. Technol. 48(6), 3120–3127 (2014). 10.1021/es403686v

45. Kameník, J., Dulaiova, H., Buesseler, K.O., Pike, S.M., Šťastná, K.: Cesium-134 and 137 activities in the central North Pacific Ocean after the Fukushima Dai-ichi nuclear power plant accident. Biogeosciences 10(9), 6045–6052 (2013). 10.5194/bg-10-6045-2013

46. Kanda, J.: Continuing ^{137}Cs release to the sea from the Fukushima Dai-ichi nuclear power plant through 2012. Biogeosciences 10(9), 6107–6113 (2013). 10.5194/bg-10-6107-2013

47. Katok, A., Hasselblatt, B.: Introduction to the Modern Theory of Dynamical Systems. Encyclopedia of Mathematics and its Applications, vol. 54. Cambridge University Press, Cambridge (1997)

48. Keating, S.R., Smith, K.S., Kramer, P.R.: Diagnosing lateral mixing in the upper ocean with virtual tracers: spatial and temporal resolution dependence. J. Phys. Oceanogr. 41(8), 1512–1534 (2011). 10.1175/2011JPO4580.1

49. Kirwan Jr., A.D.: Dynamics of "critical" trajectories. Prog. Oceanogr. 70(2–4), 448–465 (2006). 10.1016/j.pocean.2005.07.002

50. Koshel', K.V., Prants, S.V.: Chaotic advection in the ocean. Physics-Uspekhi 49(11), 1151–1178 (2006). 10.1070/PU2006v049n11ABEH006066

51. Koshel, K.V., Prants, S.V.: Chaotic Advection in the Ocean. Institute for Computer Science, Moscow (2008) [in Russian]

52. Kumamoto, Y., Aoyama, M., Hamajima, Y., Aono, T., Kouketsu, S., Murata, A., Kawano, T.: Southward spreading of the Fukushima-derived radiocesium across the Kuroshio extension in the North Pacific. Sci. Rep. 4, 1–9 (2014). 10.1038/srep04276

53. Kuznetsov, L., Toner, M., Kirwan Jr., A.D., Jones, C.K.R.T., Kantha, L.H., Choi, J.: The loop current and adjacent rings delineated by Lagrangian analysis of the near-surface flow. J. Mar. Res. 60(3), 405–429 (2002). 10.1357/002224002762231151

54. Lehahn, Y., d'Ovidio, F., Lévy, M., Heifetz, E.: Stirring of the northeast Atlantic spring bloom: a Lagrangian analysis based on multisatellite data. J. Geophys. Res. Oceans 112(C8), C08005 (2007). 10.1029/2006JC003927

55. Lekien, F., Coulliette, C., Mariano, A.J., Ryan, E.H., Shay, L.K., Haller, G., Marsden, J.: Pollution release tied to invariant manifolds: a case study for the coast of Florida. Physica D 210(1–2), 1–20 (2005). 10.1016/j.physd.2005.06.023

56. Lerman, L.M., Shil'nikov, L.P.: Homoclinical structures in nonautonomous systems: nonautonomous chaos. Chaos 2(3), 447–454 (1992). 10.1063/1.165887

57. Lichtenberg, A.J., Lieberman, M.A.: Regular and Chaotic Dynamics. Applied Mathematical Sciences, vol. 38. Springer, New York (1992). 10.1007/978-1-4757-2184-3

58. Mancho, A.M., Small, D., Wiggins, S.: Computation of hyperbolic trajectories and their stable and unstable manifolds for oceanographic flows represented as data sets. Nonlinear Process. Geophys. 11(1), 17–33 (2004). 10.5194/npg-11-17-2004

59. Mancho, A.M., Small, D., Wiggins, S.: A tutorial on dynamical systems concepts applied to Lagrangian transport in oceanic flows defined as finite time data sets: theoretical and computational issues. Phys. Rep. 437(3–4), 55–124 (2006). 10.1016/j.physrep.2006.09.005

60. Mancho, A.M., Wiggins, S., Curbelo, J., Mendoza, C.: Lagrangian descriptors: A method for revealing phase space structures of general time dependent dynamical systems. Commun. Nonlinear Sci. Numer. Simul. 18(12), 3530–3557 (2013). 10.1016/j.cnsns.2013.05.002

61. Mendoza, C., Mancho, A.M.: Hidden geometry of ocean flows. Phys. Rev. Lett. 105(3), 038501 (2010). 10.1103/PhysRevLett.105.038501

62. Mendoza, C., Mancho, A.M.: The Lagrangian description of aperiodic flows: a case study of the Kuroshio current. Nonlinear Process. Geophys. **19**(4), 449–472 (2012). 10.5194/npg-19-449-2012
63. Mendoza, C., Mancho, A.M., Rio, M.H.: The turnstile mechanism across the Kuroshio current: analysis of dynamics in altimeter velocity fields. Nonlinear Process. Geophys. **17**(2), 103–111 (2010). 10.5194/npg-17-103-2010
64. Miller, P.D., Jones, C.K.R.T., Rogerson, A.M., Pratt, L.J.: Quantifying transport in numerically generated velocity fields. Physica D **110**(1–2), 105–122 (1997). 10.1016/S0167-2789(97)00115-2
65. Mundel, R., Fredj, E., Gildor, H., Rom-Kedar, V.: New Lagrangian diagnostics for characterizing fluid flow mixing. Phys. Fluids **26**(12), 126602 (2014). 10.1063/1.4903239
66. Oikawa, S., Takata, H., Watabe, T., Misonoo, J., Kusakabe, M.: Distribution of the Fukushima-derived radionuclides in seawater in the Pacific off the coast of Miyagi, Fukushima, and Ibaraki prefectures, Japan. Biogeosciences **10**(7), 5031–5047 (2013). 10.5194/bg-10-5031-2013
67. Olascoaga, M.J., Haller, G.: Forecasting sudden changes in environmental pollution patterns. Proc. Natl. Acad. Sci. **109**(13), 4738–4743 (2012). 10.1073/pnas.1118574109
68. Olascoaga, M.J., Rypina, I.I., Brown, M.G., Beron-Vera, F.J., Koçak, H., Brand, L.E., Halliwell, G.R., Shay, L.K.: Persistent transport barrier on the West Florida shelf. Geophys. Res. Lett. **33**(22), L22603 (2006). 10.1029/2006GL027800
69. Peacock, T., Haller, G.: Lagrangian coherent structures: the hidden skeleton of fluid flows. Phys. Today **66**(2), 41–47 (2013). 10.1063/pt.3.1886
70. Pierrehumbert, R.T., Yang, H.: Global chaotic mixing on isentropic surfaces. J. Atmos. Sci. **50**(15), 2462–2480 (1993). 10.1175/1520-0469(1993)050<2462:GCMOIS>2.0.CO;2
71. Prants, S., Ponomarev, V., Budyansky, M., Uleysky, M., Fayman, P.: Lagrangian analysis of the vertical structure of eddies simulated in the Japan Basin of the Japan/East Sea. Ocean Model. **86**, 128–140 (2015). 10.1016/j.ocemod.2014.12.010
72. Prants, S.V.: Dynamical systems theory methods to study mixing and transport in the ocean. Phys. Scr. **87**(3), 038115 (2013). 10.1088/0031-8949/87/03/038115
73. Prants, S.V.: Chaotic Lagrangian transport and mixing in the ocean. Eur. Phys. J. Spec. Top. **223**(13), 2723–2743 (2014). 10.1140/epjst/e2014-02288-5
74. Prants, S.V., Andreev, A.G., Budyansky, M.V., Uleysky, M.Y.: Impact of mesoscale eddies on surface flow between the Pacific Ocean and the Bering Sea across the near strait. Ocean Model. **72**, 143–152 (2013). 10.1016/j.ocemod.2013.09.003
75. Prants, S.V., Andreev, A.G., Budyansky, M.V., Uleysky, M.Y.: Impact of the Alaskan stream flow on surface water dynamics, temperature, ice extent, plankton biomass, and walleye pollock stocks in the eastern Okhotsk Sea. J. Mar. Syst. **151**, 47–56 (2015). 10.1016/j.jmarsys.2015.07.001
76. Prants, S.V., Andreev, A.G., Uleysky, M.Y., Budyansky, M.V.: Lagrangian study of temporal changes of a surface flow through the Kamchatka Strait. Ocean Dyn. **64**(6), 771–780 (2014). 10.1007/s10236-014-0706-9
77. Prants, S.V., Budyansky, M.V., Ponomarev, V.I., Uleysky, M.Y.: Lagrangian study of transport and mixing in a mesoscale eddy street. Ocean Model. **38**(1–2), 114–125 (2011). 10.1016/j.ocemod.2011.02.008
78. Prants, S.V., Budyansky, M.V., Uleysky, M.Y.: Identifying Lagrangian fronts with favourable fishery conditions. Deep-Sea Res. I Oceanogr. Res. Pap. **90**, 27–35 (2014). 10.1016/j.dsr.2014.04.012
79. Prants, S.V., Budyansky, M.V., Uleysky, M.Y.: Lagrangian fronts in the ocean. Izv. Atmos. Oceanic Phys. **50**(3), 284–291 (2014). 10.1134/s0001433814030116
80. Prants, S.V., Budyansky, M.V., Uleysky, M.Y.: Lagrangian study of surface transport in the Kuroshio extension area based on simulation of propagation of Fukushima-derived radionuclides. Nonlinear Process. Geophys. **21**(1), 279–289 (2014). 10.5194/npg-21-279-2014
81. Prants, S.V., Budyansky, M.V., Uleysky, M.Y., Zhang, J.: Hyperbolicity in the Ocean. Discon. Nonlinearity Complex. **4**(3), 257–270 (2015). 10.5890/DNC.2015.09.004

82. Prants, S.V., Lobanov, V.B., Budyansky, M.V., Uleysky, M.Y.: Lagrangian analysis of formation, structure, evolution and splitting of anticyclonic Kuril eddies. Deep-Sea Res. I Oceanogr. Res. Pap. **109**, 61–75 (2016). 10.1016/j.dsr.2016.01.003

83. Prants, S.V., Ponomarev, V.I., Budyansky, M.V., Uleysky, M.Y., Fayman, P.A.: Lagrangian analysis of mixing and transport of water masses in the marine bays. Izv. Atmos. Oceanic Phys. **49**(1), 82–96 (2013). 10.1134/S0001433813010088

84. Prants, S.V., Uleysky, M.Y., Budyansky, M.V.: Numerical simulation of propagation of radioactive pollution in the ocean from the Fukushima Dai-ichi nuclear power plant. Dokl. Earth Sci. **439**(2), 1179–1182 (2011). 10.1134/S1028334X11080277

85. Prants, S.V., Uleysky, M.Y., Budyansky, M.V.: Lagrangian coherent structures in the ocean favorable for fishery. Dokl. Earth Sci. **447**(1), 1269–1272 (2012). 10.1134/S1028334X12110062

86. Rypina, I.I., Pratt, L.J., Pullen, J., Levin, J., Gordon, A.L.: Chaotic advection in an Archipelago. J. Phys. Oceanogr. **40**(9), 1988–2006 (2010). 10.1175/2010jpo4336.1

87. Rypina, I.I., Scott, S.E., Pratt, L.J., Brown, M.G.: Investigating the connection between complexity of isolated trajectories and Lagrangian coherent structures. Nonlinear Process. Geophys. **18**(6), 977–987 (2011). 10.5194/npg-18-977-2011

88. Samelson, R.: Lagrangian motion, coherent structures, and lines of persistent material strain. Annu. Rev. Mar. Sci. **5**(1), 137–163 (2013). 10.1146/annurev-marine-120710-100819

89. Samelson, R.M., Wiggins, S.: Lagrangian Transport in Geophysical Jets and Waves: The Dynamical Systems Approach. Interdisciplinary Applied Mathematics, vol. 31. Springer, New York (2006). 10.1007/978-0-387-46213-4

90. Shadden, S.C.: Lagrangian coherent structures. Transport and Mixing in Laminar Flows: From Microfluidics to Oceanic Currents, pp. 59–89. Wiley-Blackwell, Weinheim (2011). 10.1002/9783527639748.ch3

91. Shadden, S.C., Lekien, F., Marsden, J.E.: Definition and properties of Lagrangian coherent structures from finite-time Lyapunov exponents in two-dimensional aperiodic flows. Physica D **212**(3–4), 271–304 (2005). 10.1016/j.physd.2005.10.007

92. Shadden, S.C., Lekien, F., Paduan, J.D., Chavez, F.P., Marsden, J.E.: The correlation between surface drifters and coherent structures based on high-frequency radar data in Monterey Bay. Deep-Sea Res. II Top. Stud. Oceanogr. **56**(3–5), 161–172 (2009). 10.1016/j.dsr2.2008.08.008

93. Sulman, M.H.M., Huntley, H.S., Lipphardt, B.L., Kirwan, A.D.: Leaving flatland: Diagnostics for Lagrangian coherent structures in three-dimensional flows. Physica D **258**, 77–92 (2013). 10.1016/j.physd.2013.05.005

94. Sulman, M.H.M., Huntley, H.S., Lipphardt Jr., B.L., Jacobs, G., Hogan, P., Kirwan Jr., A.D.: Hyperbolicity in temperature and flow fields during the formation of a Loop current ring. Nonlinear Process. Geophys. **20**(5), 883–892 (2013). 10.5194/npg-20-883-2013

95. Tew Kai, E., Rossi, V., Sudre, J., Weimerskirch, H., Lopez, C., Hernandez-Garcia, E., Marsac, F., Garçon, V.: Top marine predators track Lagrangian coherent structures. Proc. Natl. Acad. Sci. **106**(20), 8245–8250 (2009). 10.1073/pnas.0811034106

96. Waugh, D.W., Abraham, E.R.: Stirring in the global surface ocean. Geophys. Res. Lett. **35**(20), L20605 (2008). 10.1029/2008gl035526

97. Wiggins, S.: Chaos in the dynamics generated by sequences of maps, with applications to chaotic advection in flows with aperiodic time dependence. Z. Angew. Math. Phys. **50**(4), 585–616 (1999). 10.1007/s000330050168

98. Zaslavsky, G.M.: The Physics of Chaos in Hamiltonian Systems, 2nd edn. World Scientific, Singapore (2007)

List of Internet Resources

[GMP] The GNU MP Bignum Library.
    http://gmplib.org

Chapter 5
Transport of Subtropical Waters in the Japan Sea

5.1 General Pattern of Circulation in the Japan Sea and Formulation of the Problem

The Japan Sea is a mid-latitude marginal sea with dimensions of 1600×900 km, the maximal depth of 3.72 km, and the mean depth of about 1.5 km. It spans conditions from subarctic to subtropical and is characterized by many of the phenomena found in the deep ocean: fronts, eddies, currents and streamers, deep water formation, convection, and subduction. It communicates with the Pacific Ocean at the south and east through the Tsushima (Korean) and Tsugaru straits, respectively. In the north it is connected with the Okhotsk Sea through the Soya (La Perouse) and Nevelskoy straits. All the four channels are shallow ones with depths not exceeding 135 m. Review of oceanography of the Japan Sea can be found in a few special journal issues (Deep-Sea Research Part II: Topical Studies in Oceanography. V. 52, is. 11–13, pp. 1359–1844, 2005. Oceanography. V. 19, is. 3, 2006. Journal of Marine Systems. V. 78, is. 2, pp. 195–316, 2009) and in the recent monographs [1, 3].

Bathymetry of the Japan Sea and its geographic and oceanographic features are shown in Fig. 5.1. Warm and saline Pacific waters enter the Tsushima (Korean) Strait and split into three currents. The Nearshore Branch of the Tsushima Current flows northward along the western coast of the Honshu Island (Japan). The Offshore Branch of the Tsushima Current with a meander-like path flows into the Yamato Basin. The East Korean Warm Current flows northward along the eastern coast of Korea to meet the North Korean Cold Current which is a prolongation of the Liman (Primorskoye) Cold Current flowing southward along the Siberian coast down to Vladivostok. One of the major large-scale features in the northern Japan Sea is a cyclonic gyre over the Japan Basin and the Tatarsky Strait. Historical and modern reviews of surface circulation in the Japan Sea can be found in [6, 7, 10, 13, 16–19, 35, 36, 39–43].

© Springer International Publishing AG 2017
S.V. Prants et al., *Lagrangian Oceanography*, Physics of Earth and Space
Environments, DOI 10.1007/978-3-319-53022-2_5

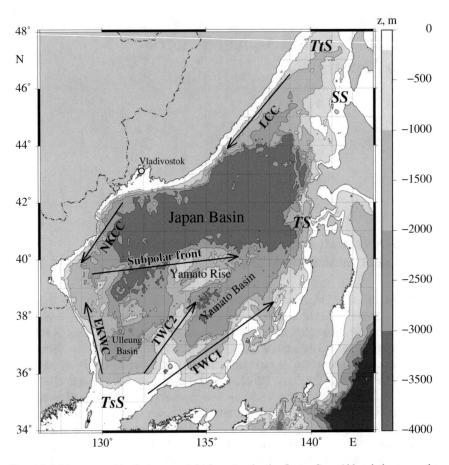

Fig. 5.1 Main geographic features and bathymetry in the Japan Sea. Abbreviations are the following: TsS (Tsushima or Korean Strait), TS (Tsugaru Strait), SS (Soya or La Perouse Strait), TtS (Tatarsky Strait), EKWC (East Korean Warm Current), NKCC (North Korean Cold Current), LCC (Liman or Primorskoye Cold Current), TWC1 and TWC2 (the first and second branches of the Tsushima Warm Current)

The Japan Sea is one of the most eddy-rich regions in the world ocean, especially, its southern part. Some of them are quasi-permanent and are manifested in the AVISO velocity field averaged for the whole simulation period in Fig. 5.2. Warm and cold mesoscale eddies are regularly observed in the Ulleung Basin. Mesoscale eddies have been found also in the area to the north of the Subpolar Front in hydrographic surveys, satellite images, and numerical simulations. It was shown that at least some of them are quasistationary and long lived, existing over the deep Japan Basin for a few months [20, 29, 37, 38].

The confluence of north-flowing warm subtropical waters with south-flowing cold subarctic ones forms one of the most remarkable features in the Sea, the distinct

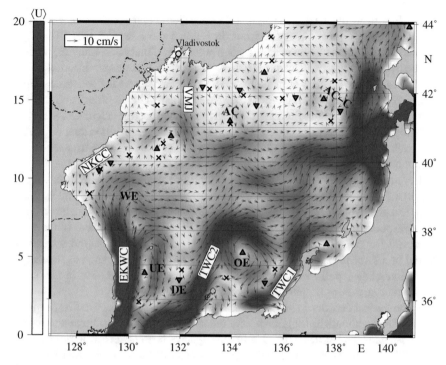

Fig. 5.2 The AVISO velocity field averaged for the period from January 2, 1993 to June 15, 2015. Elliptic and hyperbolic stationary points with zero mean velocity are indicated by *triangles* and *crosses*, respectively. Abbreviations are the following: UE (Ulleung eddy), DE (Dok eddy), OE (Oki eddy), WE (Wonsan eddy), AC-C (vortex pair near the eastern gate), AC (anticyclonic eddy over the Japan Basin), VMJ (Vladivostok meridional jet)

Subpolar Front that extends across the basin near 40° N [4, 11, 22, 28, 39, 40, 44, 45]. It is a boundary of physical and chemical properties such as temperature, salinity, dissolved oxygen, and nutrients. The Subpolar Front is clearly manifested in SST infrared satellite images, especially in winter (Fig. 3.1). Like many other hydrological fronts, the Japan Sea Subpolar Front is a highly productive zone with favorable fishery conditions. It is not a continuous curve crossing the basin with a maximal thermal gradient. It is rather a vast area between 38° N and 41° N extending across the basin from the Korea coast to the Japanese islands. The frontal area is confined by two frontal boundaries with strong gradients, the northern and southern ones, between which a zone with lower gradients is situated [45]. The northern boundary separates subarctic waters from the transformed subarctic ones. The southern one separates subtropical waters from the transformed subtropical ones. The position of the front is determined by mesoscale meanders and eddies with a comparatively short-time variability (see Fig. 3.1). Sometimes the front may be associated with an envelope of many compactly distributed eddies. Its seasonal variability dominates over interannual one.

Starting from the 1970s, the availability of satellite-derived SST data makes it possible to study position of the front on a much more regular base. The high-resolution SST images, however, are not available in cloudy and rainy days which are not rare in this area. Moreover, it is difficult to get proper location of the Japan Sea Subpolar Front with summer SST images because the seasonal heating of the sea surface reduces horizontal thermal gradients. On the other hand, satellite altimeters provide sea level anomalies which have been used to calculate and archive geostrophic velocities in the global AVISO database from 1993. It is tempting to use such rich velocity data to simulate movement of water masses in the frontal area and to find out some typical phenomena in their mixing and transport.

Understanding transport pathways of subtropical water in the Japan Sea is important for a number of reasons. Physical properties (temperature and salinity), chemical properties, pollutants, and biota (phytoplankton, zooplankton, larvae, etc.) are transported and mixed by currents and eddies. Transport of heat to the north is crucial for climatic applications. The ability to simulate transport adequately would be useful to deal with the aftermath of accidents at sea such as discharges of radionuclides, pollutants, and oil spills. It is also crucial, for instance, for understanding transport pathways for species invasions.

Since the last decades in the twentieth century, invasions of heat-loving fish (conger eel, tuna, moonfish, triggerfish, and others) and some tropical and subtropical marine organisms (turtles, sharks, and others) have been observed in the northern part of the Sea, to the coast of Russia [14]. It is natural to assume that such invasions could be caused by intrusions of subtropical waters across the Subpolar Front. They may be also one of the reasons for a prolongation of the warm period in the fall in Primorye province in Russia since the 1990s [24, 25]. From the oceanographic point of view, this transport of subtropical waters contradicts long-held beliefs on circulation in the Japan Sea. It is believed that the Subpolar Front is a transport barrier for propagation of subtropical waters across it to the north, at least, in the western and central parts of the front area (see, e.g., [7]).

In this chapter we simulate northward near-surface Lagrangian transport of subtropical waters in the central Japan Sea between 37° N and 42° N based on altimeter data. Geostrophic velocities were obtained from the AVISO database [AVISO] archived daily on a $1/4° \times 1/4°$ grid from January 2, 1993 to June 15, 2015. Advection equations (4.1) are solved for a large number of synthetic particles (tracers) advected by the AVISO velocity field. With this aim, 10^5 tracers have been launched weekly from January 2, 1993 to June 15, 2013 at the latitude 37° N from 129° E to 138° E. Trajectory of each tracer has been computed for 2 years after its launch date. We fixed the location and the instant of time where and when each tracer crossed a given latitude in the central Japan Sea between 37° N and 43° N. The first crossing of a given zonal line has been fixed, because we are interested not in a net transport but in the northward transport only. We stop to compute trajectories of those tracers which get into an AVISO cell touching the land.

To simulate and analyze transport across the frontal area we elaborate a special Lagrangian methodology consisting of a number of tasks. A complex of the numerical codes has been compiled to compute a number of Lagrangian indicators for synthetic tracers and plot the following diagrams:

(A) Meridional distribution of the number of tracers, N, crossing fixed latitudes, λ_f, in the central Japan Sea with a space step $0.1°$. The corresponding data are represented as a density map which shows by color, the density of tracks of the tracers crossed all the latitudes in the central Japan Sea from January 2, 1993 to June 15, 2015. They also can be represented as an $N(\lambda_f)$ distributions which show how many tracers reached a fixed zonal line at the longitude λ_f for the whole period of integration.

(B) Fixing initial longitudes λ_0 of launched tracers along the material line $37°$ N, we compute those final longitudes λ_f at which they cross a fixed zonal line for the whole period of integration. The results are represented as λ_0–λ_f plots.

(C) Tracking maps show where the subtropical tracers, which crossed eventually the fixed zonal line through specified meridional gates, have wandered for the whole integration period.

(D) The T–λ_f plots show when and at which longitudes those tracers crossed the zonal lines $40°$ N and $42°$ N.

(E) In order to document and visualize intrusions of subtropical waters into subarctic ones, we compute Lagrangian maps by integrating the advection equations (4.1) backward in time [30]. A subbasin in the Sea is seeded at a fixed date with a large number of tracers whose displacements are computed backward in time for a given period of time. In order to track those subtropical waters which were able to cross the Subpolar Front and reach given latitudes in the northern Japan Sea, we mark by color the tracers that reached the line $37°$ N in the past and compute how long time it took. In order to know where this or that tracer came from for a given period of time, we compute the drift maps with boundaries. The waters, that entered a given area through its southern boundary, are shown by one color on such a map, and the waters, that came through the northern boundary, are shown by another color.

5.2 Statistical Analysis of Lagrangian Transport of Subtropical Water

5.2.1 Northward Transport of Subtropical Water and Advection Velocity Field

The AVISO velocity field averaged for the period from January 2, 1993 to June 15, 2015 is shown in Fig. 5.2 with elliptic and hyperbolic stationary points imposed. This plot reflects the known features of mesoscale near-surface circulation in this part of the Sea. The Tsushima Warm Current splits into three parts. The first one (TWC1 in Fig. 5.2) is the near shore branch flowing northward along the western coast of the Honshu Island (Japan). The second one (TWC2) is the offshore meander-like branch, and the third one is the East Korean Warm Current flowing northward as a western boundary current along the eastern coast of the Korean peninsula (EKWC). It encounters the North Korean Cold Current flowing southward (NKCC). Both the currents separate from the Korean coast at about 39° N and flow to the east forming the Subpolar Front.

Some well-known quasistationary mesoscale eddy-like features are also visible in the average AVISO velocity field in Fig. 5.2. In the Ulleung Basin there are the warm Ulleung anticyclonic circulation [18, 21, 33, 34] with the center at about 37° N, 130.5° E (UE) and a cyclonic circulation called often as the cold Dok eddy (DE) [18, 21] with the center at about 36.7° N, 132° E. The flow over bottom topography around the Oki Spur in the southeastern part of the Sea generates the anticyclonic Oki Eddy (OE) with the center at about 37.5° N, 134.2° E [12, 23]. In the western part of the Sea meandering of the East Korean Warm Current produces an anticyclonic circulation called as the anticyclonic Wonsan Eddy (WE) with the center at about 39° N, 129° E [19].

Plot in Fig. 5.3a shows in accordance with the task A the density of tracks of tracers launched along 37° N and crossed all the latitudes in the central Japan Sea for the whole period of integration. The density is coded by nuances of the grey color in the logarithmic scale, $\log_{10} N_\varphi$. The magenta areas in Fig. 5.3a along the coastal line mean that the AVISO grid cells there touch the land, and we did not compute trajectories there. Uneven density of points in Fig. 5.3a means that the northward transport of subtropical waters is meridionally inhomogeneous with a kind of the gates with increased density of points. The gates are such spatial intervals along a given zonal line across which subtropical tracers prefer to cross it.

Any tracer, as a passive particle, is able to cross the fixed latitude in the northward direction if the northward component of the velocity field is nonzero at its location. In Fig. 5.3b we plot distribution of the northward component of the AVISO velocity field averaged over the whole period of integration as follows:

$$\langle v_+(\lambda, \varphi) \rangle = \frac{1}{\mathbf{n}} \sum \theta(v(\lambda, \varphi)) v(\lambda, \varphi), \tag{5.1}$$

where $v_+(\lambda, \varphi)$ is a northward (positive) component of the AVISO velocity at the point (λ, φ), $\theta(v)$ the Heaviside function, and $\mathbf{n}$ the number of days in the period from January 2, 1993 to June 15, 2015. Comparing Lagrangian representation in Fig. 5.3a with the Eulerian one in Fig. 5.3b, it is clear that areas with increased density of points in Fig. 5.3a correlate well with areas with increased average values of the northward component of the AVISO velocity field in Fig. 5.3b.

Thus, the northward transport of subtropical waters is dictated by the local advection velocity field, more precisely by local values of the northward component of the velocity. The greater is the northward component of the velocity at a given point and the longer is the period of time when it is positive the more tracers are able to cross the corresponding latitude.

The density difference in some meridional ranges in Fig. 5.3a may be very large because of the logarithmic-scale representation. There are even some places in the

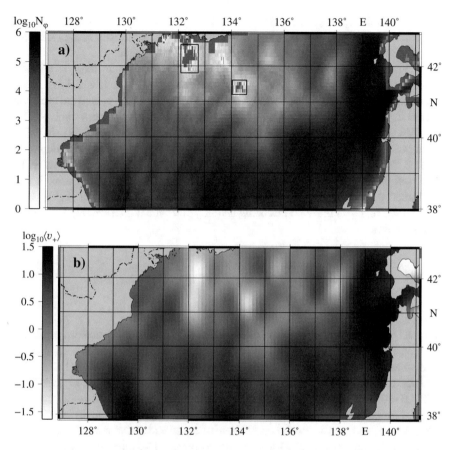

Fig. 5.3 (a) The logarithmic-scale density of tracks of the tracers crossing all the latitudes φ in the central Japan Sea, N_{φ}, from January 2, 1993 to June 15, 2015. The *rectangular magenta areas* are "forbidden zones" where the northward transport has not been observed during the whole integration period. The tracers have been launched weekly along the zonal line at $37°$ N from January 2, 1993 to June 15, 2013. (**b**) Distribution of the averaged northward component of the AVISO velocity field $\langle v_+(\lambda, \varphi) \rangle$ in the logarithmic-scale averaged over the same period

northern frontal area, the rectangular magenta areas are in Fig. 5.3a, where the northward transport has not been observed during all the simulation period, from 1993 to 2015. One "forbidden" zone is situated in the deep Japan Basin with the center at about $41.5°$ N, $134.2°$ E, and another one is situated to the south off Vladivostok from $43°$ N to $41°$ N approximately along the $132°$ E meridian. We stress that they are forbidden only to northward transport of tracers but can be and really are open to transport in other directions.

The "forbidden" zones persist due to long-term peculiarities of the advection velocity field there. The zone to the south off Vladivostok persists due to a quasi-permanent southward jet approximately along the meridian $132°$ E from $43°$ N to

40° N (VMJ in Fig. 5.2). That jet turns to the east at about 40° N and contributes to the eastward transport. In fact, the northward velocity is practically zero in this area (see Fig. 5.3b) and, therefore, the northward transport is absent.

The other "forbidden" zone persists due to two factors, the presence of a quasi-permanent anticyclonic eddy with the center at about 41.3° N, 134° E in the deep Japan Basin (AC in Fig. 5.2) and the eastward zonal jet blocking northward transport across it. Topographically constrained anticyclonic eddies with the centers at 41° N–41.5° N, 134° E–134.5° E have been regularly observed there [29, 38, 39]. For example, the current meter data from 1993 to 1996 at the moorings M3, deployed at 41.5° N and 134.3° E where the depth is about 3500 m, have shown that ACEs occurred there every year. Once an ACE is established in the Japan Basin, it will be forced to remain only in the interior region of the Basin [38]. In the next chapter we focus on those eddies using an eddy-resolved numerical regional model of circulation.

5.2.2   Gates and Barriers to the Northward Transport of Subtropical Water

Now let's look more carefully at the meridional distribution of subtropical tracers crossed the Subpolar Front for the whole period of simulation. We choose for reference four zonal lines along the AVISO grid at 42.125° N (129° E–141.24° E), 41.875° N (129° E–141.18° E), 40.125° N (128° E–140.24° E), and at 39.875° N (128° E–140.24° E). These distributions in accordance with the task A are shown in Fig. 5.4 for each zonal line by solid curves with imposed meridional distributions of the averaged northward velocity component (arrows) and of the number of crossings of those latitudes by available drifters (dashed curves). The correspondence between the peaks in the meridional distributions of the tracers and of the northward velocity component is rather good for all the chosen zonal lines confirming their direct connection.

Those peaks correlate more or less with the number of crossings of the chosen zonal lines by drifters (dashed curves in Fig. 5.4). The comparison with drifters should be taken with care because of a comparatively small number of available drifters especially in the central part of the Subpolar Front. Moreover, the drifters are not ideal passive tracers and they, of course, have not been launched at the zonal line 37° N like artificial tracers in simulation. Their launch sites have been distributed rather randomly over the basin.

The meridional distribution of tracers in Fig. 5.4 allows to distinguish the eastern, central, and western gates in the central Japan Sea, which strongly differ by the number of passing tracers. The very eastern, 138° E–140° E, and western, 129° E–131° E, gates exist mainly due to the near shore branch of the Tsushima Warm Current and the East Korean Warm Current, respectively. The central gate, 133° E–137° E, exists, probably, due to topographically constrained features over the Yamato Rise there (see Fig. 5.1). The transport through that gate will be shown to be enhanced due to a specific disposition of frontal eddies regularly observed

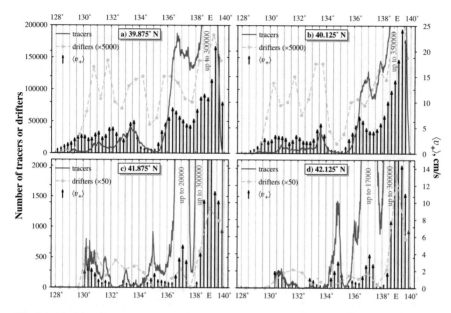

Fig. 5.4 Meridional distributions of the number of tracers which crossed indicated zonal lines (*solid curves*), of the averaged northward component of the AVISO velocity in cm s^{-1} (*arrows*) and of the number of crossings of those zonal lines by available drifters (*dashed curves*). The period of observation is from January 2, 1993 to June 15, 2015

there. The intervals between the gates may be called "conditioned barriers" because of a comparatively small number of tracers crossing zonal lines there, and because they used to "open" for a comparatively short-time intervals.

Figure 5.5a, b shows in accordance with the task B at which final longitudes λ_f the tracers, launched with the initial longitudes λ_0 at the line 37° N, reached the zonal lines 40° N and 42° N for the whole period of integration. Meridional distribution of the number of tracers with a few pronounced peaks, which crossed the zonal line 42° N for the same period, is plotted in Fig. 5.5c. This zonal line was divided for a reference in eight meridional intervals numbered by the roman numerals in Fig. 5.5b and c.

The Tsushima Warm Current contributes mainly to the eastern peak VIII at the distribution in Fig. 5.5c. The black color across all the range of initial longitudes λ_0 means that fluid particles, crossing eventually the line 42° N through the gate 138° E–140° E, could have any initial longitude λ_0 at the zonal line 37° N. They could reach that gate by different ways: either to be initially trapped by the near shore branch or to be advected by the offshore branch and then join the near shore branch. Moreover, those particles could be involved initially in the East Korean Warm Current and then move to the east along the Subpolar Front and eventually join the Tsushima Warm Current. Thus, the tracers, crossing the gate VIII, may have distinct values of some Lagrangian indicators, e.g., travelling time and distance passed.

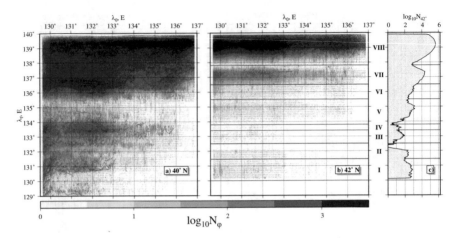

Fig. 5.5 Density plots show in the logarithmic scale how many and at which final longitudes λ_f the tracers with initial longitudes λ_0 were able to cross the zonal lines (**a**) 40° N and (**b**) 42° N for the whole simulation period. The tracers have been launched weekly at the line 37° N from January 2, 1993 to June 15, 2013. (**c**) Meridional distribution of the number of tracers, $N_{42°}$, which crossed the zonal line 42° N for the whole simulation period. This line is divided into eight intervals numbered by the roman numerals

There is a narrow barrier, the white strip in Fig. 5.5b between the gates VIII and VII, with the center at the local minimum at 137.8° E in Fig. 5.5c. A comparatively small number of tracers have been able to cross there the line 42° N for the whole simulation period. The next gate VII between 136° E and 137.8° E (see Fig. 5.5b, c) provides northward transport of subtropical tracers with the help of a quasi-permanent vortex pair located there. The corresponding transport mechanism will be described further in detail. The number of subtropical tracers, passing through this gate, is much smaller than that passing through the gate VIII (remember the logarithmic scale in Fig. 5.5). Only a small number of tracers, launched initially at the very eastern part of the zonal line 37° N, were able to cross the line 42° N through that gate because most of the eastern tracers passed through the gate VIII have been transported by the near shore branch of the Tsushima Warm Current. Most of the tracers, passed through the gate VII, came from the western and central parts of the material line at 37° N.

The number of subtropical tracers, passing through the central and western gates, is much smaller as compared with those passing through the eastern ones. We distinguish two central gates V and III at 134° E–135.5° E and 132.5° E–133.5° E, respectively, and the western gates I and II in the range 130° E–132.5° E (Fig. 5.5c). It follows from Fig. 5.5b that the western and central gates collect subtropical tracers mainly from the western part of the initial zonal line, from 129° E to 133° E. In other words, water parcels from its eastern part (133° E–137° E) practically do not pass through those gates at the latitude 42° N. Thus, the western part of the initial material line 37° N contributes to all the peaks at the distribution $N_{42°}$, whereas its eastern part contributes mainly to the Tsushima peak.

5.2.3   Transport Pathways of Subtropical Water in the Central Japan Sea

To visualize the transport pathways by which subtropical tracers, launched at the zonal line 37° N, crossed the Subpolar Front we compute in accordance with the task C so-called tracking maps. Tracking maps in Fig. 5.6 show where the subtropical tracers, which crossed eventually the zonal line 42° N, have wandered for the whole integration period. Each panel shows the tracks of those tracers which crossed that line through the corresponding interval indicated by the black strip at the top. These are the same intervals as in Fig. 5.5b, c.

The subtropical tracers, crossed eventually the latitude 42° N through the western gates 130° E–132.5° E, have been found during their travel to the north mainly in the western part of the Sea (Fig. 5.6a, b). Those ones, that crossed it through the gates at 132.5° E–133.8° E, have travelled mainly in the western and central parts of the Sea (Fig. 5.6c, d). The subtropical tracers, crossed the latitude 42° N through the central gates at 133.8° E–136.5° E, visited during their travel an enlarged area (Fig. 5.6e, f). Those ones, which crossed the latitude 42° N through the eastern gates at 136.5° E–140.2° E, visited all the southern basin (Fig. 5.6g, h) being well mixed. It is another manifestation of the fact, established before in Fig. 5.5, that the Tsushima Warm Current contains water parcels with cardinally distinct "stories."

The data, collected in Figs. 5.5 and 5.6, allow to make the conclusion that the northward transport of subtropical waters does not depend on initial spatial and temporal conditions (tracers "forget" fast their initial coordinates and launch dates), except for transport provided by the strong Tsushima and East Korean Warm currents. Therefore, it is expected that an oceanographic survey of the main outflow of the Tsushima Warm Current through the Tsugaru Strait would detect at the same place "fresh" subtropical waters, delivered by the shortest route, and transformed subtropical waters wandered in the central basin for a long time. From the ecological point of view, wherever a polluter would be situated in the central part of the Japan Sea we could found that pollution in the Tsugaru Strait.

It is interesting that there is a "hub," through which those subtropical waters pass which contribute to all the peaks in the meridional distribution of tracers along the latitude 42° N to the north off the Subpolar Front (Fig. 5.5c). This hub is the area with coordinates at about 39° N–41° N and 130° E–132° E with increased density of points in all the panels in Fig. 5.6. It is probably caused by two reasons, the presence of the quasi-permanent mesoscale Wonsan Eddy (WE in Fig. 5.2) with the center at about 39° N, 130° E and separation of the East Korean Warm Current from the coast to the east at about 39° N with formation of the Subpolar Front.

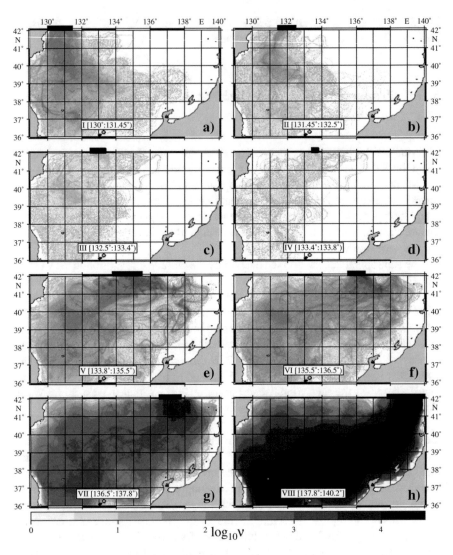

Fig. 5.6 Tracking maps show where the subtropical tracers, launched initially at the zonal line 37° N and crossed eventually the zonal line 42° N, have wandered for the whole integration period. Each panel shows the tracks of those tracers which passed through the corresponding gates indicated by *black strips* at the *top*. The track density, v, is given in the logarithmic scale

5.2.4 *Lagrangian Intrusions of Subtropical Water Across the Subpolar Front*

The T–λ_f plots in Figs. 5.7 and 5.8 show when and at which longitudes the tracers, launched weekly at the zonal line 37° N from January 2, 1993 to June 15, 2013,

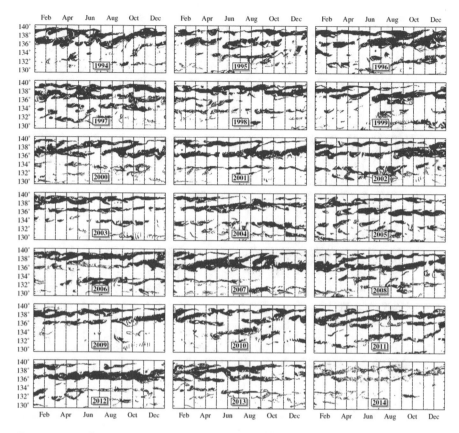

Fig. 5.7 The T–λ_f plot shows when and at which longitudes the tracers, launched at the zonal line $37°$ N, crossed eventually the zonal line $40°$ N in the period from January 1, 1994 to January 1, 2015

reached the zonal lines $40°$ N and $42°$ N, respectively. It was declared as the task D. The T–λ_f plots for two restricted time intervals are shown in Fig. 5.9. All the plots show the eastern gates VIII and VII (see Fig. 5.5) through which the subtropical tracers cross the corresponding latitudes. The locations of the central and western gates fluctuate in time, and some gates may be even closed for a while to the northward transport.

The patchiness in the plots means that subtropical tracers prefer to cross the zonal lines in the specific places (see the peaks in Figs. 5.4 and 5.5) and during specific time intervals. Any patch with a large number of tracers somewhere, say, at the central meridional gate, means that a water mass proportional to the size of this patch passed through the central gate across a given latitude during the period of time proportional to its zonal size. Thus, the northward transport of subtropical water across the Subarctic Front occurs by a portion-like manner. Specific oceanographic

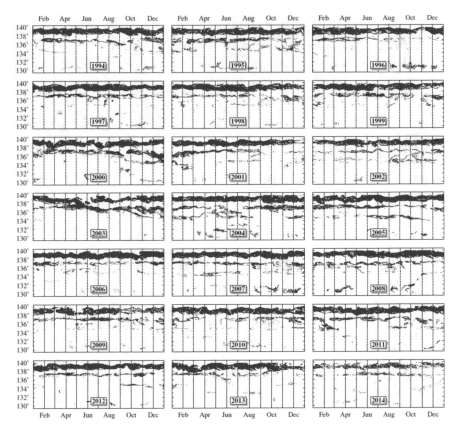

Fig. 5.8 The same as in Fig. 5.7 but for the zonal line 42° N

conditions may arise in a given area and at a given time which produce a large-scale intrusion of subtropical water to the north by means of mesoscale eddies to be present there.

One of the motivations of this study was an explanation of the fact of invasion of tropical and subtropical marine organisms in the northern part of the Sea, to the southern coast of Russia [14]. To document an intrusion of subtropical water there, we compute the backward-in-time Lagrangian drift maps. It is a realization of the task E. The basin, shown in Fig. 5.10, is seeded with a large number of tracers for each of which we compute the time required to reach its position on the map to a fixed date from the latitude 37° N. The travelling time T in days is coded by nuances of the grey color.

The maps in Fig. 5.10 illustrate two such episodes in the fall of 2005 and 2007 and a mechanism of intrusion of subtropical water to the north through the western gate. A vortex street with four anticyclones has been formed in the fall of 2005 to the north of the Subpolar Front in the western part of the Sea. Their centers are marked in Fig. 5.10a by the elliptic points (triangles) with the coordinates 39.1° N, 131.5° E; 39.3° N, 130.1° E; 40.8° N, 131.4° E, and 41.7° N, 130.8° E. Subtropical

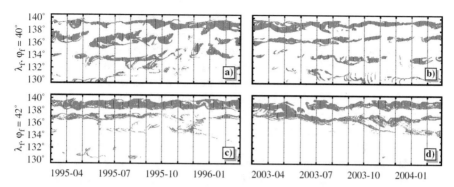

Fig. 5.9 The T–λ_f plots show when and at which longitudes the tracers, launched at the zonal line 37° N, crossed eventually the zonal lines 40° N and 42° N in the period from March 1, 1995 to March 1, 1996 ((**a**) and (**c**)) and from March 1, 2003 to March 1, 2004 ((**b**) and (**d**))

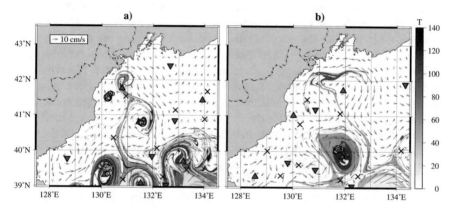

Fig. 5.10 The backward-in-time Lagrangian maps document intrusions of subtropical "grey" water to the southern coast of Russia through the western gate. Nuances of the *grey color* code travelling time T in days it took to subtropical tracers to reach their locations on the maps from the latitude 37° N to the fixed dates, (**a**) November 14, 2005 and (**b**) October 13, 2007. The AVISO velocity field is shown by *arrows*. "Instantaneous" elliptic and hyperbolic points are indicated by *triangles* and *crosses*, respectively. Locations of available drifters are shown by *full circles* for 1 day before and after the date indicated

"grey" tracers moved along the unstable manifolds of the three hyperbolic points between and around of those eddies to the north. The hyperbolic points are marked by crosses in Fig. 5.10a with the coordinates 39.2° N, 130.8° E; 40.3° N, 130.5° E, and 41.6° N, 130.9° E. Thus, the vortex street facilitates a penetration of subtropical water to the southern coast of Russia. The evidence of, at least, two anticyclones in the AVISO velocity field is confirmed by tracks of two available drifters [GDP]. Their locations are shown in Fig. 5.10a by full circles for 1 day before and after the date indicated on the map. The drifter no. 56739 has been trapped by the anticyclone with the center at 39.3° N, 130.1° E and the drifter no. 56746—by the anticyclone with the center at 40.8° N, 131.4° E.

Figure 5.10b illustrates an intrusion of subtropical water to the north in the same area in the fall of 2007. At that time, the vortex street consisted of two anticyclones with the centers at 39.8° N, 131.8° E and 41.7° N, 131.9° E and a cyclone with the center at 41.2° N, 130.8° E between them. The thin grey intrusion of subtropical water reached the Peter the Great Bay to the south off Vladivostok to October 13, 2007. The drifter no. 56727 has been trapped by the southern anticyclone that looks as the grey elliptic-like patch in Fig. 5.10b. The color coding in Fig. 5.10 provides additional information on history of subtropical tracers. The darker the color, the longer time required for a corresponding tracer to reach its location on the map. For example, the core of the anticyclone with the center at 39.8° N, 131.8° E consists of the water which is "older" than that at its periphery. Peripheries of mesoscale eddies in the ocean are known to be "vehicles" for larvae, fish, and other marine organisms [2, 26, 27, 31, 32]. In our case they could provide transport of heat-loving organisms to the southern coast of Russia [14].

The intrusions of subtropical water through the central gate across the Subpolar Front are documented in Fig. 5.11. In the beginning of September, 1995 a mesoscale cyclonic eddy (CE) to the north of the Subpolar Front with the center at about 41.5° N, 134.4° E "grabbed" some subtropical water at its southern periphery and pulled it to the north. This large-scale intrusion provided a streamer-like northward transport of subtropical water across the Subpolar Front through the central gate in the range 135° E–136° E as it is shown in Fig. 5.11a. The red and green colors on backward-in-time drift maps code the waters that entered the studied area for 2 years through its southern and northern boundaries, respectively. In the course of time, this streamer-like intrusion reached the latitude 42° N moving to the north (Fig. 5.11b).

As to the transport of subtropical waters through the eastern gate VII (see Fig. 5.5), it occurs mainly due to existence of a quasi-permanent vortex pair labeled as AC-C in the mean field in Fig. 5.2. It provides a propulsion of some subtropical tracers to the northwest whereas most of them, moving along the eastward frontal jet, join the Tsushima Warm Current and flow out to the Pacific through the Tsugaru Strait. The maps in Fig. 5.12 document the typical situation with a propulsion of

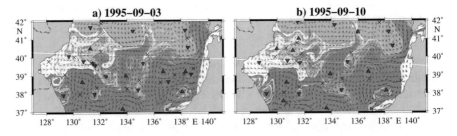

Fig. 5.11 The backward-in-time drift maps in September, 1995 document a streamer-like northward transport of subtropical waters across the Subpolar Front through a central gate with the help of the cyclone with the center at (41.5° N, 134.4° E). The *red* and *green colors* code the waters that entered the box shown for 2 years through its southern and northern boundaries, respectively. *White color* codes the tracers getting the coast

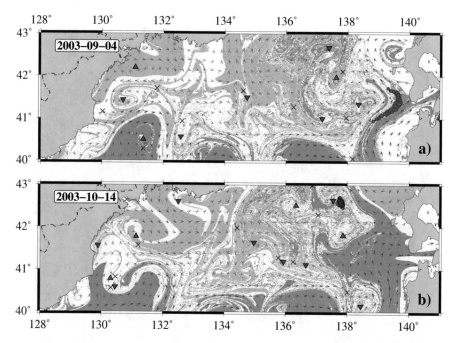

Fig. 5.12 (a) The backward-in-time drift maps in September–October, 2003 document a propulsion of subtropical tracers to the northwest with the help of the vortex pair consisting of the anticyclone with the center at 42° N, 137.7° E and the cyclone at 41.25° N, 138.35° E. Locations of the drifter no. 35660 are shown by *full circles* for 2 days before and after the day indicated

subtropical water to the northwest in September–October, 2003. The browsing and analysis of Lagrangian drift maps, computed for the whole observation period, have shown that frontal eddies have been found to facilitate northward transport of subtropical water across the Subpolar Front through the central and eastern gates.

To illustrate in more detail how this vortex mechanism works, we compute the backward-in-time D maps for September–October, 2003. The displacements D of all the tracers (4.2), distributed over the area shown in Fig. 5.13, are computed for 2 months backward in time starting from the day indicated in each panel. The values of D in km are coded by nuances of the grey color. So, black tracers displaced for the same time considerably as compared to the white ones. To validate our simulation we show in Fig. 5.13 positions of the drifter no. 35660 by full circles for 2 days before and after the date indicated in the panels with their size increasing in time. The full track of that drifter, launched on May 2, 2003 at the point 34.925° N, 129.3° E, is shown in Fig. 5.14.

In the beginning of September, 2003 (Fig. 5.13a) the vortex pair at the entrance to the gate VII consists of an anticyclone with the center at about 42° N, 137.7° E and a cyclone at about 41.25° N, 138.35° E. The cyclone winds some subtropical water from the eastward frontal jet round its northern periphery in a streamer-like manner (see the black "tongue" in Fig. 5.13a). Then this water is wound by the anticyclone

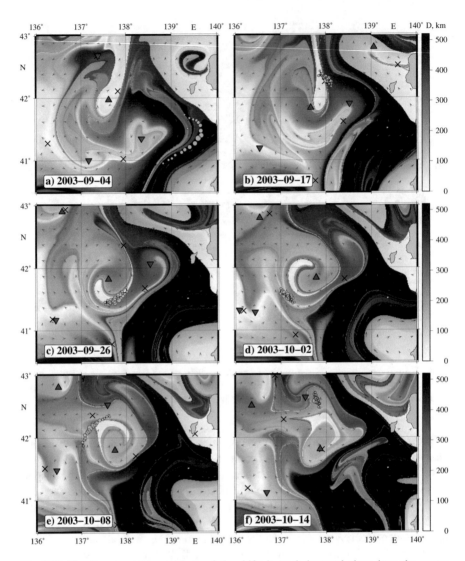

Fig. 5.13 The D maps with snapshots of the drifter's track imposed show how the vortex pair facilitates transport of subtropical tracers to the northwest through the eastern gate. The displacement of particles D in km, computed for a month backward in time from the day indicated in each panel, are coded by shades of the *grey color*. Locations of the drifter no. 35660 are shown by *full circles* for 2 days before and after the day indicated

round its southern periphery and propulsed to the northwest (Fig. 5.13b–f). This process is validated by snapshots of the track of the drifter no. 35660 for September–October, 2003. Being in the beginning of September in the main stream (Fig. 5.13a), it has drifted round the cyclone for the first half of September (Fig. 5.13b), then round the anticyclone for the second half of September (Fig. 5.13c) and in the

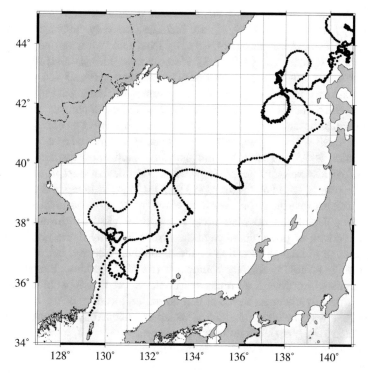

Fig. 5.14 The full track of the drifter no. 35660 launched on May 2, 2003 at the point 34.925° N, 129.3° E

beginning of October (Fig. 5.13d). Eventually, the drifter no. 35660 crossed the latitude 42° N (Fig. 5.13e) and moved to the north lugged by modified subtropical waters (Fig. 5.13f).

5.2.5   *Effect of Velocity-Field Errors on Statistical Properties of Lagrangian Transport*

It has been shown in this chapter statistically that the average northward component of the AVISO velocity field dictates preferred near-surface transport pathways of subtropical waters in the JS. The ability of satellite altimetry to accurately measure sea level anomalies has vastly improved over the last decade. However, there are still some measurement errors due to different reasons that lead to errors in the velocity field provided by the AVISO we used.

Some simulation results with an imperfect AVISO velocity field can be verified by comparing them with satellite, drifter, and in situ observations. It has been done when possible. The AVISO velocity field, averaged for the whole observation period, reproduces all the known main features of near-surface circulation in the Sea [1, 3, 7, 10, 13, 16–19, 39, 42, 43] including not only pathways of the main currents but also even locations of Ulleung, Dok, Oki, Wonsan, and other quasi-permanent mesoscale eddies (Fig. 5.2). Moreover, we have successfully compared our simulation results with available tracks of drifters (see Figs. 5.4, 5.10 and 5.13).

Nevertheless it is worth to discuss briefly some possible impact of errors in the altimetry field on our simulation results. Nobody knows, of course, "true" velocity field in the real ocean. The AVISO velocity field is imperfect as compared with an unknown "true" velocity field, and its errors can be considered as a noise $\Delta(u, v)$. The question is how reliable are our statistical simulation results based on an imperfect AVISO velocity field? To which extent one can trust to them? All the simulation results, based on the average AVISO velocity as in Fig. 5.2, are supposed to be reliable because the errors are averaged out for 22 years. As to other simulation results, they depend on possible noise Δv in the AVISO northward component v_+ which could, in principle, change the results but only if the noise would be strong enough to change direction of the meridional velocity, i.e., if $\Delta v > |v|$. If the average AVISO northward component $\langle v_+ \rangle$ is large enough as in the areas with dominated northward currents, we don't expect that it would be changed there significantly under influence of that noise. So, locations of the black-colored transport pathways in Figs. 5.3 and 5.5, which are dictated by that component, are not expected to be changed significantly.

If the average AVISO northward component $\langle v_+ \rangle$ is small, then two options are possible.

(1) It is small due to domination of a southward current somewhere, i.e., $v_- \gg \Delta v$. It is clear that possible noise would have practically no effect on northward transport in this case. For example, the forbidden zone in Fig. 5.3a to the south off Vladivostok, where the northward transport has not been observed during the whole observation period, should be located there at any realistic level of noise because it exists due to domination of a sufficiently strong southward jet (VMJ in Fig. 5.2).

(2) The average AVISO northward component $\langle v_+ \rangle$ is small due to a smallness of the absolute velocity, i.e., $\sqrt{u^2 + v^2} \sim \Delta v$. In this case northward and southward transports are equalized, and they are small if the noise is small enough. It is hardly to expect such a situation along the Subpolar Front because of a plenty of mesoscale eddies along the front where the absolute velocities are not small.

As to influence of possible errors in altimetry-derived velocity field on concrete mesoscale features, it has been analyzed in [5, 8, 9, 15] how an additional noise in advection equations, modelling unknown corrections to the AVISO velocity field, might change LCSs approximated by FTLE ridges. Strongly attracting and repelling individual LCSs in the California Current System have been shown to be robust to perturbations of the velocity field of over 20% of the maximal regional velocity [8]. Individual trajectories have been shown to be sensitive to small and moderate noisy variations in the velocity field but statistical characteristics and large-scale structures like mesoscale eddies and jets are not [5, 9, 15].

We conclude this chapter by summarizing the main results of our Lagrangian simulation and analysis of northward near-surface transport of subtropical waters in the Japan Sea for the period from January 2, 1993 to June 15, 2015.

- A Lagrangian methodology to simulate and analyze large-scale transport in the Japan Sea has been elaborated and applied.
- The northward near-surface transport of subtropical waters has been shown to be meridionally inhomogeneous with gates and barriers in the frontal zone whose locations are determined by the local advection velocity field. The very eastern and western gates exist mainly due to the Tsushima Warm Current and the East Korean Warm Current, respectively. The central gates "open" due to suitable dispositions of mesoscale eddies along the Subpolar Front facilitating propagation of subtropical waters to the north.
- The transport through the central gates has been shown to occur by a portion-like manner due to large-scale intrusions of subtropical waters round the eddies which have been documented with the help of Lagrangian maps.
- There are "forbidden" zones in the northern frontal zone where the northward transport has not been found during all the observation period. The "forbidden" zone to the south off Vladivostok exists due to a quasi-permanent southward jet there. The other "forbidden" zone exists due to the presence of a quasi-permanent topographically constrained ACE with the center at about 41.3° N, 134° E in the deep Japan Basin and the eastward zonal jet blocking northward transport there.

References

1. Akulichev, V.A. (ed.): Oceanographic studies of the Far Eastern seas and North-Western Pacific (in two books). Dalnauka, Vladivostok (2013) [in Russian]
2. Belkin, I.M., Cornillon, P.C., Sherman, K.: Fronts in large marine ecosystems. Prog. Oceanogr. **81**(1–4), 223–236 (2009). 10.1016/j.pocean.2009.04.015
3. Chang, K.I., Zhang, C.I., Park, C., Kang, D.J., Ju, S.J., Lee, S.H., Wimbush, M. (eds.): Oceanography of the East Sea (Japan Sea), 1st edn. Springer, Cham (2016). 10.1007/978-3-319-22720-7
4. Choi, B.J., Haidvogel, D.B., Cho, Y.K.: Interannual variation of the Polar Front in the Japan/East Sea from summertime hydrography and sea level data. J. Mar. Syst. **78**(3), 351–362 (2009). 10.1016/j.jmarsys.2008.11.021
5. Cotte, C., d'Ovidio, F., Chaigneau, A., Levy, M., Taupier-Letage, I., Mate, B., Guinet, C.: Scale-dependent interactions of Mediterranean whales with marine dynamics. Limnol. Oceanogr. **56**(1), 219–232 (2010). 10.4319/lo.2011.56.1.0219
6. Danchenkov, M.: Non-periodic currents of the Japan Sea. Hydrometeorology and Hydro-chemistry of Seas. The Japan Sea. Hydrometeorological Conditions, vol. 8(1), pp. 313–326. Gidrometeoizdat, Sankt-Peterburg (2003) [in Russian]
7. Danchenkov, M., Lobanov, V., Riser, S., Kim, K., Takematsu, M., Yoon, J.H.: A history of physical oceanographic research in the Japan/East Sea. Oceanography **19**(3), 18–31 (2006). 10.5670/oceanog.2006.41
8. Harrison, C.S., Glatzmaier, G.A.: Lagrangian coherent structures in the California Current System — sensitivities and limitations. Geophys. Astrophys. Fluid Dyn. **106**(1), 22–44 (2012). 10.1080/03091929.2010.532793

9. Hernández-Carrasco, I., López, C., Hernández-García, E., Turiel, A.: How reliable are finite-size Lyapunov exponents for the assessment of ocean dynamics? Ocean Model. **36**(3–4), 208–218 (2011). 10.1016/j.ocemod.2010.12.006

10. Hirose, N., Fukumori, I., Kim, C.H., Yoon, J.H.: Numerical simulation and satellite altimeter data assimilation of the Japan Sea circulation. Deep-Sea Res. II Top. Stud. Oceanogr. **52**(11–13), 1443–1463 (2005). 10.1016/j.dsr2.2004.09.034

11. Isoda, Y.: Interannual SST variations to the north and south of the polar front in the Japan Sea. La Mer **32**(4), 285–293 (1994). http://www.sfjo-lamer.org/la_mer/32-4/32-4-9.pdf

12. Isoda, Y.: Warm eddy movements in the eastern Japan Sea. J. Oceanogr. **50**(1), 1–15 (1994). 10.1007/bf02233852

13. Ito, M., Morimoto, A., Watanabe, T., Katoh, O., Takikawa, T.: Tsushima Warm Current paths in the southwestern part of the Japan Sea. Prog. Oceanogr. **121**, 83–93 (2014). 10.1016/j.pocean.2013.10.007

14. Ivankova, V.N., Samuilov, A.E.: New fish species for the USSR waters and an invasion of heat-loving fauna in the north-western part of the Japan Sea. Voprosy Ihtiologii **19**(3), 449–550 (1979) [in Russian]

15. Keating, S.R., Smith, K.S., Kramer, P.R.: Diagnosing lateral mixing in the upper ocean with virtual tracers: spatial and temporal resolution dependence. J. Phys. Oceanogr. **41**(8), 1512–1534 (2011). 10.1175/2011JPO4580.1

16. Kim, K., Chang, K.I., Kang, D.J., Kim, Y.H., Lee, J.H.: Review of recent findings on the water masses and circulation in the East Sea (Sea of Japan). J. Oceanogr. **64**(5), 721–735 (2008). 10.1007/s10872-008-0061-x

17. Kim, T., Yoon, J.H.: Seasonal variation of upper layer circulation in the northern part of the East/Japan Sea. Cont. Shelf Res. **30**(12), 1283–1301 (2010). 10.1016/j.csr.2010.04.006

18. Lee, D.K., Niiler, P.: Eddies in the southwestern East/Japan Sea. Deep-Sea Res. I Oceanogr. Res. Pap. **57**(10), 1233–1242 (2010). 10.1016/j.dsr.2010.06.002

19. Lee, D.K., Niiler, P.P.: The energetic surface circulation patterns of the Japan/East Sea. Deep-Sea Res. II Top. Stud. Oceanogr. **52**(11–13), 1547–1563 (2005). 10.1016/j.dsr2.2003.08.008

20. Lobanov, V., Ponomarev, V., Tishchenko, P., Talley, L., Mosyagina, S., Sagalaev, S., Salyuk, A., Sosnin, V.: Evolution of the anticyclonic eddies in the northwestern Japan/East Sea. In: Proceedings of the 11th PAMS/JECSS Workshop, April 11–13, 2001, Cheju, Korea, pp. 37–40 (2001)

21. Mitchell, D.A., Teague, W.J., Wimbush, M., Watts, D.R., Sutyrin, G.G.: The Dok cold eddy. J. Phys. Oceanogr. **35**(3), 273–288 (2005). 10.1175/jpo-2684.1

22. Mooers, C., Kang, H., Bang, I., Snowden, D.: Some lessons learned from comparisons of numerical simulations and observations of the JES circulation. Oceanography **19**(3), 86–95 (2006). 10.5670/oceanog.2006.46

23. Morimoto, A., Yanagi, T., Kaneko, A.: Eddy field in the Japan Sea derived from satellite altimetric data. J. Oceanogr. **56**(4), 449–462 (2000). 10.1023/A:1011184523983

24. Nikitin, A.A., Danchenkov, M.A., Lobanov, V.B., Yurasov, G.I.: Surface circulation and synoptic eddies in the Japan Sea. Izv. TINRO **157**, 158–167 (2002) [in Russian]

25. Nikitin, A.A., Lobanov, V.B., Danchenkov, M.A.: Possible pathways for transport of warm subtropical waters to the area of the Far Eastern marine reserve. Izv. TINRO **131**, 41–53 (2002) [in Russian]

26. Olson, D., Hitchcock, G., Mariano, A., Ashjian, C., Peng, G., Nero, R., Podesta, G.: Life on the edge: marine life and fronts. Oceanography **7**(2), 52–60 (1994). 10.5670/oceanog.1994.03

27. Owen, R.W.: Fronts and eddies in the sea: mechanisms, interactions and biological effects. In: Longhurst, A.R. (ed.) Analysis of Marine Ecosystems, pp. 197–233. Academic Press, London (1981)

28. Park, K.A., Chung, J.Y., Kim, K.: Sea surface temperature fronts in the East (Japan) Sea and temporal variations. Geophys. Res. Lett. **31**(7), L07304 (2004). 10.1029/2004gl019424

29. Prants, S., Ponomarev, V., Budyansky, M., Uleysky, M., Fayman, P.: Lagrangian analysis of the vertical structure of eddies simulated in the Japan Basin of the Japan/East Sea. Ocean Model. **86**, 128–140 (2015). 10.1016/j.ocemod.2014.12.010

30. Prants, S.V.: Backward-in-time methods to simulate large-scale transport and mixing in the ocean. Phys. Scr. **90**(7), 074054 (2015). 10.1088/0031-8949/90/7/074054
31. Prants, S.V., Budyansky, M.V., Uleysky, M.Y.: Identifying Lagrangian fronts with favourable fishery conditions. Deep-Sea Res. I Oceanogr. Res. Pap. **90**, 27–35 (2014). 10.1016/j.dsr.2014.04.012
32. Prants, S.V., Budyansky, M.V., Uleysky, M.Y.: Lagrangian fronts in the ocean. Izv. Atmos. Oceanic Phys. **50**(3), 284–291 (2014). 10.1134/s0001433814030116
33. Shin, C.W.: Characteristics of a warm eddy observed in the Ulleung Basin in July 2005. Ocean Polar Res. **31**(4), 283–296 (2009). 10.4217/opr.2009.31.4.283
34. Shin, H.R., Shin, C.W., Kim, C., Byun, S.K., Hwang, S.C.: Movement and structural variation of warm eddy WE92 for three years in the Western East/Japan Sea. Deep-Sea Res. II Top. Stud. Oceanogr. **52**(11–13), 1742–1762 (2005). 10.1016/j.dsr2.2004.10.004
35. Shrenk, L.: On the currents of the Okhotsk, Japan and adjacent seas. Mem. Emperor Acad. Sci. **23**(Suppl. 3), 1–112 (1874) [in Russian]
36. Stepanov, D.V., Diansky, N.A., Novotryasov, V.V.: Numerical simulation of water circulation in the central part of the Sea of Japan and study of its long-term variability in 1958–2006. Izv. Atmos. Oceanic Phys. **50**(1), 73–84 (2014). 10.1134/s0001433813050149
37. Sugimoto, T., Tameishi, H.: Warm-core rings, streamers and their role on the fishing ground formation around Japan. Deep Sea Res. Part A **39**, S183–S201 (1992). 10.1016/S0198-0149(11)80011-7
38. Takematsu, M., Ostrovski, A.G., Nagano, Z.: Observations of eddies in the Japan basin interior. J. Oceanogr. **55**(2), 237–246 (1999). 10.1023/a:1007846114165
39. Talley, L., Min, D.H., Lobanov, V., Luchin, V., Ponomarev, V., Salyuk, A., Shcherbina, A., Tishchenko, P., Zhabin, I.: Japan/East Sea water masses and their relation to the sea's circulation. Oceanography **19**(3), 32–49 (2006). 10.5670/oceanog.2006.42
40. Trusenkova, O., Nikitin, A., Lobanov, V.: Circulation features in the Japan/East Sea related to statistically obtained wind patterns in the warm season. J. Mar. Syst. **78**(2), 214–225 (2009). 10.1016/j.jmarsys.2009.02.019
41. Uda, M.: The results of simultaneous oceanographical investigations in the Japan Sea and its adjacent waters in May and June 1932. J. Imp. Fisheries Exp. Station **5**, 57–190 (1934)
42. Yoon, J.H.: Numerical experiment on the circulation in the Japan Sea. Part I. Formation of the East Korean warm current. J. Oceanogr. Soc. Japan **38**(2), 43–51 (1982). http://www.terrapub.co.jp/journals/JO/JOSJ/pdf/3802/38020043.pdf
43. Yoon, J.H., Kim, Y.J.: Review on the seasonal variation of the surface circulation in the Japan/East Sea. J. Mar. Syst. **78**(2), 226–236 (2009). 10.1016/j.jmarsys.2009.03.003
44. Zhao, N., Manda, A., Han, Z.: Frontogenesis and frontolysis of the subpolar front in the surface mixed layer of the Japan Sea. J. Geophys. Res. Oceans **119**(2), 1498–1509 (2014). 10.1002/2013jc009419
45. Zuenko, Y.I.: Interannual variations of location of the polar front in the nortwestern Japan Sea. Izv. TINRO **127**, 37–49 (2000) [in Russian]

List of Internet Resources

[AVISO] Archiving, Validation and Interpretation of Satellite Oceanographic (AVISO).
  http://www.aviso.altimetry.fr
[GDP] The Global Drifter Program.
  http://www.aoml.noaa.gov/phod/dac

Chapter 6
Dynamics of Eddies in the Ocean

6.1 Eddies in the Ocean

Eddies, which are energetic, swirling, robust, and long-lived features, have been found almost everywhere in the ocean [7, 10, 11]. Their size varies with latitude, region, bottom topography, and other factors. They can be characterized by a horizontal scale to be submesoscale eddies with the diameter <10–20 km and mesoscale eddies with larger diameters. Most of the kinetic energy of ocean circulation is contained in eddies. Eddies are important also because they transport momentum, heat, nutrients and provide mixing of waters with different properties.

Mesoscale eddies are particularly apparent along the paths of intense western boundary ocean current like the Gulf Stream, the Kuroshio, the Agulhas Current, and others. They also can be found in the mid-ocean both at the high and low latitudes. There are a number of mechanisms of generation of such eddies. A class of eddies, which are formed from the western boundary currents through the pinch-off of meanders, are called "rings" [22]. They may transport water of the parent current over thousands of kilometers and may exist for a few years. Free eddies in the mid-ocean are formed mainly due to a baroclinic instability of surface currents. Some eddies can be generated by winds or cooling at the sea surface. Flows over sea mounts and other prominent features of the sea floor, past islands, and capes generate topographically constrained eddies. Submesoscale and mesoscale eddies are often formed near the coasts and over the continental shelf due to a side friction and the corresponding instability.

By their shape and structure, the vortex features can be divided into monopole, dipole (mushroom currents), tripole, and multipole ones. Eddies with the same polarity can form vortex pairs exchanging by water with each other. Satellite images sometimes show vortex streets with submesoscale or mesoscale eddies moving over a trench, along the continental shelf or a coastline. The vortex configuration may be rather complex, for example, with a large mesoscale eddy or a ring surrounded by smaller-size eddies with opposite polarity. A specific class of eddies with the

© Springer International Publishing AG 2017
S.V. Prants et al., *Lagrangian Oceanography*, Physics of Earth and Space
Environments, DOI 10.1007/978-3-319-53022-2_6

scale smaller than 30 km (typically around 10–20 km) are frequently observed at the sea surface both from satellites and space shuttles. They are spiral in form, overwhelmingly rotate cyclonically, and extend to 300 m below the surface. They have been observed in the regions where currents produce no or a little horizontal stress [41].

As to the vertical structure, most of the large-scale eddies are cone-like "bags" of water reaching sometimes a few km downward. Moreover, there are intrathermo-clinic eddies which are manifested not at the sea surface but inside the thermocline [9, 12, 14, 43, 45]. There also exist hetons which are eddies of opposite polarity one over another [42, 44, 45]. The eddies move either being advected by a background current or due to interaction with other eddies, streamers, intrusions, etc. The β-effect is not large enough to influence mesoscale eddies. However, it is sufficiently large to drag mesoscale eddies and rings asymptotically to the west in the northern hemisphere. Decay and death of eddies may be caused by different reasons: small-scale turbulent diffusion, radiation of planetary waves, interaction with background currents, interaction with other eddies, etc.

Before the satellite era it was difficult to detect eddies in the mid-ocean, just like to search for a needle in a haystack. Sparse distribution of in situ measurements limited our knowledge of eddies in the ocean. As to rings around the Gulf Stream and Kuroshio, they have been known to sailors and fishermen long ago. Mesoscale eddies in the mid-ocean have been discovered only in the late 1960s. Satellite SST sensors and altimeters changed the situation cardinally. The current satellite missions provide global real-time data for measurements of SST and the SSH above the ocean rest state with high precision and high space and time resolution.

The FTLE diagnostic enables to identify boundaries of mesoscale eddies in a given velocity field. As an example, we demonstrate in Fig. 6.1a, a NASA satellite image of the sea color in the Oyashio–Kuroshio frontal zone in the North Western Pacific Ocean to the east of the Hokkaido Island (Japan) in May 2009 during the period of a spring bloom of phytoplankton. The phytoplankton becomes concentrated along the boundaries of the mesoscale eddy tracing out the motion of water. Computing altimetry-based daily FTLEs and other Lagrangian indicators we were able to trace out the birth and history of that eddy. It was found to be a Kuroshio warm-core ring to be split off the Kuroshio Extension Current, the powerful eastward jet, that transports warm subtropical waters approximately along 35° N–37° N latitudes (see Figs. 4.1 and 4.3). That ring slowly migrated to the north and reached to May 2009 its place in Fig. 6.1. It should be stressed that its core contains mainly warm and salty Kuroshio water that is strongly different from more cold and fresh ambient Oyashio water. The boundaries of such rings are known to be attractive for phyto- and zooplankton, and, therefore, fish and other marine organisms may accumulate there for feeding [28, 29, 33].

In simulation we distribute a large number of tracers in the area and advect them forward and backward in time for 30 days in the AVISO velocity field starting from May 21, 2009. We integrate Eq. (4.1) for each pair of initially closed tracers and compute the FTLE in accordance with (4.12). Those forward- and backward-in-time values, Λ_+ and Λ_-, then are coded by color and plotted together on the geographic

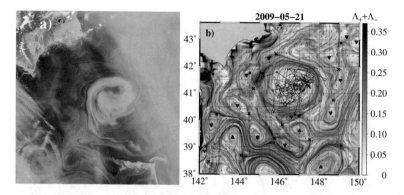

Fig. 6.1 (a) NASA image of *ocean color* illustrating spring bloom of phytoplankton near Hokkaido (Japan) in May 2009. Phytoplankton concentrates along the boundaries of the mesoscale eddy tracing out the vortex motion of water. (b) Combined forward- and backward-in-time (Λ_+ and Λ_-, respectively) FTLE maps computed with the AVISO velocity field superimposed by *arrows*. Λ is in days^{-1}. Track of the Argo float no. 2900946, launched at the western periphery of that eddy on January 27, 2009, is shown by the *green stars* from the launching date up to the end of its mission, on January 16, 2010. The *largest star marks* its position on May 21, 2009 and the other ones show its positions each 5 days before and after that date. The *crosses* and *triangles mark* "instantaneous" hyperbolic and elliptic stationary points on May 21, 2009, respectively

map in Fig. 6.1b. We recall that the physical meaning of the FTLE is the rate of divergence of initially closed particles for a given period of time.

Track of the Argo float no. 2900946 (for reference, see [ARGODB]), launched at the western periphery of that ring on January 27, 2009, is shown in Fig. 6.1b by the green stars. The float has been captured by the ring and provided valuable information on vertical profiles of temperature, density, and salinity up to January 16, 2010, the end of its mission. That Hokkaido ring was a quasistationary feature staying practically at the same place for a year but gradually decreasing in diameter. Large mesoscale ACEs appear often to the east off the Hokkaido Island. Sometimes they are Kuroshio rings, and sometimes not.

6.2 Altimetry-Based Lagrangian Analysis of Formation, Structure, Evolution, and Splitting of Mesoscale Kuril Eddies

6.2.1 Mesoscale Kuril Eddies

Mesoscale ACEs have been regularly observed on the oceanic side off the Kuril Islands near the Bussol' Strait in the North Western subarctic Pacific to the south of the Kamchatka Peninsula (see Fig. 6.2) [1, 5, 6, 19, 35–39, 52]. It is the deepest Kuril strait through which water inflows to the Okhotsk Sea and outflows from it

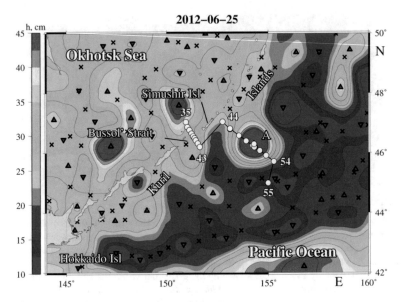

Fig. 6.2 The mesoscale Bussol' ACE A in the SSH field by satellite altimetry. Locations of the CTD transect in the R/V *Professor Gagarinskiy* cruise in 2012 are shown by the *white circles* with some stations numbered. "Instantaneous" elliptic and hyperbolic points in the AVISO velocity field, to be present in the area on June 25, 2012, are indicated by the *triangles* and *crosses*, respectively

[34, 50]. Such eddies are observed in this area every year in the SSH-anomaly field and infrared satellite images. They are one of the most energetic and prominent features in the western subarctic gyre in the North Pacific which require detailed study of their hydrographic structure and understanding the processes of their origin, dynamics, transformation, and decay. They are supposed to control significantly transport of the East Kamchatka and Oyashio currents and modify the properties of these source waters of the North Pacific Intermediate Water [52].

Extensive hydrographic observations of those eddies [6, 19, 35–39, 52], especially in the last decade of the twentieth century, allowed to sample their vertical structure, dynamics, and kinematics. They are observed along the entire chain of the Kuril Islands from the Hokkaido Island up to the Kamchatka Peninsula and thus control water exchange at the western boundary of the subarctic gyre. However, the origin of those eddies is not clear. Are they advected from the north (south) or do they form locally?

Based on satellite altimetry data and current meter moorings, it was demonstrated that the eddies in the southern and central Kuril area move typically to the northeast while the eddies in the northern Kurils area move southwestward [15]. At least, some Kuril ACEs are believed to be modifications of warm-core Kuroshio rings that moved northward for several years. One of them, called WCR86B, split off from the Kuroshio Extension in 1986 and reached the Bussol' Strait at the latitude 46.5° N in September 1990 when it has been sampled [19, 35, 37, 53]. This eddy moved along the Kuril–Kamchatka Trench and disappeared in the end of 1991. Even so far away from its origin, it had a warm and high-salinity core in the upper layer and a secondary low salinity core at the intermediate depth (250–600 m). Other large warm-core eddy has been sampled off the Bussol' Strait in summer of 1995 [52]. This eddy can be tracked back in time when it was a Kuroshio warm-core ring with its center located around 46.5° N and 146.5° E in summer of 1994, and then it moved northeastward with decreasing size for a year and disappeared in February 1996. During the evolution of the WCR86B, its warm core has been observed to shrink with subsequent formation and enlargement of its secondary cold core. This allowed to propose that during its evolution in the subarctic waters the Kuroshio warm-core rings eventually change their structure to cold-core rings typical for Kuril eddies [19]. Another suggestion is that the Kuril ACEs are formed locally by intrusion of cold fresh water outflow from the Okhotsk Sea [37, 52]. Even if an eddy was not formed locally but migrated from the south, an interaction with the Okhotsk Sea water outflow should impact significantly on its water mass structure and dynamics.

It was shown that regular appearance of large ACEs to the east off the Kuril Islands are important for water dynamics and its modification in the area. Approaching the coast, such a migrating eddy could block the coastal flow and thus intensify the offshore branches of the East Kamchatka and Oyashio currents. The Kuril eddies affect the distributions and vertical fluxes of the dissolved oxygen, nutrients, and dissolved inorganic carbon in the Oyashio region and thus affect the plankton bloom there with an impact on marine biota. Using in situ observations and satellite data, it has been found that boundaries of the Kuril eddies were composed of productive coastal Oyashio water which was wrapped around the eddy core creating a high productive belt [18].

A recent interest to those eddies has been motivated by the accident at the FNPP in March 2011. Measurements of the ^{137}Cs and ^{134}Cs radioactivity along with hydrographic sampling of a number of mesoscale eddies in the broad area to the east off Japan and the Kuril Islands have been conducted in the R/V *Professor Gagarinskiy* cruise (June 13–July 10, 2012) [4] (see Chap. 7). In particular, a cold-core Kuril ACE, called the eddy A in [4], has been sampled in the end of June 2012 with the center at that time at 46.19° N, 154.33° E. It was suggested that ACEs could accumulate water with high concentration of radioisotopes because of convergence and subduction processes and then transport this water while migrating northward. Even if it was actually proved for the Kuroshio warm-core rings it might not be so for the Kuril eddies. The observed concentrations of ^{134}Cs and ^{137}Cs across the eddy A at different depths have not exceeded the background level in the ocean [4].

The eddies consist of a core and a periphery. The core is a uniform water mass which conserves for a while its physicochemical properties. Periphery is a water mass involved in a vortex motion around the eddy center. Strong potential vorticity gradient at the core boundary allows fluid particles in theory to enter and leave the core only due to diffusion. The eddy periphery deforms and exchanges water with its surroundings much more effectively than the core. However, peripheral water masses may differ by its properties from ambient waters. In spite of the tremendous progress in studying oceanic eddies, mechanisms of their formation, propagation, and decay are not well understood and may be different in different regions. Even eddy identification is not a simple task (see, e.g., [7, 8, 17]). Water exchange between the eddy core and the surroundings is also not well understood. It is important as well to know how eddies gain, retain, and release water.

We focus in this section on the Kuril anticyclonic eddy A or the Bussol' eddy, that has been sampled in the cruise R/V *Professor Gagarinskiy* in June 2012, and on its parent eddy B. We apply a Lagrangian approach to study formation, structure, evolution, splitting, and merging of those eddies based on the AVISO velocity field. To be more concrete we will deal with the following issues:

- Identification of eddies and their boundaries in the altimetric velocity field.
- Documenting deformation, interaction, and splitting of eddies.
- Tracking water exchange between eddies and its surroundings: how and when they gain, retain, and release water.
- Tracking the origin of core and peripheral waters: which waters fill the eddy core and its periphery and how they do that.
- How fluid parcels leave the eddy core and its periphery and where they are headed for.

Our simulations are verified in part by tracks of available surface drifters and Argo floats [GDP, ARGO].

6.2.2   CTD Sampling of the Bussol' Eddy A

The cruise of the R/V *Professor Gagarinskiy* was conducted by the Pacific Oceanological Institute of the Far Eastern Branch of the Russian Academy of Sciences in June and July 2012 with the aim to collect data on distribution of Fukushima-derived ^{134}Cs and ^{137}Cs isotopes in some areas of the Japan and Okhotsk seas and the adjacent area in the North Western Pacific [4]. Part of the cruise included observations in the area of the central Kuril Islands with a CTD transect from the Okhotsk Sea to the Pacific Ocean through the Bussol' Strait crossing a large mesoscale Bussol' eddy A (Fig. 6.2). That eddy was identified by satellite altimetry data. Its center and boundaries have been identified with the help of Lagrangian diagnostic maps operationally computed and transmitted to the board by radio. Special attention was paid to cross the eddy as close as possible to its center. With this aim a few stations have been done at the expected location of its center obtained by computing the corresponding elliptic point in the AVISO velocity field. Observations were implemented from surface down to the depth of 2000 m with the CTD SBE 911plus equipped with two sets of temperature and conductivity sensors, as well as sensors of dissolved oxygen, turbidity, and fluorescence. Observations across the Bussol' Strait have been done by the CTD SBE 19plus down to 600 m.

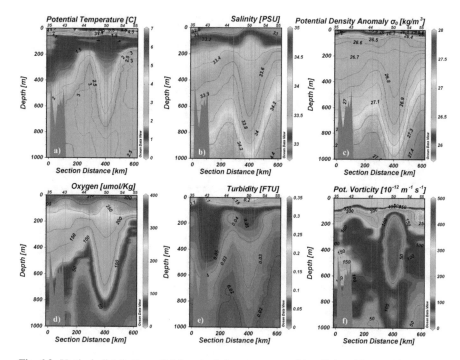

Fig. 6.3 Vertical distribution of (**a**) potential temperature, (**b**) salinity, (**c**) potential density anomaly, (**d**) dissolved oxygen, (**e**) turbidity, and (**f**) potential vorticity along the section from station 35 in the Okhotsk Sea (*left*) to station 55 in the Pacific Ocean (*right*) across the Bussol' eddy A by CTD observations in the R/V *Professor Gagarinskiy* cruise from June 23 to 27, 2012. The numbers of stations are indicated on the *top* of each panel

In total, 13 stations down to 2000 m and 8 stations down to 600 m have been done in June 23–27, 2012 for the following analysis of the eddy structure.

The SSH field in Fig. 6.2 shows that the eddy A located east of the Bussol' Strait and the Simushir Island is the most prominent and intense dynamic feature in the region with the positive SSH in its center exceeding 40 cm. Distribution of water properties along the CTD transect from the Okhotsk Sea to the Pacific Ocean, crossing the eddy center, is presented in Fig. 6.3. The eddy had a well-developed vertically uniform core located between 100 and 700 m. Eddy core water had lower temperature and salinity and higher concentration of dissolved oxygen and turbidity. The core water mass characteristics are similar to the Okhotsk Sea water. However, the core of the eddy has much lower stratification than the Okhotsk Sea water which is clearly seen at the distribution of potential vorticity (Fig. 6.3f). This suggests that water has been well mixed vertically by convergence process in the eddy core and possible convection in winter. High content of dissolved oxygen may prove intense vertical convective mixing in this layer.

The uplifting of the isolines in the upper layer of the eddy indicates a relative cyclonic shear above 200 m depth (Figs. 6.3 and 6.4). Thus we may suggest that

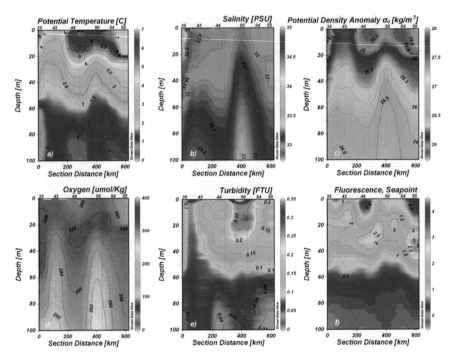

Fig. 6.4 The same as in Fig. 6.3 but in the upper 100 m layer

the eddy has its main core located between 100 and 700 m, and the upper part of toroidal shape is clearly seen in the surface layer above 40 m (Fig. 6.4c). There is a colder and higher salinity and thus higher density water in the center of the eddy and lower density water at its periphery in the layer above 200 m. This water has a little different properties than that in the Okhotsk Sea. It has a little higher temperature and salinity.

It is interesting to note that in the center of the eddy surface water has much higher turbidity up to 0.25–0.33 FTU (Fig. 6.4e). It could be caused by plankton activity. There is a high chlorophyll content (a high fluorescence) at the layer corresponding to seasonal pycnocline with maximum at 10–20 m in the Okhotsk Sea and the Bussol' Strait, around 20–40 m in the eddy, and 30–50 m in the Pacific to the south (Fig. 6.4f). It results in the high content of dissolved oxygen in this layer because of photosynthesis (Fig. 6.4d). However, there is no indication to high plankton activity in the center of the eddy where a high turbidity water has been observed. It is not clear why this water has a much higher turbidity than surrounding waters. One of the mechanisms could be convergence of water in the center of the eddy. Thus, water in the eddy center has been trapped there for a long time, while water at its periphery is a result of involvement of surrounding water masses, and its lifetime in the eddy is much shorter. In the next section we will examine the water exchange in the eddy using the Lagrangian approach.

6.2.3 Lagrangian Analysis of the Sampled Bussol' Eddy A

6.2.3.1 Evolution and Metamorphoses of the Kuril Eddies in 2012

In this section we focus on the first two issues mentioned in Sect. 6.2.1. Lagrangian tools are applied to identify a Kuril anticyclone, which we call the eddy B, and to document its deformation, splitting, and genetic coupling with the eddy A that has been sampled in the R/V *Professor Gagarinskiy* cruise in the end of June 2012. To do that let's compute backward-in-time Lagrangian drift maps and plot particle's displacements D (4.2) for 30 days in the AVISO velocity field. Such a map on February 1, 2012 in Fig. 6.5a clearly shows the eddy B with the elliptic point at its center 46.3° N, 153.3° E. The boundary of that eddy can be identified approximately as a closed curve with the maximal (locally) gradient of D that passes through the hyperbolic point at 47.2° N, 153.7° E. That curve separates the waters, involved in the rotational motion around the vortex center, from ambient waters. Displacements of fluid particles for the former ones are much smaller than for the particles outside the eddy ("dark" waters in Fig. 6.5a). The spirality of the eddy B on the map is a typical feature for drift maps with eddies (see, e.g., [26, 27]). The spirality just means that water parcels inside an eddy rotate with different angular velocities, and the distances, D, between final and initial positions of particles differ in bands of different color.

In the beginning of 2012, the Kuril mesoscale ACE B was observed near the coast of the Simushir Island to the east of the Bussol' Strait (Fig. 6.5a). The East Kamchatka Current rounds it from the east and south (the black band around that eddy in Fig. 6.5a). The eddy B started to deform in February due to an intensification of the East Kamchatka Current and development of a meander. Its elliptic point has been shifted during a month to the south by almost 1° (Fig. 6.5b). An elliptic point at 46° N, 153° E appeared on April 14 at the center of the newly formed eddy A. Simultaneously a hyperbolic point appeared at 46.2° N, 152.5° E between the centers of the eddies B and A (Fig. 6.5d). The blue patch, placed on March 24 at the center of the eddy B (Fig. 6.5c), stretched strongly to the middle of April due to splitting of the eddy B and formation of the eddy A. The blue filament encircles the young eddy A around its elliptic point.

Splitting of the eddy B and birth of the eddy A in the middle of April are confirmed by a track of the drifter no. 42949 coming from the north (the string with the red circles in Fig. 6.5d–f). The circles increase in size through time and show locations of the drifter every 6 h, i.e., Fig. 6.5d shows a fragment of the drifter's trajectory from 0:00 GMT on April 12 to 0:00 GMT on April 14. It encircles the eddy A anticyclonically in the second half of April confirming our simulation (Fig. 6.5e, f). Water from the core of the eddy B formed the core of the eddy A (look at evolution of the blue patch). The periphery of the eddy A consists mainly of waters from the north (look at evolution of the red patch in Fig. 6.5c placed on March 24 to the north of the eddy B).

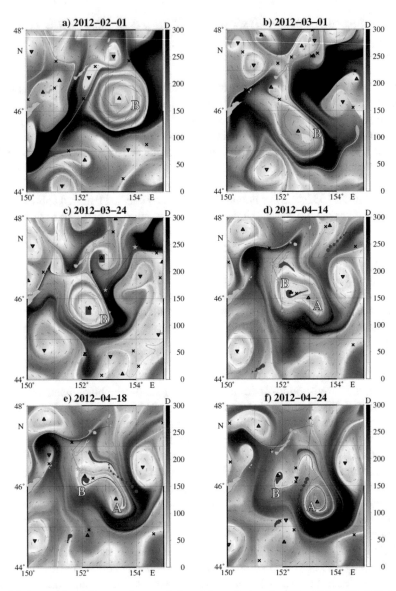

Fig. 6.5 Lagrangian drift maps show evolution, deformation, and splitting of the eddy B that gave birth to the eddy A. The displacement of particles D (in km), computed for a month backward in time from the day indicated in each panel, are coded by nuances of the *grey color*. "Instantaneous" elliptic and hyperbolic points, to be present in the area on fixed days, are indicated by *triangles* and *crosses*, respectively. Tracks of two drifters are shown by *full circles* for 2 days before the day indicated with the size of the circles increasing in time. Tracks of the Argo float no. 4900939, approached the eddy B in the end of March, are shown by the *stars*. The *largest star* corresponds to the day indicated and the other ones show its location each 5 days before and after that date. The *colored squares* are patches with tracers whose deformation is shown

The eddies A and B form an anticyclonic vortex pair with a hyperbolic point arising just after splitting of the eddy B. It is shown on April 14 by the cross in Fig. 6.5e at 46.2° N, 152.5° E. By definition, each hyperbolic point has contracting (stable **manifold**) and expanding (unstable **manifold**) directions along which water parcels in the course of time converge to it and diverge away from it, respectively. Figure 6.6 with the drifter tracks clarifies the role of hyperbolic points and their manifolds in organizing oceanic flows around.

Two drifters nos. 42949 and 42970 in Fig. 6.6a on April 28 approach that point, located now at 46° N, 152.3° E, along its stable manifolds from the south and north. When approaching the hyperbolic point, they slow down (compare Figs. 6.5e, f and 6.6a, b) because of the presence of a dynamical saddle trap in a vicinity of hyperbolic (saddle) points [51] (see Sect. 2.2.3). Then the drifters begin to move away from the hyperbolic point to the west and east in the course of time along the unstable manifolds of that point (Fig. 6.6b–d on April 29, May 2 and 3). The southern drifter no. 42970 moves away from the hyperbolic point and eventually enters the Okhotsk Sea via the Bussol' Strait, whereas the northern one no. 42949 encircles the eddy A once more before going away. The eddy A progressively grows in size during April, May, and June gaining surrounding water (look at the red filaments encircling its core in Fig. 6.6a–d). The eddy A has practically a circular form by the days of sampling with the elliptic point on June 24 at 46.3° N, 154.3° E and the size of $\simeq 140 \times 135$ km in the meridional and zonal directions, respectively.

The FTLE maps provide a complementary tool to visualize eddies and confirm the results obtained with the D maps. We compute forward-in-time (Λ_+) and backward-in-time (Λ_-) FTLE fields by the method presented in Sect. 4.3.1 [27] for 15 days in the future and in the past, respectively, starting from June 25. The combined map in Fig. 6.6f clearly shows the eddy A as a closed region bounded by black "ridges" intersecting at two hyperbolic points. The FTLE "ridges" are locations of particles on a given day with maximal (locally) values of the FTLE. Particles on both sides of a "ridge" diverge maximally in the future and in the past. It means that the water parcels on one side of the "ridge" are involved in the vortex motion around an elliptic point, whereas the ones on the other side move away from the eddy. Thus, intersecting "ridges" on the FTLE map approximate the eddy boundary.

In Fig. 6.7 snapshots of the SSH field on March 24, April 18, and April 24, 2012 show, respectively, the eddy B with the center at 45.7° N, 152.3° E (the left panel), its splitting (the central panel) and formation of the eddy A with the center at 45.6° N, 153.3° E (the right panel).

6.2.3.2   Structure of the Kuril Eddies A and B and Water Exchange

Using specific backward-in-time Lagrangian maps we study here horizontal structure of the eddy A, the origin of its core, and peripheral waters and how they are situated within that eddy. The other task, connected with the first one, is to track by

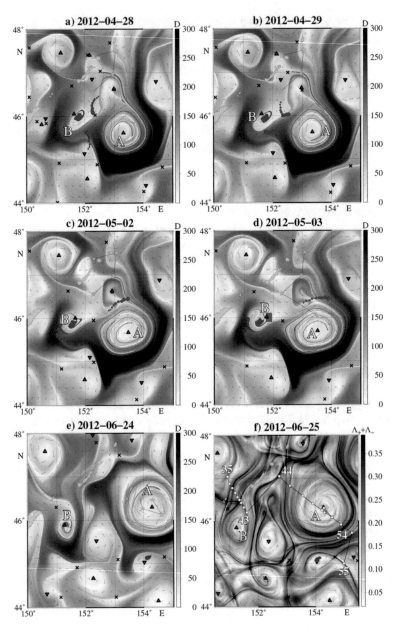

Fig. 6.6 (**a**)–(**e**) Lagrangian drift maps show evolution of the eddy A till the sampling days in the end of June 2012. Tracks of the drifters are indicated by *full circles* for 2 days before the day indicated with the size of the circles increasing in time. Tracks of the Argo float no. 4900939, approached the eddy A in the end of April and be trapped by it from May to October, are shown by the *stars*. The *largest star* corresponds to the day indicated and the other ones show its location each 5 days before and after that date. (**f**) Combined Lyapunov map of forward (Λ_+) and backward-in-time (Λ_-) FTLEs on June 25, 2012. Λ is in days^{-1}. Locations and numbers of sampling stations across the eddy A are shown

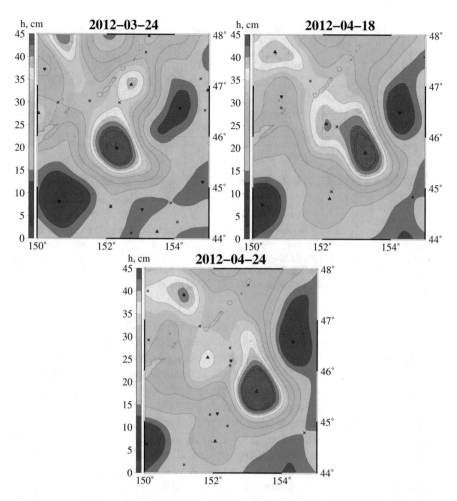

Fig. 6.7 Snapshots of the SSH field on March 24, April 18, and April 24, 2012 which show, respectively, the eddy B with the center at 45.7° N, 152.3° E (*the left panel*), its splitting (*the central panel*), and formation of the eddy A with the center at 45.6° N, 153.3° E (*the right panel*)

means of forward-in-time Lagrangian maps by which ways fluid parcels leave the parent eddy B to form the core of the newborn eddy A and its periphery.

The tracking maps in Fig. 6.8 show in the logarithmic scale where the tracers placed on the material line along the R/V *Professor Gagarinskiy* cruise track, crossing the eddy A, have been found for the period in the past from June 25 to April 11, 2012 (Fig. 6.8a) and for a much longer period from June 25, 2012 to the day of the accident at the FNPP, March 11, 2011 (Fig. 6.8b). The map in Fig. 6.8a, computed throughout life of the eddy A, confirms that some tracers of that line came to their locations on the dates of sampling from the Okhotsk Sea, whereas the most part of them has been advected by the East Kamchatka Current

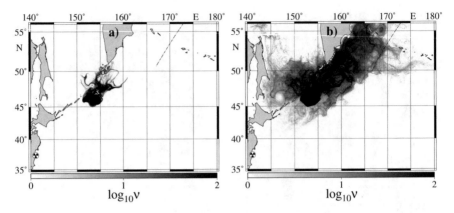

Fig. 6.8 Simulated backward-in-time tracking maps for the tracers distributed on the material line along the R/V *Professor Gagarinskiy* cruise track crossing the eddy A. The maps show where those tracers have been from (**a**) June 25 to April 11, 2012 and from (**b**) June 25, 2012 to the day of the accident at the Fukushima Nuclear Power Plant, March 11, 2011. The density of tracers, ν, is in the logarithmic scale. Radioactivity sign shows location of the FNPP

from the north. The longer history of the eddy A waters in Fig. 6.8b shows a larger area visited by the eddy A tracers for 15 months after the Fukushima accident. This map demonstrates that the risk of a radioactive contamination of the eddy A surface waters is minimal. Observed ^{137}Cs and ^{134}Cs concentrations inside the eddy A in the R/V *Professor Gagarinskiy* cruise in the end of June 2012 have been found to be at the background level not only in surface waters but also in the depth [4].

A large number of tracers were uniformly distributed in the box [153° E–155.5° E; 45.5° N–47° N], including the eddy A, and integrated backward in time from June 25 to April 11, 2012 in the regional AVISO velocity field. Coding by different colors the particles which entered the area through its geographical borders for that period of time, we plot in Fig. 6.9a the drift Lagrangian map that shows where water masses, constituting that eddy, came from. Yellow color means that the corresponding particles entered the box shown in Fig. 6.9a through its western boundary, green, blue, and red ones—through its eastern, northern, and southern boundaries, respectively. "Grey" particles in the inner core of the eddy A are those ones which were present in the box for the integration period.

To reveal horizontal structure of the eddy A we fix the dates in the past when the particles entered the box. The box shown in Fig. 6.9 is seeded with a large number of tracers on June 25. The advection equations (4.1) are solved backward in time from June 25 to April 11. The days, when tracers entered the box, T, are fixed and coded by nuances of the grey color. The entrance-time map in Fig. 6.9b shows how many days ago a tracer with given initial coordinates crossed one of the box boundaries. The colored patch at the eddy center means that the corresponding tracers have been there before April 11. The rings in Fig. 6.9b of different color mean that the eddy has captured and wound water by portions. It is an analogue of tree rings. The eddy core can be identified there as a boundary of contrast colors. The rotation map in

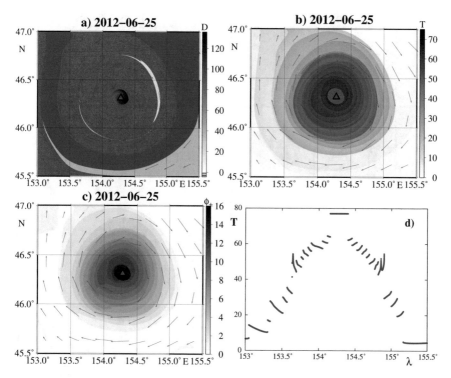

Fig. 6.9 (**a**) Backward-in-time drift map shows where the waters of the eddy A came from for the integration period from June 25 to April 11: *yellow*—from the west, *green*—the east, *blue*—the north, *red*—the south, *grey color* codes the waters that have been in the box during all that period in the past. The absolute displacement, *D*, for the grey particles is in km. (**b**) Vortex rings revealed by the entrance-time map. The day *T*, when particles entered the eddy A in the past, is coded by the *grey color*. (**c**) The rotation map shows how many times particles have rotated around the eddy center from April 11 to June 25, 2012. (**d**) Dependence of the entrance day *T* on longitude λ of particles distributed initially on the zonal material line along 46.3° N

Fig. 6.9c shows how many times the particles rotated around the eddy center from April 11 to June 25, 2012. This plot has a ring-like structure as well.

The process of winding of water onto the eddy by portions becomes evident with the plot of dependence of some particle's characteristics inside the eddy on their coordinate. We solve advection equations (4.1) backward in time from June 25 to April 11 for the particles distributed initially on the zonal material line crossing the eddy along 46.3° N and fix the day, *T*, when particles with given longitudes λ entered the vortex area (Fig. 6.9d). The plot in Fig. 6.9d consists of a hierarchy of segments with "quantized" portions of the zonal material line with close entrance dates. As expected, the entrance dates for particles in the outer parts of each segment are older than for those in the inner parts. The straight line with the maximal values of *T* in Fig. 6.9d corresponds to the patches in the inner core in Fig. 6.9a, b.

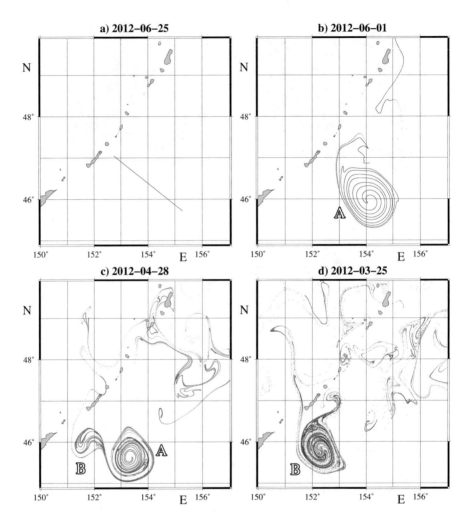

Fig. 6.10 Backward-in-time evolution of a material line placed along the R/V *Professor Gagarin-skiy* cruise track crossing the eddy A in the end of June 2012

A direct evidence of subsequent winding of surrounding water onto the eddy A is provided by backward-in-time evolution of a material line placed along the R/V *Professor Gagarinskiy* cruise track crossing the eddy A. The particles of that line have been integrated backward in time from June 25 (Fig. 6.10a). Figure 6.10b demonstrates how the eddy has wound water for June. The process of splitting of the eddy B is clearly seen in Fig. 6.10c plotted on April 28. Figures 6.5 and 6.6 illustrate a genetic connection between the eddies B and A.

In order to track the origin of the eddy A by a complementary way we compute forward-in-time Lagrangian maps for the eddy B. The box with that eddy in Fig. 6.11a is seeded with a large number of tracers which are integrated forward

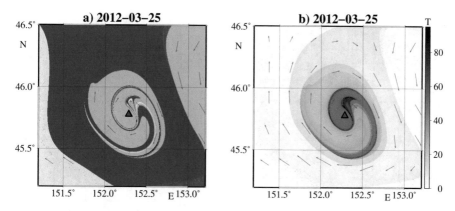

Fig. 6.11 (**a**) Forward-in-time drift map shows where the waters of the eddy B headed for over the integration period: *yellow*—to the west, *green*—to the east, *blue*—to the north, *red*—to the south, and *grey color* codes the waters that did not leave the box for that period of time. (**b**) The residence time map. The day T, when the particles left the vortex area, is coded by the *grey color*

in time from March 25 to June 25, 2012. Coding by different colors the particles which left the box through its geographical borders for that period of time, we get the drift Lagrangian map that shows where water masses of the eddy headed for (Fig. 6.11a). Yellow color means that the corresponding particles crossed the western boundary, green, blue, and red ones—the eastern, northern, and southern boundaries, respectively. The "grey" particles in the inner core are those ones which did not leave the box for the integration period. "Green" waters in the eddy B core is a source of eddy A waters. In fact, the eddy B split into two parts, as shown in Figs. 6.5 and 6.6, with the eastern part to be formed initially by the core waters of the eddy B. Then surrounding waters wound onto that split part of the eddy B by portions as shown in Figs. 6.9 and 6.10 and eventually formed the eddy A which has been sampled in the end of June 2012.

Two $0.25° \times 0.25°$-size patches with tracers were placed on June 25, 2012 at the locations of stations 45 (blue patch in Fig. 6.12a) and 49 (red patch in Fig. 6.12a) at a periphery of the eddy A and near its center, respectively, and advected backward in time. Tracking the history of those patches, we see that on March 10 the red tracers were concentrated mainly in the eddy B (Fig. 6.12d). However, some of them originated from the Okhotsk Sea and the East Kamchatka Current. On April 28, they were compactly distributed among two eddies, inside a remnant of the eddy B and in the newborn eddy A (Fig. 6.12c). As to the blue tracers, they were distributed over a large area, mainly to the east of the Kuril Islands with a small "tail" in the Okhotsk Sea. Figure 6.12b plotted on June 1 shows the red tracers distributed in the eddy A core and the blue ones winding around that core.

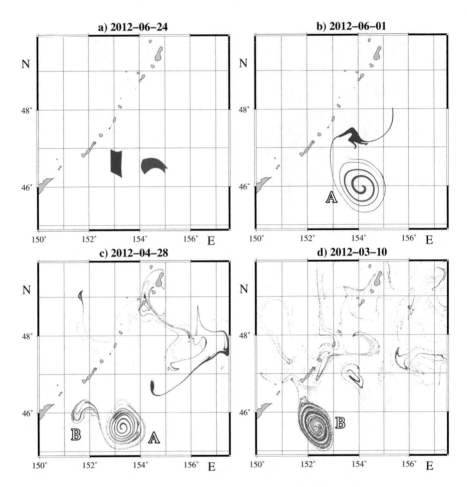

Fig. 6.12 Backward-in-time evolution of the patches with tracers distributed at the locations of (**a**) station 45 at a periphery of the eddy A and (**b**) of station 49 near its center in the R/V *Professor Gagarinskiy* cruise

6.2.4 Vertical Profiles of Temperature and Salinity by the Argo Floats

The eddy A was born as a result of splitting of another Kuril eddy B in the area east of the central Kuril Islands which happened in middle April of 2012. We have observed evolution of the eddy B since the beginning of 2012 (Fig. 6.5). In early February the eddy had a circular form around 165–170 km in diameter and was located just to the east of the Simushir Island. Then during February–March the eddy was moving to the southwest by around 100 km down to the area off the Bussol' Strait. We explain this transition of the eddy by an intensification of the

East Kamchatka Current. Its branch is clearly seen in the D field as an area with the highest displacement's values just upstream of the eddy (Fig. 6.5a and b). The SSH snapshots in Fig. 6.7 clearly show the eddy B, its splitting and formation of the eddy A in accordance with Fig. 6.5c, e, and f.

It may be suggested that the impact of the East Kamchatka Current resulted in changes of the eddy form from a circular to elliptic one and beginning of its splitting in the middle of April. The splitting of mesoscale eddies could be traced by satellite altimetry charts or infrared images. However, Lagrangian drift and FTLE maps show more detailed pattern of this process in space and time. They demonstrate clear boundaries between the eddies identifying water particles belonging to one or another vortices (Figs. 6.6 and 6.9). The eddy B splitting and formation of the eddy A from its southeastern part were also confirmed by satellite-tracked drifters and Argo floats. After its formation, the newborn eddy A was quickly increasing in size by involving surrounding water and was moving to the north. By the end of June when its CTD sampling was implemented by the R/V *Professor Gagarinskiy*, the eddy was located just to the east of the Simushir Island with its center at 46.3° N, 154.3° E and had a size of $\simeq 140 \times 135$ km (Fig. 6.6e, f).

The center of the eddy A was identified in the cruise to be at the point 46.197° N, 154.334° E on June 26, 2012. It is station 50 (see Table 7.1 in the end of Chap. 7). The corresponding simulated elliptic point of that eddy had coordinates 46.3° N, 154.3° E on the same days (see Fig. 6.6e, f). The simulated and measured locations of the center of the eddy A coincide with the accuracy of 7–10 km. As to comparison of simulated and measured boundaries of the eddy A, it follows from the potential vorticity plot in Fig. 6.4 that the eddy core was located on June 25–26, 2012 between stations 46 (46.5° N, 153.5° E) and 53 (46° N, 155° E). Positive SSH values in Fig. 6.2 were located between the points with coordinates 45.7° N and 47° N and 153.5° E and 155° E. The boundary of the eddy core on the simulated D map in Fig. 6.6e can be identified as a closed outer contour with a maximal local gradient of D between 45.7° N and 47° N and between 153.3° E and 155.2° E. The black "ridge" on the combined Lyapunov map in Fig. 6.6f, delineating approximately the boundary of the eddy core, crosses the cruise transect between stations 53 and 54. It delineates approximately the same boundary of the eddy A as that on the potential vorticity plot, D and SSH maps.

Previous studies of the eddies in the area off the central Kuril Islands [5, 15, 19] suggested that their migration to the northeast was caused by the mean flow. This corresponds to our observations of the drift of the eddies B and A. Southwestward motion of the eddy B and northeastward motion of the eddy A represent seasonal variations of the mean flow in the area east of the Kuril Islands, e.g., the winter intensification of the East Kamchatka Current and the summer intensification of the flow from the subarctic front area to the north.

The Argo float no. 4900939 was deployed by the Japan Agency for Marine-Earth Science and Technology (JAMSTEC) on August 31, 2009 at the western Aleutian Islands. Its parking depth was around 500 m and a frequency of profiling was 5 days [ARGODB]. The float travelled through the Bering Sea more than 2 years and then came out to the North Western Pacific with the East Kamchatka Current. Drifting

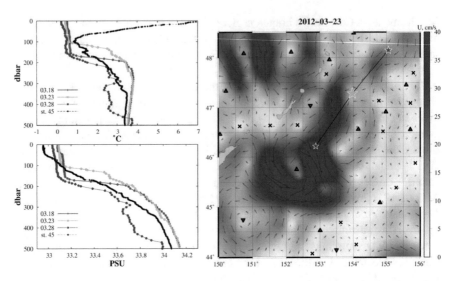

Fig. 6.13 (*Left*) Vertical profiles of temperature T and salinity S obtained by the Argo float no. 4900939 in the vicinity of the eddy B on March 18, 23, and 28, 2012. T and S profiles by CTD observations of R/V *Professor Gagarinskiy* on June 25 at the periphery of the eddy A (st. 45) are shown for a comparison. (*Right*) The velocity field on March 23 and the floats locations on March 18, 23, and 28 in the vicinity of the eddy B

to the southwest along the Kuril Islands, the float has been captured by the eddy B at its northeastern periphery in the end of March 2012. Profiles of temperature and salinity taken by the float in March represent a typical winter modification of North Western Pacific Subarctic water mass structure with cold mixed upper layer (0–150 m) of 0.5–0.6 °C and 33.09–33.14 PSU and warm intermediate layer between 200 and 400 m with temperature increasing up to 3.5–3.7 °C (Fig. 6.13).

When the float came into the area of the eddy B on March 28, 2012, its profiles were changed presenting more colder and lower salinity water between 200 and 450 m. Similar profiles were obtained at the southern and western periphery of the eddy on April 2, 7, and 12 when the float left the eddy and drifted into the Okhotsk Sea (Fig. 6.14). These profiles coincide with the T and S profiles obtained on June 25, 2012 by the CTD observations of R/V *Professor Gagarinskiy* at the periphery of the eddy A at station 45 (Figs. 6.3, 6.4, 6.13, and 6.14). The comparison suggests that the float has not entered the core of the eddy B, but it was travelling along its periphery clockwise. Slight difference of profiles for April 7 and 12 from those of March 28, April 2 at station 45, and similarity of the first ones with the profile taken in the mixed water area in the Bussol' Strait on April 17 may suggest that the float was leaving the eddy that time.

By May 12, 2012, the float has entered the eddy A from the northwest (Fig. 6.15). It seems that this time the float finally came into the core of the eddy and was trapped there by October. Vertical profiles of T and S inside the eddy and in the area around the Kuril Islands are similar and demonstrate mixed cold and low salinity water of

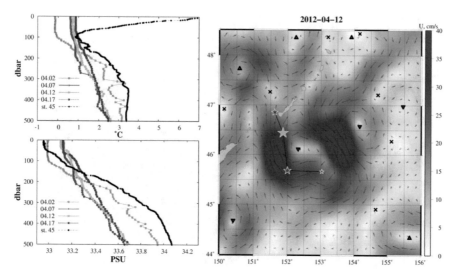

Fig. 6.14 (*Left*) Vertical profiles of temperature T and salinity S obtained by the Argo float no. 4900939 at the periphery of the eddy B on April 2, 7, 12, and 17, 2012. T and S profiles by CTD observations of R/V *Professor Gagarinskiy* on June 25 at the periphery of the eddy A (st. 45) are shown for a comparison. (*Right*) The velocity field on April 12 and locations on April 2, 7, 12, and 17 of the float moving clockwise around the center of the eddy B

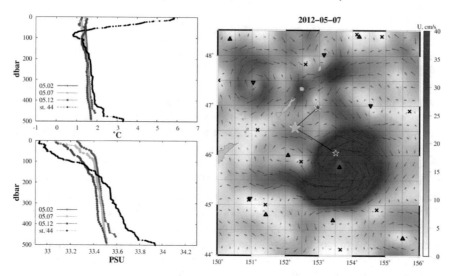

Fig. 6.15 (*Left*) Vertical profiles of temperature T and salinity S obtained by the Argo float no. 4900939 at the periphery of the eddy A on May 2, 7, and 12, 2012. T and S profiles by CTD observations of R/V *Professor Gagarinskiy* on June 24 at the periphery of the eddy A (st. 44) are shown for a comparison. (*Right*) The velocity field on May 7 and the floats locations on May 2, 7, and 12 in the vicinity of the eddy A

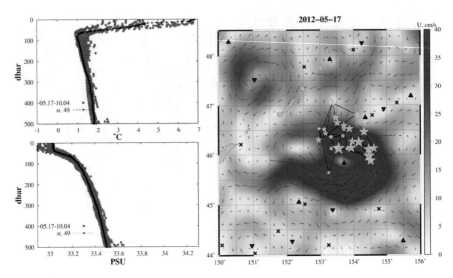

Fig. 6.16 (*Left*) Accumulated vertical profiles of temperature T and salinity S obtained by the Argo float no. 4900939 in the eddy A from May 17 to October 4, 2012. T and S profiles by CTD observations of R/V *Professor Gagarinskiy* on June 26 in the core of the eddy A (st. 49) are shown for a comparison. (*Right*) The velocity field on May 17 and locations of the float trapped by the eddy A for the period from May 17 to October 4

1.2–1.8 °C and 33.1–33.5 PSU between 100 and 450 m. The same vertical structure was also observed by the CTD profiling close to the Kuril Islands by R/V *Professor Gagarinskiy* on June 25 (st. 44). Similarity of the profiles and the fact that the float has penetrated into the center of the eddy (not just taken along its periphery) may suggest an intrusion of the mixed Okhotsk Sea water into the core of the eddy A. The float has circulated in the core of the eddy A for the next 4.5 months. It sent the last signal on October 4, 2012 from the center of the eddy after more than 3 years of operation. All 27 profiles, taken in the eddy since May 17, show no changes in the structure of the eddy core and demonstrate good coincidence with the ship CTD observations taken in the end of June (Fig. 6.16). The fact that the float was trapped inside the eddy core suggests its uniform vertical structure by the depth 500 m, otherwise the float could hardly stay inside the eddy for such a long time. It is difficult to estimate currents inside the eddy by float track data. We may estimate a time of full rotation of the eddy to be 6–7 days which is close to the period of the float profiling (5 days). So we do not see a loop-like track of the float in Fig. 6.16. Instead, one can note a northward expansion of the float track which is caused by the northward drift of the eddy over May–October period.

Observations by the float show a similarity of vertical structure of the eddies B and A. Thus, along with our Lagrangian simulations, this confirms a genetic connection of the studied eddies. The fact that the float came into the center of the eddy A and a similarity of its profiles, taken along the Kuril Islands with those in the core of the eddy, suggest an intrusion of the Okhotsk Sea water into the eddy A during or just after its separation from eddy B in late April. This mechanism of the Kuril eddies formation was discussed by Yasuda et al. [52].

The CTD observations across the eddy A by the R/V *Professor Gagarinskiy* in June 26–27, 2012 showed its vertical structure to be quite typical for the Kuril eddies with a core of relatively low temperature and low salinity water at the intermediate depth [5, 39, 52]. This very uniform core of the eddy A with an extremely low potential vorticity layer had a very large vertical size extending from 100 down to 700 m (Fig. 6.3). As to the other mesoscale ACEs in the ocean, the uniform core forms during eddy evolution by entrainment of surrounding water and mixing inside the eddy core (see, e.g., [22]). This process of vertically uniform core formation takes time from a few months to a few years. So it is hardly possible that this core, containing such a large volume of well mixed water, was formed in the eddy A since the eddy formation in middle April during 2 months before the CTD sampling. It looks more probable that this core is a remnant of the eddy B. Lagrangian simulations, presented in the section, were started in the beginning of 2012 and continued 6 months by the period of ship sampling. However, we expect the age of this deep core of the eddies B and A to be much longer because the eddy B was presented in the area long before our observations. In fact, altimetric observation and simulation showed that it was born in the area in June 2009 with the elliptic point to be at that time at 45° N, 153° E.

Based on the CTD observations, we may suggest that the core of the eddy A consisted of its deep part located between 100 and 700 m and upper part above 100 m (Figs. 6.3 and 6.4). While the deep core was quite isolated from surrounding water, the upper core was influenced by intrusions of surface water from the adjacent areas. Lagrangian simulations provide tracing the origin of the entrained water. First, we confirm that the main part of water, forming the core of the eddy A, came from the eddy B (Figs. 6.5 and 6.6). Entrainment of surrounding water into the eddy occurred by portions and was associated with development of streamers transported water mainly from the north and winding onto the eddy forming its spiral-like structure (Figs. 6.9 and 6.10). While water at the eddy periphery was renewing quickly, water at the very central part of the eddy was quite isolated. It was captured inside the eddy for a long time and originated from the eddy B which was influenced, in turn, by the Okhotsk Sea water (Figs. 6.8, 6.12 and 6.10). This may explain different properties of water in the center of the eddy observed by the CTD sampling, in particular, a high turbidity in the upper layer (Fig. 6.4). Water at the eddy periphery originated mostly from the East Kamchatka Current and Okhotsk Sea (Figs. 6.8 and 6.12). There was no water entrainment into the eddy from the southern areas during the study period. This result is very important to estimate a role of this eddy in transporting water contaminated by radionuclides after the Fukushima accident in March 2011. Our results confirmed that even the eddy had a slightly increased concentrations of radiocesium isotopes in June 2012 it was not a result of direct water transport from Tohoku area while it could be a result of atmospheric transport and deposition [4].

6.3   Lagrangian Analysis of the Vertical Structure of Numerically Simulated Eddies in the Japan Sea

6.3.1   Topographically Constrained Frontal Eddies in the Japan Basin

When simulating mean near-surface circulation in the Japan Sea based on multi-year altimetry data in Chap. 5, we have found a quasi-permanent anticyclonic feature to the north of the central Polar Front, labeled as AC in the averaged AVISO velocity field in Fig. 5.2. This anticyclonic eddy with the center at about 41.3° N, 134° E has been shown to play an important role in existence there a "forbidden" zone where northward transport of water has not been observed for the period of integration from 1993 to 2015. Topographically constrained anticyclonic eddies have been really regularly observed there [24, 47, 48].

As part of an international cooperative program "Circulation Research of the East Asian Marginal Seas" (CREAMS), long-term moored current measurements have been carried out at seven sites in the Japan Basin. One of the moorings, M3, was deployed at 41.5° N and 134.3° E where the Japan Basin is 3500 m deep. It was equipped with three current meters at about 1000, 2000, and 3000 m depths which made the measurements for 3 years, from August 1993 to July 1996, with the data sampling period of 1 h. The current meter data of 3-year duration have shown that deep-sea ACEs with the orbital speeds of the order of 0.1 m/s occurred every year in the deep layers [47]. Available time series of SST images and World Ocean Circulation Experiment (WOCE) drifter tracks well correlated to that finding. The currents at 1000, 2000, 3000 m have been found [47] to be highly coherent throughout the observational period. They have observed intensification of the current in fall and winter. The eddies observed in the Japan Basin did not exhibit any definite direction of propagation. It has been noted in [47] that the effect of the bottom geometry may be important. In fact, the eddy currents were observed only at M3, but not in the rim area of the Basin at M1, M2, M4, M6, and M7 stations [47]. During the oceanographic CTD-hydrochemical survey in summer 1999 [48, 49], the mesoscale ACE with the center approximately at the site of the M3 mooring has been found in temperature, salinity, dissolved oxygen, and NO_3 sections along 134° E and 41.25° N.

In this section we use the output from the eddy-resolved multilayered circulation Marine Hydrophysical Institute Model (MHIM) [40] to analyze from a Lagrangian perspective, the vertical structure of simulated deep-sea ACEs in the Japan Basin constrained by bottom topography. We focus on the ACE, generated in the MHIM approximately at the place of the M3 mooring [47] and the hydrographic sections [48, 49], where such eddies have been regularly observed in different years (1993–1997, 1999–2001).

Lagrangian tools have been shown in the preceding section to be useful in identifying 2D structure of eddies, their cores and boundaries, and pathways by which they gain and expel water (see also [2, 21, 26, 27, 30, 31, 46]). The challenge is in identification of the vertical structure of eddies. In order to quantify properties of the 3D eddy structure (for example, the volume of eddies), it is often the eddy's surface edges are simply extended to a given depth along the vertical. It is well known, however, that most eddies do not have a cylinder-like form. Moreover, the intriguing problem is changed in the structure of eddies at different depths in the course of time.

The analysis in this section will be performed using two Lagrangian indicators, the FTLE [23] and displacements for a large number of tracers [25, 30, 32]. The Lagrangian diagnostics we use seem to be more appropriate to identify eddies than the commonly used techniques, because FTLE and drift maps are imprints of history of water mass involved in the vortex motion whereas vorticity, the Okubo–Weiss parameter, and similar indicators are "instantaneous" snapshots. That is why one can see eddies more clearly on the Lagrangian maps. We will show how our modeled ACE evolves from the eddy without any signs of rotation motion at the sea surface in summer into a one reaching the surface in fall. In order to demonstrate that we implement a quasi-3D computation of those Lagrangian indicators. We use the velocity field from the output of the MHIM and compute the Lagrangian maps of the FTLE and particle's displacements in each model layer.

Quasi-3D Lagrangian approach has been applied recently [3] for diagnostics of 3D LCSs around a particular CE pinched off from a Benguela upwelling front. Three-dimensional LCSs were extracted as "ridges" of the calculated fields of the FTLE obtained from an output of the Regional Ocean Modelling System (ROMS). They have been found [3] to be quasi-vertical surfaces. Another eddy feature, a ring of the Loop Current in the Gulf of Mexico has been studied [46] by the similar method, using the data-assimilating HYbrid Coordinate Ocean Model (HYCOM). Those authors have studied the location of relevant transport barriers during the formation of the Eddy "Franklin" in 2010 at several depths from the surface down to 200 m.

6.3.2 Regional Circulation Marine Hydrophysical Institute Model

We introduce briefly the quasi-isopycnic layered MHIM [20, 40] with a free surface boundary condition incorporating the horizontally inhomogeneous upper mixed layer. The model is based on a system of primitive equations integrated within each quasi-isopycnic layer. All layers are assumed to be horizontally inhomogeneous, however, the density in each thermocline layer changes within the limits determined *a priori* by the prescribed basic stratification.

It is assumed that the layers may outcrop. The layer outcropping is similar to the isopycnal model applied by Hogan and Hurlburt [13] to the Japan Sea. The interfaces of the inner model layers can climb to the upper mixed layer in the frontal zones. The horizontally inhomogeneous upper mixed layer model includes a parametrization of turbulent heat, salt and momentum fluxes, drift current in that layer, entrainment and subduction processes at the bottom of the layer [20, 40]. The basic equations for momentum, temperature, and salinity in the upper mixed layer are similar to the vertically integrated equations for inner layers of the model. The commonly used convective adjustment scheme is applied to simulate vertical convection. According to our previous studies [27], the MHIM successfully simulates the mesoscale eddy dynamics, interaction between eddies over the shelf edge and steep continental slope of the Japan Basin, as well as, mesoscale eddies and currents, mixing and transport of water masses in the Peter the Great Bay in the Japan Sea [31].

The numerical experiment with the MHIM is focused in this section on simulation of the mesoscale circulation over the deep Japan Basin, its continental slope and shelf during summer and fall. The model domain is the closed sea area [39° N–44° N; 129° E–139° E] with the horizontal resolution 2.5 km [27].

Both biharmonic and harmonic viscosities are used in the momentum equation of the sea circulation model, and only harmonic horizontal diffusion is used in the equation for the temperature and salinity transfer. The coefficients of biharmonic horizontal viscosity and diffusion are set to be constant in space. The coefficient of harmonic horizontal viscosity is a constant in space only during model spin up. After the model spin up, the harmonic horizontal viscosity is taken into account only near the boundary of the model domain. The model is integrated in time from June to November to simulate relatively stable mesoscale ACEs in the Japan Basin. As horizontal boundary conditions, we set zero values for the current velocity and its second derivative along a normal to the boundary, as well as zero values for the first derivatives of temperature and salinity. Therefore, we consider a closed model domain. Due to the realistic initial conditions, basic density stratification below seasonal pycnocline and main features of the large-scale cyclonic circulation over the Japan Basin do not significantly change during this 6 months period. Formation and evolution of the mesoscale eddies in the area studied (Fig. 6.17b) do not substantially depend on the boundary conditions in the model domain from early summer to late fall.

The model has ten quasi-isopycnic layers with the first one to be a horizontally inhomogeneous upper mixed layer. The first nine layers are located inside the main pycnocline with the lower boundary of the ninth layer to be the lower boundary of the main pycnocline which is not deeper than 250 m in the area studied. The lower tenth layer includes deep and bottom waters of the Japan Sea. The realistic bottom topography (Fig. 6.17), obtained from ETOPO2 (2-Minute Gridded Global Relief Data), is one of the most important factors in simulation of the large-scale and mesoscale circulations.

The initial conditions for summer isopycnal interfaces in the model layers, temperature, and salinity distribution include only large-scale features of the model variables. They have been taken from oceanographic survey in 1999 [48, 49] with a substantial smoothing. After smoothing, there were no any mesoscale structures in the initial conditions. The MHIM has been integrated with the time step of 4 m in from the initial condition under realistic meteorological situations. The first

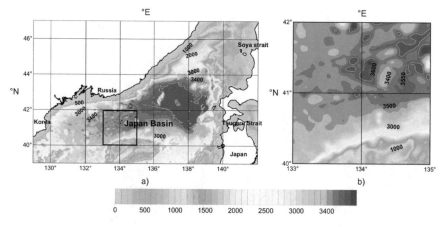

Fig. 6.17 The bottom topography in (**a**) the northern part of the Japan Sea and (**b**) the studied area

month with June meteorological conditions is a time interval of the model spin up. The wind stress, short wave radiation, near-surface air temperature, humidity, and wind speed have been taken from daily mean National Centers for Environmental Prediction/National Center for Atmospheric Research (NCEP/NCAR) Reanalysis. The heat fluxes at the sea surface, being depended on the mixed layer temperature and meteorological characteristics, are calculated in the model.

In the model spin up (30 days), the coefficients of quasi-isopycnic biharmonic viscosity, harmonic viscosity, and diffusion used in the momentum and heat/salt transfer equations have been decreased correspondingly from $10^{17}\,\mathrm{cm^4\,s^{-1}}$, $10^7\,\mathrm{cm^2\,s^{-1}}$, and $0.4 \cdot 10^7\,\mathrm{cm^2\,s^{-1}}$ to $2 \cdot 10^{16}\,\mathrm{cm^4\,s^{-1}}$, $0.5 \cdot 10^6\,\mathrm{cm^2\,s^{-1}}$, and $0.2 \cdot 10^6\,\mathrm{cm^2\,s^{-1}}$. After spin up, the harmonic viscosity is applied only in two grid lines adjacent to the contour of the model domain with the coefficient $0.5 \cdot 10^6\,\mathrm{cm^2\,s^{-1}}$. In other offshore grid points, this coefficient is set to be zero and only biharmonic viscosity in the momentum equations is taken into account with the coefficient $2 \cdot 10^{16}\,\mathrm{cm^4\,s^{-1}}$. The coefficient of temperature, salinity diffusion is $0.2 \cdot 10^6\,\mathrm{cm^2\,s^{-1}}$ in all the grid points after spin up. We basically simulate the nonlinear large-scale and mesoscale circulation over the Japan Basin, continental slope, and shelf taking into account daily mean external atmospheric forcing from July 1 to December 1. The numerical experiments with minimized coefficients of the horizontal and vertical viscosity show intensive mesoscale dynamics. In particular, they demonstrate variability of the mesoscale ACEs and CEs over the shelf, the shelf break and base of the continental slope, and in the central deep area of the Japan Basin as well. The anticyclonic eddies, generated over the shelf break and continental slope, move usually downstream with prevailing phase velocity of about 6–8 cm/s [27]. Some quasistationary mesoscale eddies in the central Japan Basin area, including the eddies studied in this section, are generated in the thermocline and deep Japan Basin water over the mesoscale bottom troughs and sea mounts.

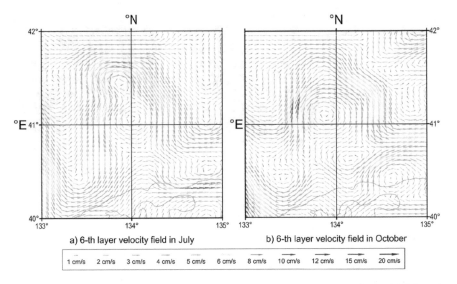

Fig. 6.18 Simulated velocity field in the studied area in the 6th layer averaged over (**a**) July and (**b**) October of a typical year

Figure 6.18 demonstrates the monthly mean velocity fields in July and October in the 6th model layer in the deep Japan Basin area (see Fig. 6.17a). The system of ACEs and CEs in the velocity field varies from July to October–November. The most stable one is an ACE simulated in the central region of the studied area over a mesoscale bottom trough. It is a trough between the steepest slope of the Yamato Rise near the southern–southeastern edge of the area and high sea mounts of the Tarasov Rise, the Vasilkovsky Ridge up to 1700 m (42° N, 134° E) and the Bersenev Ridge up to 2200 m (42° N, 133.71° E) in the northern part [16]. This trough is a small southwestern part of the Japan Basin (see Fig. 6.17). The mid-scale cyclonic circulation is formed during model run over the trough. The mesoscale eddies are supposed to be formed in the central area of the trough (and in the center of the mid-scale cyclonic circulation) due to baroclinic instability of sea currents and bottom topography effects. This quasistationary ACE is situated over the mesoscale bottom saddle between bottom depression in the southern and northern edges of the saddle and between sea mounts in the eastern and western parts. Its center is shifting with time during the model run from July to November in the vicinity of the point with coordinates 41.15° N and 134° E where similar topographically controlled ACEs have been often observed [47–49]. The studied topographically controlled mesoscale ACE looks like an eddy separated off the current in the cyclonic circulation and stagnating over the complicated bottom topography between sea mounts and bottom depressions. Moreover, it also interacts with surrounding mesoscale CEs.

6.3.3 Three-Dimensional Structure and Evolution of Eddies in the Japan Basin

The drift map in Fig. 6.19a clearly demonstrates a vortex pair with two ACEs in the 9th layer with elliptic points at their centers 41° N, 134° E and 41.4° N, 133.9° E and a hyperbolic point in between at 41.3° N, 134° E. The boundary of the northern eddy can be identified as a closed curve with the maximal local gradient of the tracer's displacement, D, that separates waters involved in the rotational motion around the vortex center from ambient waters. Magnitude of the absolute displacements for the former ones are much smaller than that for the particles outside the eddy.

The FTLE field provides complementary information on the horizontal eddy structure. We computed forward (Λ_+) and backward-in-time (Λ_-) FTLE maps by the method described in Sect. 4.3.1 for 30 days in the future and in the past, respectively, starting from July 23. The combined map in Fig. 6.19b demonstrates the same vortex pair as in Fig. 6.19a. The eddy may be identified in the combined FTLE field as a closed region bounded by the black ridges intersecting at two hyperbolic points. Water parcels on one side of the ridge are involved in the vortex motion around the elliptic point, whereas the ones on the other side move downstream away from the eddy. For comparison, we compute the Okubo–Weiss

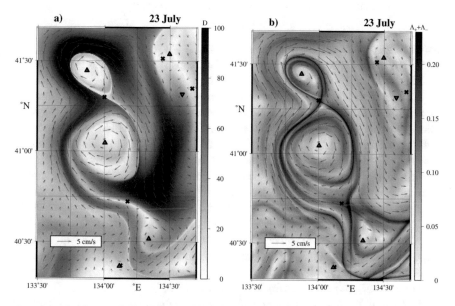

Fig. 6.19 Manifestation of modeled eddies on July 23 in the ninth layer on (**a**) the drift map (D in km) and (**b**) the combined forward-in-time (Λ_+) and backward-in-time (Λ_-) FTLE maps with the velocity field imposed. Λ is in days^{-1}. "Instantaneous" elliptic and hyperbolic points, to be present in the area on July 23, are indicated by the *triangles* and *crosses*, respectively

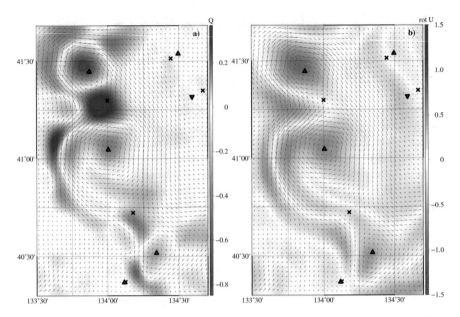

Fig. 6.20 (**a**) Manifestations of the simulated eddies on July 23 in the ninth layer in the field of the Okubo–Weiss parameter, Q. Vortex motion (*red color*) occurs if the vorticity prevails over the deformation, i.e., if $W < 0$. (**b**) The relative vorticity, rot U, in the area

parameter and vorticity field in the same area. It is evident that the eddies can be identified more clearly on the drift and FTLE maps (Fig. 6.19) than in those fields (see Fig. 6.20).

The southern eddy has a more complicated structure because of its intensive interaction with ambient waters during the month of integration. The drift maps allow to delineate the transport pathways by which eddies could exchange water with their surroundings. They look like dark "tongues" in Fig. 6.19a encircling that eddy. The origin of those waters can be traced out by computing tracer's displacements in zonal and meridional directions backward in time for the month, from July 23 to June 23. The corresponding zonal and meridional drift maps shown in Fig. 6.21 allow to visualize where the waters in the ninth layer came from. Blue color of the water tongues around that eddy mean that it captured water from the south (Fig. 6.21a) and east (Fig. 6.21b). Complementary backward-in-time Lagrangian longitudinal (Fig. 6.22a) and latitudinal (Fig. 6.22b) drift maps show by color the longitudes and latitudes, respectively, from which tracers, initialized in the area on June 23, came to their final positions on July 23.

The sharp boundary between waters with high gradients of a Lagrangian indicator (e.g., the absolute displacement D) was called a "Lagrangian front" in [25, 29] (see Chap. 8). The Lagrangian fronts, encircling each of the eddies in the pair in Fig. 6.19a, can be identified by a narrow white stripe demarcating the curve with the maximal gradient of D. White color means that the corresponding particles

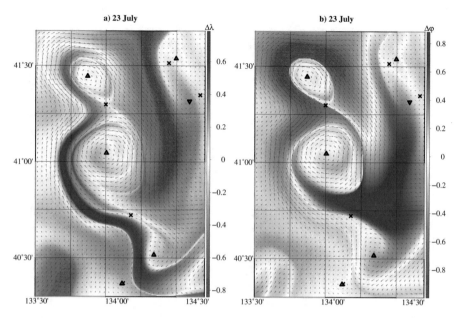

Fig. 6.21 (a) Backward-in-time meridional drift map shows tracer's displacements in the area with the simulated eddies in the meridional direction for the month, from July 23 to June 23. The *color* codes the displacement from the south to north (*blue*) and vice versa (*red*). (b) Backward-in-time zonal drift maps show tracer's displacements in the zonal direction for the same period. The *color* codes the displacement from the east to west (*blue*) and vice versa (*red*). Both the displacements, $\Delta\lambda$ and $\Delta\phi$, are computed in the ninth layer of the model in geographic degrees

have experienced very small displacements over the period of integration because they rotated around the eddy's centers. The sizes of the southern and northern simulated eddies are estimated to be $\simeq 35 \times 45$ km and $\simeq 20 \times 20$ km, respectively.

Trajectories of fluid parcels and stationary points in a velocity field can gain and lose hyperbolicity over time. It means, in particular, that hyperbolic stationary points may appear and disappear in the course of time. Only those ones, which exist on July 23, are shown in Fig. 6.19. As to the southern eddy, it is confined from the east and south by the S-like unstable manifold of the hyperbolic point located between the eddies in the vortex pair. Any unstable manifold influences strongly on adjacent fluid parcels. It is illustrated in Fig. 6.23, where we placed blue and rose patches with tracers near the S-like unstable manifold, and advected them forward in time for three and half months. Both the patches elongate along that manifold. The red patch was chosen in the center of the northern eddy and is shown in Fig. 6.23 to deform slightly in the course of time until the northern eddy begins to break down to the end of summer. The southern eddy is confined from the west by the unstable manifold of a hyperbolic point which disappeared to July 23 but existed before. The complicated pattern of the black ridges around the southern eddy in Fig. 6.19b confirms the conclusion about its intensive interaction with ambient waters.

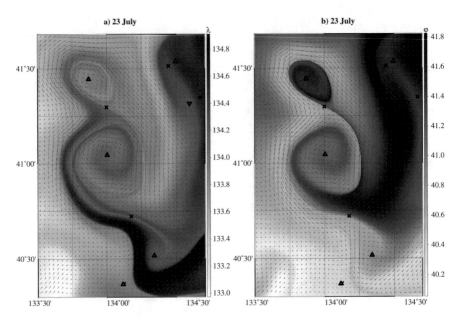

Fig. 6.22 (a) Backward-in-time longitudinal drift map in the area with the simulated eddies (the ninth layer) with the color coding the longitudes, λ, from which the tracers, initialized on June 23, came to their final positions on July 23. (b) Backward-in-time latitudinal drift map with the color coding the latitudes, ϕ, from which the tracers, initialized on June 23, came to their final positions on July 23

To illustrate the vertical structure of the vortex pair and its evolution we show in Fig. 6.24 the summer FTLE maps in the first, third, fifth, and ninth layers. On July 23 the pair with two ACEs, elliptic points at their centers, and a hyperbolic point in between are clearly seen in the lower layers below the 4th one. The vortex pair is especially prominent in the 9th layer, that is a lower boundary of the main pycnocline. It is not recognizable in the third layer and above. The elliptic points are absent at the surface where there are no signs of vortex motion. In the course of time, the pattern changes. The pair gradually decays in the sense that the northern eddy merges with the southern one (compare the map on August 7 with the map on August 22 when the northern elliptic point disappears in the 9th layer). As to the other layers, it is seen that the northern elliptic point disappears earlier. It is possible to recognize on August 22 a single ACE with the size $\simeq 40 \times 50$ km in the 5th layer and below. We see the ACE that has not reached the surface to the end of summer.

Changes in the vertical vortex structure in September and October are shown in Fig. 6.25. The eddy in fall is clearly visible in the 5th layer and below as a single ACE of the same size as in August. As to the surface layers, a prominent eddy structure becomes visible there only to the end of October. In the beginning of September, the elliptic point appears in the first layer at the place where the eddy is visible in the 5th layer and below, but the eddy cannot be clearly detectable on

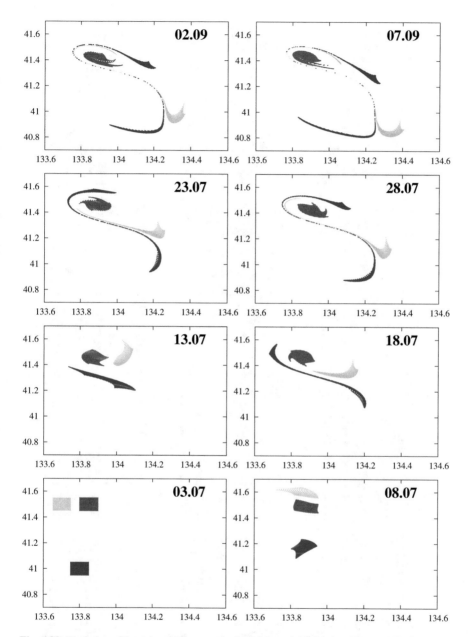

Fig. 6.23 Evolution of 3 patches with tracers in the 9th layer initialized on July 3 in the area with the simulated vortex pair (see Figs. 6.21 and 6.22 above). The *red patch* was chosen in the center of the northern eddy, the *blue* and *rose* ones—on a S-like unstable manifold somewhere at the black ridge encircling the eddies. The *blue* and *rose patches* elongated from the beginning along the same unstable manifolds. The *red patch* deforms slightly in the course of time until the northern eddy begins to break down to the end of summer. After that, it elongates along an unstable manifold of the survived southern eddy

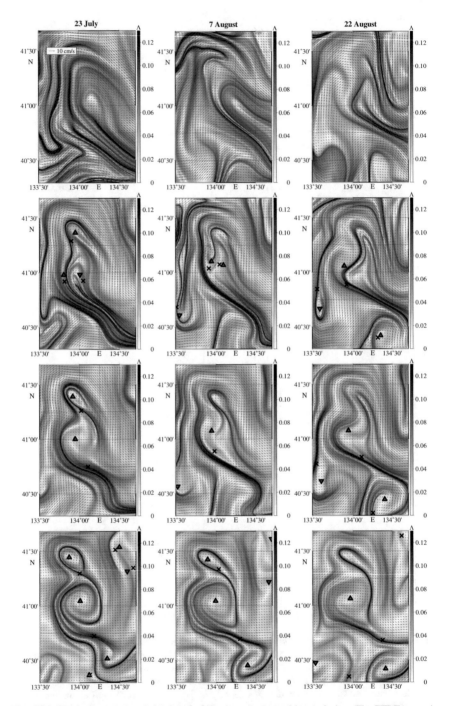

Fig. 6.24 Vertical structure of simulated eddies in summer and its evolution. The FTLE maps in the 1st, 3rd, 5th, and 9th layers are shown on July 23, August 7 and 22 *from the top to the bottom*, respectively. The vortex pair, seen in the lower layers, evolves gradually with time in a single eddy which, however, is not visible in the surface layers

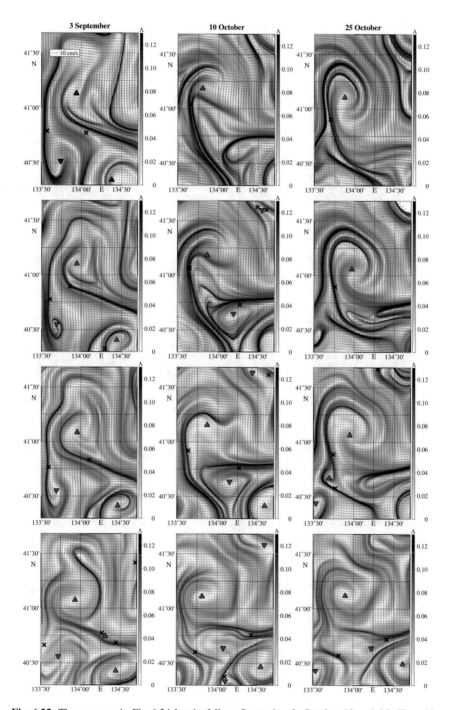

Fig. 6.25 The same as in Fig. 6.24 but in fall on September 3, October 10 and 25. The eddy appears at the surface to the end of October

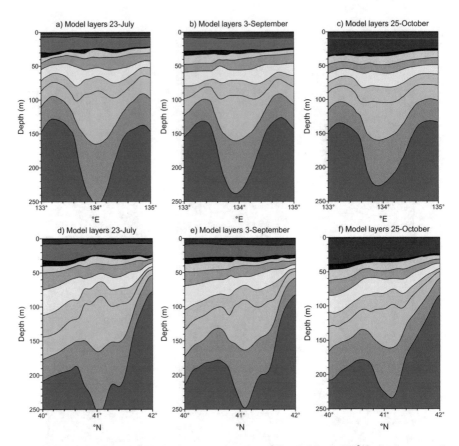

Fig. 6.26 Zonal (along 41° N, the *upper row*) and meridional (along 134° E, the *lower row*) sections of the interfaces between modeled eddy layers on July 23, September 3, and October 25. Each quasi-isopycnic layer is shown by its own color

the corresponding FTLE maps. Thus, the bowl-shaped eddy is formed to the end of October. It penetrates from the surface to the bottom gradually decaying to the end of November.

Ten layers have been used in the MHIM adopted to the Japan Sea. In order to illustrate transformation of the studied vortex structure with time, we compute vertical zonal and meridional sections across the simulated eddy. Figures 6.26 and 6.27 show evolution of its vertical structure from late July to late October in terms of sections of quasi-isopycnic layer interfaces (Fig. 6.26) and water temperature in the layers (Fig. 6.27). The studied anticyclonic eddies have been formed over the mesoscale bottom trough in early July, at first, in the main pycnocline and underlying layers and were present within the layers from the 4th one and below all the time after its formation.

The Lagrangian maps on July 23 in Fig. 6.19 clearly show the vortex pair oriented in the meridional direction. The zonal sections in Fig. 6.26 along the latitude 41° N cross the southern ACE which is manifested as a depression of the layers from the 3rd to 9th ones around the elliptic point. The vortex pair evolves to September 3 to a single ACE (see Fig. 6.25) whose zonal cross-section is shown in Fig. 6.26b. It still does not extend to the surface. To October 25 the eddy reaches the surface (see Figs. 6.25 and 6.26c). The northern ACE of the vortex pair, seen in Fig. 6.19 on July 23, is hardly visible in the meridional cross-section as a small depression in the lower layers to the right of the main depression (Fig. 6.26d and e).

From July to early September, the ACE is present only occasionally in the numerical solution in the upper mixed layer (the layer 1) and in the seasonal pycnocline (the layers 2–4). In this period of warm season, when the upper mixed layer is comparatively thin (see Fig. 6.26a, b, d, and e) and the seasonal pycnocline is very strong (see Fig. 6.27a, b, d, and e), the simulated ACE is unstable in the upper layers. Its lifetime in these layers does not exceed a few days. During summer, it episodically appears and breaks down into smaller submesoscale eddies in the upper layers. During October–November, when the thickness of the upper mixed layer increases from 10 to 15–30 m (Fig. 6.26c and f) and the seasonal pycnocline is weak, the ACE becomes as stable in the upper layers as in the underlying ones. It is also manifested in the zonal temperature section in Fig. 6.27c as the closed isotherm at the sea surface.

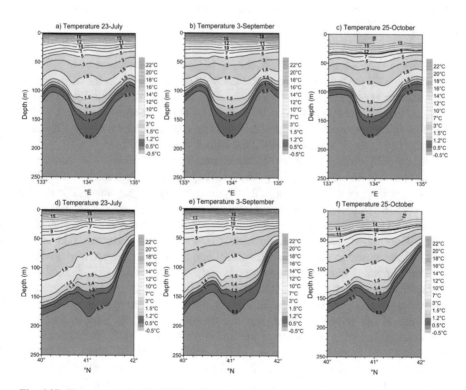

Fig. 6.27 The same as in Fig. 6.26 but for temperature sections across the simulated eddy on July 23, September 3, and October 25

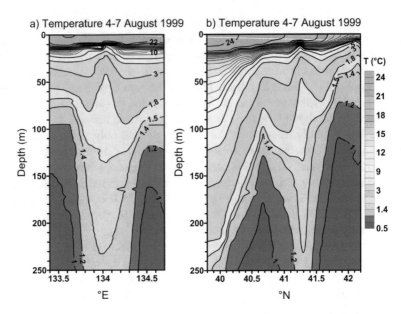

Fig. 6.28 (**a**) Zonal (along 41° N) and (**b**) meridional (along 134° E) observed sections of temperature across the ACE to be found during the oceanographic CTD-hydrochemical survey in early August 1999 [49]

During the oceanographic CTD-hydrochemical survey in summer 1999 [49], the mesoscale ACE (with the center approximately at the same place as the M3 mooring [47] and our simulated eddy) has been observed in temperature, salinity, dissolved oxygen, and NO_3 sections along 134° E and 41.25° N. The warm fresh core of the eddy with high gradients in temperature, salinity, and dissolved oxygen at its edge was situated in the thermocline within the layer from 50 to 150 m.

Figure 6.28 shows water temperature structure of that ACE in zonal and meridional sections of the oceanographic survey in early August 1999 [49]. The density gradient in this eddy basically depends on the temperature gradient. The eddy core, surrounded by maximal temperature and density gradients, looks like a lens. The eddy occupies the water column below the seasonal pycnocline (30–50 m). The anticyclonic eddy was not clearly observed in the upper mixed layer and in the seasonal pycnocline both in the zonal and meridional temperature sections. It is an important feature of the observed eddy closely related to the simulation results discussed above. Eddies with similar vertical structure in temperature and density cross-sections, named as *intrathermocline eddies*, have been observed to the south of the Subarctic Front over the western side of the Yamato Rise and within quasistationary meanders of the Tsushima Current [12]. They have been successfully simulated in this area [13] using an isopycnal ocean circulation model. The observed eddy in Fig. 6.28 is situated over the mesoscale bottom trough surrounded by sea mounts in the western area of the Japan Basin (see Fig. 6.17a and b) practically in the same area as our simulated eddy. In summer, the position and the vertical structure of the simulated ACE in Figs. 6.26 and 6.27 are similar to those for the observed ACE in Fig. 6.28. Both the simulated and observed ACEs have the similar eddy core, the relief of layer interfaces and isotherms.

The observed ACE is stronger than the simulated one with a much more strong temperature/density front situated to the south of the ACE and a stronger vertical stratification. That difference may be explained by the fact that the observation has been made in the warm climatic period and warm year (1999), whereas the simulation has been performed under daily meteorological conditions averaged over 25-years period from 1976 to 2000. Moreover, we did not take into account meridional heat advection from the southern sea area and the southern boundary of the model domain due to no-slip boundary condition for the current velocity.

To visualize the origin and fate of water masses inside the eddy, a large number of tracers (250,000) were distributed on September 1 in each layer around the eddy center inside the patch 7×11 km [133.87° E–133.97° E; 41° N–41.1° N]. They have been advected for 2 months forward and backward in time by the velocity field in each layer. The tracking maps are computed as follows. The area under study [131° E–136° E; 40° N–43.5° N] is divided into 500×500 cells. We compute how many tracers have visited each cell from September 1 to November 1 (forward-in-time tracking maps) and from September 1 to July 1 (backward-in-time tracking maps).

The results are shown in Fig. 6.29 and may be interpreted as follows. The forward-in-time tracking maps in Fig. 6.29a show where the corresponding tracers in each chosen layer were walking from September 1 to November 1. In this period, the tracers from the patches in the lower layers have rotated around the eddy center with an insignificant flow outside in the 5th layer. The "tail" on the upper plane means that at the surface those tracers have been transported by a current to the southeast, i.e., the eddy did not exist at the surface most part of that period.

The backward-in-time tracking maps in Fig. 6.29b show that the tracers in the layers from the 7th to 10th have rotated around the eddy center preserving the eddy core in these layers for 2 months in the past. It means that the eddy existed in those layers all that time at the same place without exchanging by the core water with its surroundings. The "tail" in the 5th layer means that the eddy in this layer gained the water from the south, but its core has been at the same place for the 2 months in the past. As to the surface layer, the tracers from the initial patch in the 1st layer came to their positions on September 1 from the west. So the eddy did not exist at the surface in summer. Those conclusions are confirmed by the FTLE maps in Figs. 6.24 and 6.25.

The main ACE under study is an eddy-like feature in the region of the Japan Basin to the north of the Yamato Rise (see Fig. 6.17a). That eddy has been found to be practically stationary for a half-year integration period including summer and fall. It is seen from the lower panels in Figs. 6.24 and 6.25 that displacement of its elliptic point for 3 months did not exceed 10 km. Inspection of the AVISO velocity field at the sea surface for 1992–2014 [AVISO] has shown that surface eddy-like anticyclonic features often appeared in the area around 134° E, 41° N in cold seasons and disappeared in warm ones. No significant directed transport of such

a)   b)

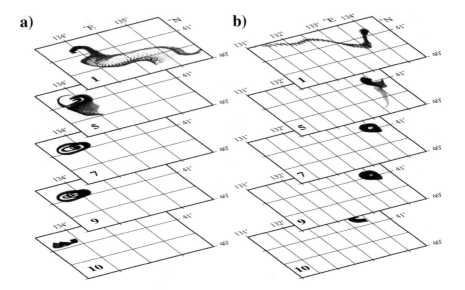

Fig. 6.29 Tracking maps for the tracers distributed over the eddy's core on September 1 in the 1st, 5th, 7th, 9th, and 10th layers numbered in the left lower corners. (**a**) The forward-in-time map shows where the corresponding tracers were walking from September 1 to November 1. (**b**) The backward-in-time map shows where they were walking from September 1 to July 1. The density of traces is in the logarithmic scale

eddies has been found in those altimetric data. Taking into account these findings, complicated bottom topography in the area with underwater seamounts and trenches (see Fig. 6.17b) and observations [47, 48], it is reasonable to suppose that we deal with the ACE constrained by bottom topography.

Computations of the FTLEs and particle's displacements in each depth layer clearly show that the simulated ACE evolves from the eddy that does not reach the surface in summer into a one reaching the surface in fall. The corresponding elliptic points, demarcating the eddy's center, are absent in the upper layers in summer (Fig. 6.24) and appear in fall (Fig. 6.25). This result is confirmed by computing deformation of the model layers and the temperature cross-sections. Whether the eddy reaches the surface or not depends on the stratification measure in the thermocline, topographic, and other parameters. In summer, when the upper mixed layer is comparatively thin and the stratification of seasonal pycnocline is very strong, the simulated eddy is unstable in the upper layers. In fall, when the stratification of seasonal pycnocline is much weaker than in summer, the eddy becomes as stable in the upper layers as in the underlying ones.

The eddy must be sufficiently nonlinear to exist as a stable entity. The measure of the nonlinearity is the so-called quasigeostrophic nonlinearity parameter Q_β, which is the ratio of the relative vorticity advection to the planetary vorticity advection [7] defined as $Q_\beta = U/\beta L^2$, where U is the maximum rotational speed, L the eddy radius, and β is the planetary vorticity gradient. To estimate the quasigeostrophic nonlinearity parameter of our simulated eddy in the Japan Basin, we take $U = 0.05–0.1\,\mathrm{m\,s^{-1}}$, $L = 30{,}000\,\mathrm{m}$, and $\beta = 1.73 \cdot 10^{-11}\,\mathrm{m\,s^{-1}}$ to get $Q_\beta = 3.3–6.6$. This range of values means that the relative vorticity dominates and suggests that our simulated eddy is nonlinear. We may conclude that the quasistationary simulated eddy persists as a stable entity during the simulation period. That conclusion is based not only on the Q_β criterion, but it is also confirmed by different Lagrangian diagnostics including daily FTLE and drift maps (see Figs. 6.19, 6.24, and 6.25) and existence of elliptic points in the eddy's center in the lower layers during the simulation period (see Figs. 6.19, 6.24, and 6.25).

References

1. Andreev, A.G., Zhabin, I.A.: Origin of the mesoscale eddies and year-to-year changes of the chlorophyll-*a* concentration in the Kuril Basin of the Okhotsk Sea. In: PICES-2012 Program and Abstracts, Hiroshima, Japan (2012). https://www.pices.int/publications/presentations/PICES-2012/2012-POC/POC-P-1540-Andreev.pdf

2. Beron-Vera, F.J., Wang, Y., Olascoaga, M.J., Goni, G.J., Haller, G.: Objective detection of oceanic eddies and the Agulhas leakage. J. Phys. Oceanogr. **43**(7), 1426–1438 (2013). 10.1175/JPO-D-12-0171.1

3. Bettencourt, J.H., Lopez, C., Hernandez-Garcia, E.: Oceanic three-dimensional Lagrangian coherent structures: a study of a mesoscale eddy in the Benguela upwelling region. Ocean Model. **51**, 73–83 (2012). 10.1016/j.ocemod.2012.04.004

4. Budyansky, M.V., Goryachev, V.A., Kaplunenko, D.D., Lobanov, V.B., Prants, S.V., Sergeev, A.F., Shlyk, N.V., Uleysky, M.Y.: Role of mesoscale eddies in transport of Fukushima-derived cesium isotopes in the ocean. Deep-Sea Res. I Oceanogr. Res. Pap. **96**, 15–27 (2015). 10.1016/j.dsr.2014.09.007

5. Bulatov, N.V., Lobanov, V.B.: Investigation of mesoscale eddies east of Kuril islands on the basis of meteorological satellite data. Sov. J. Remote. Sens. **3**, 40–47 (1983) [in Russian]

6. Bulatov, N.V., Lobanov, V.B.: Influence of the Kuroshio warm-core rings on hydrographic and fishery conditions off Southern Kuril Islands. In: Proceedings of the PORSEC-92 in Okinawa, pp. 1127–1131 (1992)

7. Chelton, D.B., Schlax, M.G., Samelson, R.M.: Global observations of nonlinear mesoscale eddies. Prog. Oceanogr. **91**(2), 167–216 (2011). 10.1016/j.pocean.2011.01.002

8. Dong, C., Nencioli, F., Liu, Y., McWilliams, J.C.: An automated approach to detect oceanic eddies from satellite remotely sensed sea surface temperature data. IEEE Geosci. Remote Sens. Lett. **8**(6), 1055–1059 (2011). 10.1109/lgrs.2011.2155029

9. Dugan, J.P., Mied, R.P., Mignerey, P.C., Schuetz, A.F.: Compact, intrathermocline eddies in the Sargasso Sea. J. Geophys. Res. Oceans **87**(C1), 385–393 (1982). 10.1029/jc087ic01p00385

10. Early, J.J., Samelson, R.M., Chelton, D.B.: The evolution and propagation of quasigeostrophic ocean eddies. J. Phys. Oceanogr. **41**(8), 1535–1555 (2011). 10.1175/2011jpo4601.1

11. Fu, L.L.: Pattern and velocity of propagation of the global ocean eddy variability. J. Geophys. Res. Oceans **114**(C11), C11017 (2009). 10.1029/2009jc005349

12. Gordon, A.L., Giulivi, C.F., Lee, C.M., Furey, H.H., Bower, A., Talley, L.: Japan/East Sea intrathermocline eddies. J. Phys. Oceanogr. **32**(6), 1960–1974 (2002). 10.1175/1520-0485(2002)032<1960:jesie>2.0.co;2

13. Hogan, P., Hurlburt, H.: Why do intrathermocline eddies form in the Japan/East Sea? a modeling perspective. Oceanography **19**(3), 134–143 (2006). 10.5670/oceanog.2006.50

14. Hormazabal, S., Combes, V., Morales, C.E., Correa-Ramirez, M.A., Lorenzo, E.D., Nuñez, S.: Intrathermocline eddies in the coastal transition zone off central Chile (31 − 41° S). J. Geophys. Res. Oceans **118**(10), 4811–4821 (2013). 10.1002/jgrc.20337

15. Isoguchi, O., Kawamura, H.: Eddies advected by time-dependent Sverdrup circulation in the western boundary of the subarctic North Pacific. Geophys. Res. Lett. **30**(15), 1794 (2003). 10.1029/2003GL017652

16. Karnaukh, V.N., Karp, B.Y., Tsoy, I.B.: Basement structure and sedimentary cover seis-mostratigraphy in the northern part of the Japan Basin in the region of the Tarasov Rise (Sea of Japan). Oceanology **47**(5), 691–704 (2007). 10.1134/s0001437007050128

17. Karrasch, D., Huhn, F., Haller, G.: Automated detection of coherent Lagrangian vortices in two-dimensional unsteady flows. Proc. R. Soc. A Math. Phys. Eng. Sci. **471**(2173), 20140639–20140639 (2014). 10.1098/rspa.2014.0639

18. Kusakabe, M., Andreev, A., Lobanov, V., Zhabin, I., Kumamoto, Y., Murata, A.: Effects of the anticyclonic eddies on water masses, chemical parameters and chlorophyll distributions in the Oyashio current region. J. Oceanogr. **58**(5), 691–701 (2002). 10.1023/a:1022846407495

19. Lobanov, V.B., Rogachev, K.A., Bulatov, N.V., Lomakin, A.F., Tolmachev, K.P.: Long-term evolution of the Kuroshio warm eddy. Dokl. USSR Akad. Sci. **317**(4), 984–988 (1991). [in Russian]

20. Mikhailova, E.N., Shapiro, N.B.: Quasi-isopycnic layer model for large-scale ocean circula-tion. Phys. Oceanogr. **4**(4), 251–261 (1993). 10.1007/bf02197624

21. Miller, P.D., Jones, C.K.R.T., Rogerson, A.M., Pratt, L.J.: Quantifying transport in numerically generated velocity fields. Physica D **110**(1–2), 105–122 (1997). 10.1016/S0167-2789(97)00115-2

22. Olson, D.B.: Rings in the ocean. Annu. Rev. Earth Planet. Sci. **19**(1), 283–311 (1991). 10.1146/annurev.ea.19.050191.001435

23. Pierrehumbert, R.T., Yang, H.: Global chaotic mixing on isentropic surfaces. J. Atmos. Sci. **50**(15), 2462–2480 (1993). 10.1175/1520-0469(1993)050<2462:GCMOIS>2.0.CO;2

24. Prants, S., Ponomarev, V., Budyansky, M., Uleysky, M., Fayman, P.: Lagrangian analysis of the vertical structure of eddies simulated in the Japan Basin of the Japan/East Sea. Ocean Model. **86**, 128–140 (2015). 10.1016/j.ocemod.2014.12.010

25. Prants, S.V.: Dynamical systems theory methods to study mixing and transport in the ocean. Phys. Scr. **87**(3), 038115 (2013). 10.1088/0031-8949/87/03/038115

26. Prants, S.V., Andreev, A.G., Budyansky, M.V., Uleysky, M.Y.: Impact of mesoscale eddies on surface flow between the Pacific Ocean and the Bering Sea across the near strait. Ocean Model. **72**, 143–152 (2013). 10.1016/j.ocemod.2013.09.003

27. Prants, S.V., Budyansky, M.V., Ponomarev, V.I., Uleysky, M.Y.: Lagrangian study of trans-port and mixing in a mesoscale eddy street. Ocean Model. **38**(1–2), 114–125 (2011). 10.1016/j.ocemod.2011.02.008

28. Prants, S.V., Budyansky, M.V., Uleysky, M.Y.: Identifying Lagrangian fronts with favourable fishery conditions. Deep-Sea Res. I Oceanogr. Res. Pap. **90**, 27–35 (2014). 10.1016/j.dsr.2014.04.012

29. Prants, S.V., Budyansky, M.V., Uleysky, M.Y.: Lagrangian fronts in the ocean. Izv. Atmos. Oceanic Phys. **50**(3), 284–291 (2014). 10.1134/s0001433814030116

30. Prants, S.V., Budyansky, M.V., Uleysky, M.Y.: Lagrangian study of surface transport in the Kuroshio extension area based on simulation of propagation of Fukushima-derived radionu-clides. Nonlinear Process. Geophys. **21**(1), 279–289 (2014). 10.5194/npg-21-279-2014

31. Prants, S.V., Ponomarev, V.I., Budyansky, M.V., Uleysky, M.Y., Fayman, P.A.: Lagrangian analysis of mixing and transport of water masses in the marine bays. Izv. Atmos. Oceanic Phys. **49**(1), 82–96 (2013). 10.1134/S0001433813010088

32. Prants, S.V., Uleysky, M.Y., Budyansky, M.V.: Numerical simulation of propagation of radioactive pollution in the ocean from the Fukushima Dai-ichi nuclear power plant. Dokl. Earth Sci. **439**(2), 1179–1182 (2011). 10.1134/S1028334X11080277

33. Prants, S.V., Uleysky, M.Y., Budyansky, M.V.: Lagrangian coherent structures in the ocean favorable for fishery. Dokl. Earth Sci. **447**(1), 1269–1272 (2012). 10.1134/S1028334X12110062

34. Qiu, B.: Kuroshio and Oyashio currents. In: Steele, J.H. (ed.) Encyclopedia of Ocean Sciences, 2nd edn., pp. 358–369. Academic Press, Oxford (2001). 10.1016/B978-012374473-9.00350-7

35. Rabinovich, A.B., Thomson, R.E., Bograd, S.J.: Drifter observations of anticyclonic eddies near Bussol' Strait, the Kuril Islands. J. Oceanogr. **58**(5), 661–671 (2002). 10.1023/a:1022890222516

36. Rogachev, K.A.: Rapid thermohaline transition in the Pacific western subarctic and Oyashio fresh core eddies. J. Geophys. Res. Oceans **105**(C4), 8513–8526 (2000). 10.1029/1999jc900330

37. Rogachev, K.A.: Recent variability in the Pacific western subarctic boundary currents and Sea of Okhotsk. Prog. Oceanogr. **47**(2–4), 299–336 (2000). 10.1016/s0079-6611(00)00040-9

38. Rogachev, K.A., Carmack, E.C.: Evidence for the trapping and amplification of near-inertial motions in a large anticyclonic ring in the Oyashio. J. Oceanogr. **58**(5), 673–682 (2002). 10.1023/A:1022842306586

39. Rogachev, K.A., Tishchenko, P.Y., Pavlova, G.Y., Bychkov, A.S., Carmack, E.C., Wong, C.S., Yurasov, G.I.: The influence of fresh-core rings on chemical concentrations (CO_2, PO_4, O_2, alkalinity, and pH) in the western subarctic Pacific Ocean. J. Geophys. Res. Oceans **101**(C1), 999–1010 (1996). 10.1029/95jc02924

40. Shapiro, N.: Formation of a circulation in the quasiisopycnic model of the Black Sea taking into account the stochastic nature of the wind stress. Phys. Oceanogr. **10**, 513–531 (2000). 10.1007/BF02519258

41. Shen, C.Y., Evans, T.E.: Inertial instability and sea spirals. Geophys. Res. Lett. **29**(23), 2124 (2002). 10.1029/2002gl015701

42. Sokolovskiy, M.A., Carton, X.J.: Baroclinic multipole formation from heton interaction. Fluid Dyn. Res. **42**(4), 045501 (2010). 10.1088/0169-5983/42/4/045501

43. Sokolovskiy, M.A., Filyushkin, B.N., Carton, X.J.: Dynamics of intrathermocline vortices in a gyre flow over a seamount chain. Ocean Dyn. **63**(7), 741–760 (2013). 10.1007/s10236-013-0628-y

44. Sokolovskiy, M.A., Verron, J.: Finite-core hetons: stability and interactions. J. Fluid Mech. **423**, 127–154 (2000). 10.1017/s0022112000001816

45. Sokolovskiy, M.A., Verron, J.: Dynamics of Vortex Structures in a Stratified Rotating Fluid. Atmospheric and Oceanographic Sciences Library, vol. 47. Springer, New York (2014). 10.1007/978-3-319-00789-2

46. Sulman, M.H.M., Huntley, H.S., Lipphardt Jr., B.L., Jacobs, G., Hogan, P., Kirwan Jr., A.D.: Hyperbolicity in temperature and flow fields during the formation of a Loop current ring. Nonlinear Process. Geophys. **20**(5), 883–892 (2013). 10.5194/npg-20-883-2013

47. Takematsu, M., Ostrovski, A.G., Nagano, Z.: Observations of eddies in the Japan basin interior. J. Oceanogr. **55**(2), 237–246 (1999). 10.1023/a:1007846114165

48. Talley, L., Min, D.H., Lobanov, V., Luchin, V., Ponomarev, V., Salyuk, A., Shcherbina, A., Tishchenko, P., Zhabin, I.: Japan/East Sea water masses and their relation to the sea's circulation. Oceanography **19**(3), 32–49 (2006). 10.5670/oceanog.2006.42

49. Talley, L.D., Lobanov, V.B., Tishchenko, P.Y., Ponomarev, V.I., Sherbinin, A.F., Luchin, V.A.: Hydrographic observations in the Japan/East Sea in winter, 2000, with some results from summer, 1999. In: Danchenkov, M.A. (ed.) Oceanography of the Japan Sea. Proceedings of CREAMS'2000, pp. 25–32. Dalnauka Publishing House, Vladivostok (2001). http://www.ferhri.ru/science/conference/creams2000/CREAMS.proceedings.shtml

50. Talley, L.D., Nagata, Y. (eds.): The Okhotsk sea and Oyashio region. No. 2 in PICES Scientific Reports. PICES (1995). https://www.pices.int/publications/scientific_reports/Report2/Rpt2.pdf

51. Uleysky, M.Y., Budyansky, M.V., Prants, S.V.: Effect of dynamical traps on chaotic transport in a meandering jet flow. Chaos **17**(4), 043105 (2007). 10.1063/1.2783258
52. Yasuda, I., Ito, S.I., Shimizu, Y., Ichikawa, K., Ueda, K.I., Honma, T., Uchiyama, M., Watanabe, K., Sunou, N., Tanaka, K., Koizumi, K.: Cold-core anticyclonic eddies south of the Bussol' Strait in the Northwestern Subarctic Pacific.   J. Phys. Oceanogr. **30**(6), 1137–1157 (2000). 10.1175/1520-0485(2000)030<1137:CCAESO>2.0.CO;2
53. Yasuda, I., Okuda, K., Hirai, M.: Evolution of a Kuroshio warm-core ring — variability of the hydrographic structure.   Deep Sea Res. Part A **39**(Supplement 1), S131–S161 (1992). 10.1016/s0198-0149(11)80009-9

List of Internet Resources

[ARGO] International Argo Program Homepage.
    http://www.argo.net/
[ARGODB] WHOI Argo Atlas & Database.
    http://argoweb.whoi.edu/
[AVISO] Archiving, Validation and Interpretation of Satellite Oceanographic (AVISO).
    http://www.aviso.altimetry.fr
[GDP] The Global Drifter Program.
    http://www.aoml.noaa.gov/phod/dac

Chapter 7
Fukushima-Derived Cesium Isotopes in the North Western Pacific: Direct Observation and Altimetry-Based Simulation of Propagation

7.1 Transport of Cesium Isotopes in the Kuroshio Extension Area Just After the Accident

The great Tohoku earthquake of magnitude 9.0 on March 11, 2011 followed by the tsunami inflicted heavy damage on the Fukushima Nuclear Power Plant (FNPP). Due to lack of electricity it was not possible to cool nuclear reactors and the fuel storage pools that caused numerous explosions at the FNPP (for details, see [32]). The Fukushima accident was classified at the maximum level of 7, similar to the Chernobyl accident which happened in 1986 in the former Soviet Union.

The radioactivity databases at the International Atomic Energy Agency and at the Meteorological Research Institute in Tsukuba were developed. Radionuclides were released from the FNPP through two major pathways, direct discharges of radioactive water, and atmospheric deposition onto the North Pacific Ocean. That deposition was indirectly estimated to be in the range 6.4–35 PBq [22]. Large amount of radioactive water leaked directly into the ocean [18, 42]. The total amount of ^{137}Cs isotope released was estimated to be 3.6 ± 0.7 PBq by the end of May [43].

Radioactive cesium isotopes with 30.17 yr half-life for ^{137}Cs and 2.06 yr half-life for ^{134}Cs have been detected over a broad area in the North Western Pacific in 2011 and 2012 [2, 6, 10–12, 15–18, 22, 31]. ^{137}Cs is a passive tracer in seawater which can be used to study long-term circulation and ventilation of water masses in the global ocean. In particular, distribution of Fukushima-derived ^{137}Cs in the ocean would help to verify numerical circulation models and their parameters. Different numerical circulation models have been used to simulate propagation of the radioactive pollution in the ocean [4, 8, 10, 25, 28, 29, 38, 39, 42, 43]. The results of in situ research and impact of the accident on ocean and coastal ecosystems to the end of 2013 have been collected in monographs [30, 32].

© Springer International Publishing AG 2017
S.V. Prants et al., *Lagrangian Oceanography*, Physics of Earth and Space Environments, DOI 10.1007/978-3-319-53022-2_7

Before March 2011, ^{137}Cs concentration levels off Japan were 1–2 Bq m^{-3} $\simeq$ 0.001–0.002 Bq kg^{-1}, while ^{134}Cs was not detectable. Because of a comparatively short half-life time, any measured concentrations of ^{134}Cs could only be Fukushima derived. Concentrations at the FNPP discharge channels in early April 2011 were more than 50 million times greater than the preexisting ocean level of ^{137}Cs [6].

One month after the accident, seawater, suspended solids, and zooplankton samples were collected from the surface mixed layer and subsurface layers at a number of stations, 200–2000 km offshore from the FNPP [10]. In surface water, ^{137}Cs concentrations were ranged from several times to two orders of magnitude higher than that before the accident. ^{134}Cs isotope was also detected with the ratio ^{134}Cs/^{137}Cs to be about 1. The highest concentrations, from $\simeq$150 Bq m^{-3} to $\simeq$350 Bq m^{-3}, have been found off the FNPP ($\simeq$200 km from the nuclear power plant) and Miyagi (the earthquake source). ^{137}Cs concentrations to the east of this region in the area [146° E–147° E; 37° N–38° N] were also high (50–60 Bq m^{-3}). The ^{137}Cs concentrations in the Kuroshio Extension, <10 Bq m^{-3}, were unexpectedly low, because it was considered to be the main potential pathway for contaminated water to the open ocean.

The expedition of the Russian Hydrometeorological Service on R/V *Pavel Gordienko* in April 24–May 6, 2011 [19] proved increased concentration of both ^{137}Cs and ^{134}Cs in surface water along the whole Kuril Island chain (2.2–3.6 Bq m^{-3} and 1.2–2.9 Bq m^{-3}, correspondingly) but not at the southern part of the Kamchatka Peninsula, where those concentrations were 1.4 and 0.4 Bq m^{-3}, correspondingly. Such distribution of radionuclides could be explained by atmospheric transport. They also found high concentration of radionuclides in the area about 350 km east of Tohoku, where ^{137}Cs and ^{134}Cs contents were found to be up to 24 and 21 Bq m^{-3}, correspondingly.

The R/V *Ka'imikai-o-Kanaloa* cruise has been conducted in June 4–18, 2011 [6] to investigate the distribution of Fukushima-derived radionuclides in seawater, zooplankton, and micronectonic fish 30–600 km offshore from the FNPP. Activities up to 325 Bq m^{-3} were found more than 600 km offshore. As to ^{137}Cs, the highest level (except for the discharge channels), 600–800 Bq m^{-3}, has been detected 30 km offshore. In June, Fukushima-derived cesium did not generally penetrate below 100–200 m. Over time, it is expected to find deeper penetration proving a means to study the rates of vertical mixing processes in the Pacific. Fukushima-derived isotopes have also been detected in zooplankton (with the maximal level about $5 \cdot 10^4$ Bq m^{-3} dry weight comparable with the recommended value of $4 \cdot 10^4$ Bq m^{-3}) and jellyfish but not in micronectonic fish. In June 2011, the highest surface water concentrations for both the isotopes, $3.9 \cdot 10^3$ Bq m^{-3}, have been detected in a semipermanent mesoscale eddy centered at 142.5° E, 37° N, not the nearest location to the nuclear power plant.

Results of direct observation of ^{134}Cs and ^{137}Cs in surface seawater collected from R/V *Kaiun Maru* in a broad area in the western and central North Pacific in July, October 2011 and July 2012 have been reported in Ref. [15]. In particular, seawater samples were collected at their stations C43–C55 (July 26–29, 2011) located from $35°$ N to $41°$ N along the $144°$ E transect with its southern edge crossing a crest of the Kuroshio Extension meander and the northern edge crossing partly the Tohoku mesoscale ACE centered at that time at $144°$ E, $38°$ N. It is a semipermanent eddy to be present in the region before and after the accident. It is clearly seen in an earlier simulation of Fukushima-derived radionuclides propagation (see Fig. 3b in Ref. [34]). After the accident, that eddy has interacted with a number of adjacent eddies and streamers promoting transport of contaminant water to the north, south, and east. The measured ^{137}Cs concentrations at stations C43–C55 [15] have been varied from the background level of $1.9 \pm 0.4\,Bq\ m^{-3}$ (station C52) to $153 \pm 6.8\,Bq\ m^{-3}$ (station C47). The ratio $^{134}Cs/^{137}Cs$ was close to 1.

7.1.1 The Kuroshio Rings and Near-Surface Cross-Jet Transport

The area east of Japan is known as the Kuroshio–Oyashio confluence zone [20] or subarctic frontal area. The Kuroshio Extension prolongs the Kuroshio Current, a western boundary current in the North Western Pacific, when the latter separates from the continental shelf of the Japanese island Honshu at Cape Inubo about $35°42'$ N (see Fig. 4.1). It flows eastward from this point as a strong unstable meandering jet constituting a front separating the warm subtropical and cold subpolar waters of the North Pacific Ocean. There are cyclonic and anticyclonic recirculation gyres on the northern and southern flanks of the jet. The main features of the Kuroshio Extension are described, for example, in Refs. [13, 35] and other papers. The Kuroshio and the Kuroshio Extension transport a large amount of heat and release that to the atmosphere strongly affecting climate. It is a region with one of the most intense air-sea heat exchange and the highest eddy kinetic energy level. It is also a region with commercial fishing grounds of Pacific saury, tuna, squid, Japanese sardine, and other species.

The Kuroshio–Oyashio confluence zone is populated with a plenty of mesoscale and submesoscale eddies that transfer heat, salt, nutrients, carbon, pollutants, and other tracers across the ocean (see Fig. 4.3). They originate, besides from the Kuroshio Extension, from the Tsugaru Warm Current, flowing between the Honshu and Hokkaido islands, and from the cold Oyashio Current flowing out of the Arctic along the Kamchatka Peninsula and the Kuril Islands (see Fig. 6.2). Those eddies may persist for the periods ranging from a few weeks to a few years and have a strong influence on the local climate, hydrography, and fishery.

A study of Kuroshio Extension rings and their interaction with the mean flow is important by many reasons. They act to transfer energy to the mean currents, influence on the Kuroshio Extension jet dynamics, and drive the recirculation gyres. They transport for a long distance water masses with biophysical properties different from ambient waters that may have a great impact on living organisms. The strongest mesoscale eddies of both polarities are generated along the Kuroshio Extension jet.

The warm-core anticyclonic rings are pinched off from the meandering Kuroshio Extension mainly to the north whereas the cold-core cyclonic ones—to the south of it. The occurrence, distribution, and behavior of the Kuroshio Extension rings, moving generally westward due to the planetary β-effect, have been studied in a number of papers via hydrographic observations, infrared imaging, and altimetry data (see, e.g., [9, 13, 41, 44]). However, the process of their separation from the parent jet is not fully understood.

Lagrangian tools have been successfully used before to obtain a detailed description of different advective transport phenomena in the ocean and atmosphere. As to the problem of eddy separation from strong jet currents and a CJT, there are papers on Lagrangian approach to the Loop Current eddy separation in the Gulf of Mexico [1, 23] and on Lagrangian description of CJT in the Kuroshio Current [26, 27]. Near-surface velocity fields from numerical models of circulation in the Gulf of Mexico have been used to study the eddy separation process by computing effective invariant manifolds [23] and FTLEs [1]. It has been shown that the Lagrangian methods are a useful supplement to traditional approaches as they reveal flow details not easily extracted from Eulerian point of view. In Chap. 2 we applied a dynamical system theory approach to study mechanisms of chaotic zonal and CJT for kinematic and dynamical analytic models of meandering jets.

In this section we use the AVISO velocity field to study the process of interaction of cold-core cyclonic rings with the Kuroshio Extension main current, the events of their separation from the parent jet, and their role in near-surface CJT. The special task is to know whether it was possible for Fukushima-derived radionuclides to cross the Kuroshio Extension jet which is supposed to be an impenetrable barrier. We apply here different Lagrangian tools to trace the origin of water parcels with measured levels of concentrations of Fukushima-derived cesium isotopes collected in two research vessel (R/V) cruises in June and July 2011 in the large area of the North Western Pacific [6, 15]. The results of simulation are supported by tracks of the surface drifters which were deployed in the area.

Satellite altimetry and hydrographic surveys demonstrate clearly that the Kuroshio Extension alternates between two dominant states: one with two quasistationary meanders and another one when the meanders are not especially prominent [35, 40]. During the meandering state, steep troughs develop stimulating ring pinchoff events on the both flanks of the jet. We will here focus on that state with increased eddy activity. A sketch of the meandering Kuroshio Extension state in Fig. 7.1 shows the eastward jet current with two crests near 143° E and 151° E which are anticyclonic side of the jet and two troughs near 147° E and 153° E with cyclonic rotations. In reality, the Kuroshio Extension jet is highly unstable, the meander's amplitude may change in the course of time, and locations of the crests and troughs may fluctuate strongly both in the meridional and zonal directions.

Documenting separation of rings from jets, merging with jets, and tracking their propagation are longstanding problems in oceanography. The AVISO velocity field is shown in Fig. 7.2 on fixed days. Figure 7.2 illustrates the process of formation of a cyclonic ring (which we denote as the ring 1) from the Kuroshio Extension meander in June 2011 when the Kuroshio Extension was in the state with two prominent

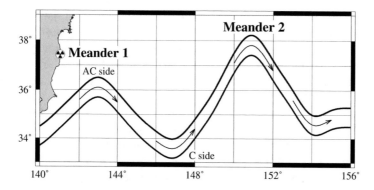

Fig. 7.1 Schematic view of the Kuroshio Extension state with two quasistationary meanders. Location of the FNPP is shown

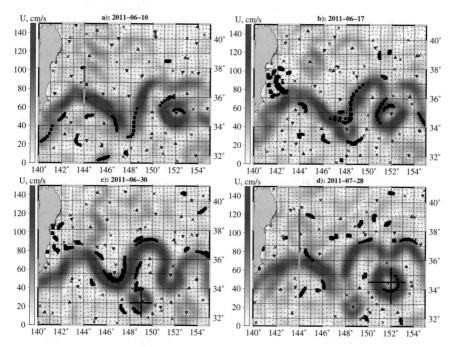

Fig. 7.2 (**a**), (**b**), and (**c**) Metamorphoses of the AVISO velocity field in the process of formation of the cyclonic ring 1 on June 10, 17, and 30, 2011, respectively, with tracks of the drifters imposed. (**d**) Velocity field on July 28, 2011 with the cyclonic ring 2 separated from the jet. Nuances of the *blue color* measure the modulus of the linear velocity U in cm/s. The straights are material lines that have been evolved backward in time to trace origin of the corresponding tracers. The perpendicular *black lines* cross the cyclonic ring 1 (in panel (**c**)) and cyclonic ring 2 (in panel (**d**)). The *colored lines* in (**a**), (**d**) are placed along the transects where seawater samples have been collected in June 10 and 11, 2011 in the R/V *Ka'imikai-o-Kanaloa* cruise [6] and in July 26–29, 2011 in the R/V *Kaiun Maru* cruise [15], respectively

quasistationary meanders. In the beginning of June, the trough of the first meander started to steepen with its edges becoming day by day closer and closer to each other (Fig. 7.2a). The edges merged eventually closing a volume of water with a cyclonic rotation to be connected with the parent jet by an arch (Fig. 7.2b). An elliptic point in its center and a hyperbolic point in the neck of the meander appeared. Approximately 5 days later, the velocity field bifurcated, the ring 1 was separated from the jet, and the meander amplitude decreased correspondingly (Fig. 7.2c). Tracks of the two drifters, encircling partly the ring, are shown in Fig. 7.2c.

There are almost always a number of eddies around the Kuroshio Extension. Here we focus on the ring 1 (Fig. 7.2b, c) separated from the jet in the middle of June 2011, the cyclonic ring 2 (Fig. 7.2d) separated from the jet on July 28, 2011, and the Tohoku ACE, visible in all the panels in Fig. 7.2, with the center on July 28, 2011 to be at $144°$ E, $38°$ N. The material line technique, presented in Chap. 4, is applied now to trace the origin and history of water masses inside the rings 1 and 2 crossing them by the black perpendicular lines shown in Fig. 7.2c, d, respectively. In the next section we apply the same technique to trace the origin and history of tracers placed on colored segments along the transects where seawater samples have been collected on June 10 and 11, 2011 in the R/V *Ka'imikai-o-Kanaloa* cruise [6] (Fig. 7.2a) and on July 26–29, 2011 in the R/V *Kaiun Maru* cruise [15] (Fig. 7.2d).

Transport of water masses across strong jet currents, like the Gulf Stream and the Kuroshio, is important because they separate waters with distinct bio-physicochemical properties (see Chap. 2). It may cause heating and freshing of waters with a great impact on the weather and living organisms. The question whether the stable Kuroshio Extension jet between $141°$ E and $153°$ E is an impenetrable barrier for CJT is still open. The new aspect of that problem arised suddenly after the Fukushima accident on March 11, 2011. By the common opinion (see, e.g., [6]), it was difficult to expect observation of Fukushima-derived radionuclides on the southern side of the Kuroshio Extension jet. Could contaminant waters from the Fukushima area cross the Kuroshio Extension jet and appear on the southern side of the jet or not?

In order to document directly the CJT of Fukushima-derived radionuclides, we compute latitudinal Lagrangian maps (see Chap. 4) by integrating the advection equations (4.1) backward in time from a fixed date till the day of the accident for a large number of particles distributed in the studied area. The latitude, φ, from which each particle came to its final position, is coded by color. The latitudinal map on June 30, 2011 in Fig. 7.3a demonstrates that "red" waters crossed the latitude $31°$ N from the south, "green" waters crossed the latitude of the FNPP ($141°05'$ E, $37°25'$ N) from the north, whereas nuances of the grey color code the particles originated from the latitudes between $31°$ N and $37°$ N. Zoom in Fig. 7.3b shows the cyclonic ring 1 with "green" water in its core (originated from the latitudes $>37°$ N) that may contain increased concentration of Fukushima-derived radionuclides.

The simulation results are confirmed by tracks of two surface drifters which were captured by the ring 1 (Fig. 7.3b). The southern track belongs to the drifter no. 98875 released on December 2010 at the point $126.784°$ E, $15.778°$ N [GDP]. It was transported by the Kuroshio Current from the southwest. The northern track belongs to the drifter no. 36473 released on June 11 in the R/V *Ka'imikai-o-Kanaloa*

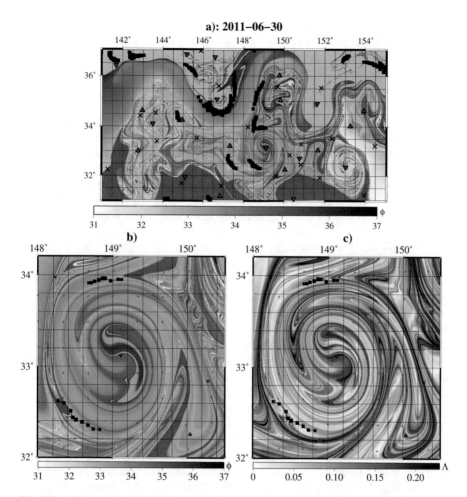

Fig. 7.3 (**a**) Lagrangian latitudinal map on June 30, 2011 computed from that date to the day of the accident, March 11, 2011, with the "*red*" and "*green*" particles originated from the latitudes <31° N and >37° N, respectively. Nuances of the *grey color* code in geographic degrees the particles originated from the latitudes between 31° N and 37° N. (**b**) Zoom of the area with the cyclonic ring 1 demonstrates a small transport across the Kuroshio Extension jet (the "*green*" particles). Tracks of two drifters, captured by the cyclonic ring 1, are shown as the *black squares*. (**c**) FTLE map of the same area as in (**b**).

cruise [6] at the point 144.087° E, 35.901° N. It crossed the Kuroshio Extension jet and was captured by the ring 1. It is a "green" drifter in Fig. 2 in Ref. [6]. Some of their "red" drifters in that Fig. 2 have been trapped by the meander trough that formed the cyclonic ring 2 in July 2011.

The map in Fig. 7.3b is an evidence of transport of water across the Kuroshio Extension jet. To check that finding we initialized a material line crossing the ring 1 core along 31°06′ N (the horizontal black line in Fig. 7.2c) and evolved it backward in time till the day of the accident. It has been found that the fragments of that line, containing "green" particles, really came from the area nearby the location of the FNPP, whereas the other particles of that line came from the west, mainly along the Kuroshio Extension jet. However, the amount of potentially dangerous Fukushima waters in the core of the ring 1 is comparatively small.

Figure 7.3c is a plot of the FTLE field Λ. That quantity has been computed backward in time from June 30 to May 30 by the method of the singular-value decomposition of an evolution matrix [33] (see Sect. 4.3.1). The FTLE map demonstrates the same large-scale structures as the latitudinal map. It provides also a typical fine-scale pattern of mixing in the ring 1 with alternating black ridges and white bands. The black bands are locus of those particles that separated from each other for 1 month in the past at a maximal rate. The boundaries between particles of different origin in Fig. 7.3b correspond to the black ridges with high values of Λ in Fig. 7.3c.

Now we apply the material line technique to trace the origin of water masses in cores of the rings 1 and 2. The ring 2 was born after a separation of the trough of the second meander from the parent jet. In June–July 2011 that meander was deforming strongly and eventually produced a ring-like feature with a diameter of about 300 km that has been detached to the south from the parent jet and then reabsorbed in a short time. That ring-like feature with the center near 152° E and 34°30′ N is seen in Fig. 7.2d in the AVISO velocity field on July 28, 2011.

Coming back to the question, whether the Kuroshio Extension jet is an impenetrable barrier for Fukushima-derived radionuclides, let us compute tracking maps. The region under study [130° E–170° E, 25° N–50° N] is divided into 500 × 500 cells. We advect particles of the perpendicular black material lines crossing the cyclonic rings 1 (Fig. 7.2c) and 2 (Fig. 7.2d) backward in time from June 30 (ring 1) and from July 28 (ring 2) until the day of the accident, March 11, 2011. Then one fixes each day how many times particles visited each cell during the month after the accident, from March 11 to April 10, when the maximal leakage of radionuclides from the FNPP directly into the ocean and their atmospheric fallout on the ocean surface were registered. Summarizing those numbers, we get a total number of visits in each cell to be proportional to the density of particle's traces, v, which is coded by color in the logarithmic scale. Probability that the ring 1 contains contaminated water is comparatively small because the density of points in the region around the FNPP is small (Fig. 7.4a). This map confirms the result of direct calculation of CJT in Fig. 7.3b where only a small amount of potentially contaminated "green" water, originated from the latitudes >37° N, is visible. It is clear from Fig. 7.4a that the core of the cyclonic ring 1 consists mainly of Kuroshio water.

This method provides a direct evidence of a CJT in difference from forward-in-time simulation of a patch of particles initially located around the Fukushima area. Previous simulations in Refs. [6] (see Fig. S3 in a supplementary material to that paper) and [15] (see their Fig. 2) have found some particles from the initial patch on the southern Kuroshio Extension flank. However, day by day inspection of simulated propagation has shown that most of those particles, in fact, did not cross the jet.

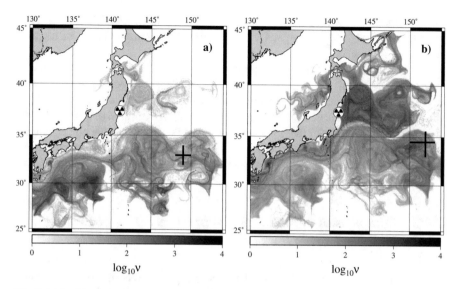

Fig. 7.4 Tracking maps for the tracers placed on the perpendicular material lines crossing (**a**) the cyclonic ring 1 (Fig. 7.2c) and (**b**) the cyclonic ring 2 (Fig. 7.2d). The maps show where the corresponding tracers were walking after the FNPP accident, from March 11 to April 10, 2011. The density of traces is in the logarithmic scale

They have propagated to the east along the northern Kuroshio Extension side and after reaching the longitude about 155° E, where the Kuroshio Extension becomes unstable and even may break down for a while, part of them were advected to the south and then to the west propagating along the southern Kuroshio Extension flank. It is not a CJT. In order to document the real CJT it is necessary to select candidate particles on the southern Kuroshio Extension flank and compute the corresponding longitudinal and tracking maps backward in time.

As to particles in the core of the ring 2, the density of their traces in the area, that is supposed to be contaminated, is much higher as compared to the ring 1 case (Fig. 7.4b). It means that the probability to observe higher concentrations of Fukushima-derived radionuclides in surface waters of the ring 2 was expected to be comparatively large. The periphery of the ring 2 contained water parcels which moved during the month after the accident around the mesoscale eddies present to the north and east from the FNPP location. On the day of the accident, there was an eddy system with the large Tohoku ACE with the center around 144° E and 39° N, a small ACE to the north of it and a medium CE at the traverse of the Tsugaru Strait. It has been shown in our paper [34] that namely this eddy system has governed mixing and transport of radioactive water, part of which has been captured by the Tohoku ACE and advected around the adjacent eddies to the north. The concentration of radionuclides around those eddies might be comparatively large because they were able to capture ambient contaminated water [34]. The influence of the Tohoku ACE is evident on both the tracking maps as a patch with increased density of traces at its place.

We conclude this section by emphasizing that the material line technique may be useful for finding the surface areas in the ocean that are potentially dangerous due to the risk of radioactive or other contamination. Before choosing the track of a planed R/V cruise, it is instructive to make a simulation by initializing backward-in-time evolution of material lines, crossing eddies in the region visible on Lagrangian maps. The corresponding tracking maps computed would help to know where one could expect higher or lower concentrations of contamination in this or that eddy.

7.1.2 Comparison of Simulation with Observation of Cesium Isotopes During the Cruises in June and July 2011

In this section we apply the material line technique to trace the origin of water parcels with measured levels of ^{134}Cs and ^{137}Cs concentrations collected in the two R/V cruises in June and July 2011 [6, 15]. Starting from the dates of sampling, we evolve backward-in-time material lines placed along the transects, where stations with collected surface water samples were located. Results of direct observation of radioactive cesium in surface seawater collected from R/V *Kaiun Maru* in a broad area in the western and central North Pacific in July, October 2011 and July 2012 have been reported in Ref. [15]. In this study, we focus on the results of measurements to be carried out at their stations C43–C55 from July 26 to July 29, 2011 along the 144° E meridian from 35° N to 41° N. That transect is shown in Fig. 7.2d. The measured ^{137}Cs concentrations at the stations C43–C55 have been found to be in the range from the background level 1.9 ± 0.4 mBq/kg (station C52) to 153 ± 6.8 mBq/kg (station C47). The ratio ^{134}Cs/^{137}Cs was close to 1. The level of the concentration of the cesium isotopes in the sea surface waters in the North Pacific before the accident did not exceed 2–3 mBq/kg.

A material line was placed along that transect and divided into four colored segments: (1) the orange segment, 35° N–36°30′ N, crossing the first meander's crest with initialization on July 26, (2) the red segment, 36°30′ N–30° N, crossing the southern part of the Tohoku ACE with initialization on July 27, (3) the blue segment, 38° N–39°30′ N, crossing its northern part with initialization on July 28, and (4) the green segment, 39°30′ N–30° N with initialization on July 29.

The tracking maps in Fig. 7.5, colored as the corresponding segments, show where particles were walking from March 11 to April 10, 2011. Particles from the orange segment, as expected, were advected to their places in the end of July mainly by the Kuroshio Current from the southwest and did not cross the latitude of the segment's northern end, 36°30′ N (Fig. 7.5a). The risk of their radioactive contamination is small. It was confirmed by the measured ^{137}Cs concentrations at the stations C52–C55 of that segment to be 2–5 mBq/kg [15], which was slightly higher than the background level.

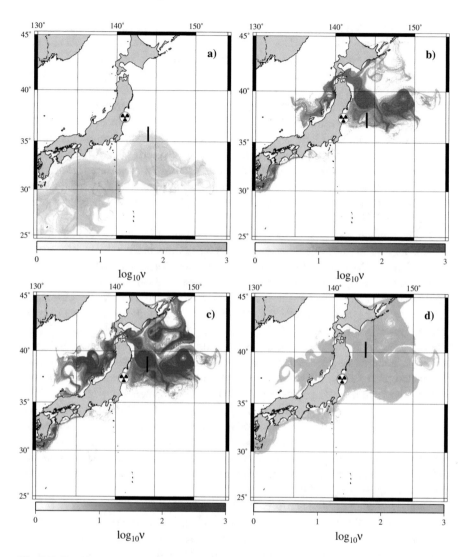

Fig. 7.5 Tracking maps for the particles placed on four material line segments along the transect where seawater samples have been collected in July 26–29, 2011 [15]. The maps are colored as the corresponding segments in Fig. 7.2d and show where particles were walking from March 11 to April 10, 2011. Particles were initially placed along the 144° E meridian at the segments: (**a**) 35° N–36° 30′ N, (**b**) 36° 30′ N–30° N, (**c**) 38° N–39° 30′ N, and (**d**) 39° 30′ N–30° N

Particles from the red segment have been found walking mainly in the area to the north from the latitude 36° 30′ N (Fig. 7.5b). Part of that segment crossed the Tohoku ACE existing in the area from the day of the accident to the days of in situ measurements. That is why we see increased density of points in the area where it has been located from March 11 to April 10. A comparatively high level of Fukushima-derived cesium isotopes was expected in the corresponding water samples. It is really the case. The ^{137}Cs concentrations at the stations C49 and C50 of that segment were measured to be 36 ± 3.3 and 50 ± 3.6 mBq/kg [15]. The high cesium concentration levels up to 153 ± 6.8 mBq/kg [15] were measured at the stations C46, C47, and C48 (blue segment) and C43, C44, and C45 (green segment). The tracking maps in Fig. 7.5c, d show clearly an increased density of traces of the corresponding particles in the area where the maximal leakage of radionuclides from the FNPP directly into the ocean and their atmospheric fallout on the ocean surface were registered from March 11 to April 10, 2011. Those maps also demonstrate strong mixing in the Kuroshio–Oyashio frontal zone. Traces of the particles, initialized at the end of July at a comparatively compact material line, have been found from March 11 to April 10 in an extended area including some parts of the Japan and Okhotsk seas.

Fukushima-derived ^{134}Cs and ^{137}Cs were measured in surface and subsurface waters, as well as in zooplankton and fish, at 50 stations in June 4–18, 2011 during the R/V *Ka'imikai-o-Kanaloa* cruise [6]. We initialized a material line as shown in Fig. 7.2a where ^{137}Cs concentrations were measured on June 10 and 11 at 25 stations in the range from 1.4 ± 0.2 mBq/kg (station 13) to 173.6 ± 9.9 mBq/kg (station 10). The ratio ^{134}Cs/^{137}Cs was close to 1. Particles were placed on the green segment, 35°30′ N–36°30′ N, crossing the first meander's crest (June 11), and the red one, 37° N–38° N, to the north from the meander's crest (June 10). Traces of the particles of the green segment have been found on both sides of the Kuroshio Extension jet, whereas traces of the particles of the red one were on the northern side of the jet only. The southern part of the green segment crossed the jet itself but its northern part was outside of it (Fig. 7.2a). That is why the tracking map in Fig. 7.6a consists of two disconnected domains, one is to the south of the jet and another one is to the north. The measured ^{137}Cs concentrations at the stations 13 and 14, situated in the green segment, were at the background level, in the range 1.4– 3.6 mBq/kg [6], because the corresponding particles were advected by the Kuroshio Extension current (Fig. 7.6a).

Density of traces of the particles from the red segment is comparatively high in the area around the FNPP (Fig. 7.6b). This finding is confirmed by measurements at stations 10, 11, and 12 [6] where the concentrations of Fukushima-derived ^{134}Cs and ^{137}Cs were in the range 21.9–173.6 mBq/kg. The red segment crossed partly the Tohoku ACE, which is visible in Fig. 7.2a, and the traces of its particles are dense at the place of that ring. Some particles were advected to their places on the initial segment by the Tsushima Current from the Japan Sea to the Pacific Ocean through the Tsugaru Strait.

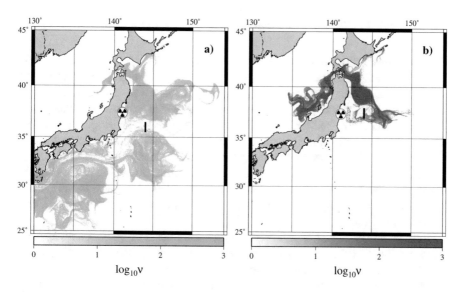

Fig. 7.6 Tracking maps for the particles placed on two material line segments along the transect where seawater samples have been collected on June 10 and 11, 2011 [6]. The maps are colored as the corresponding segments in Fig. 7.2a and show where particles were walking from March 11 to April 10, 2011. Particles were initially placed along the 144° E meridian at the segments: (**a**) 35°30′ N–36°30′ N and (**b**) 37° N–38° N

7.2 Role of Mesoscale Eddies in Transport of Cesium Isotopes

It is known that Kuroshio warm-core rings could move northeastward towards the Kuril Islands [7, 13, 24, 45] and thus transport water with a higher cesium content to the north of the Subarctic Front. To prove this hypothesis we have performed a numerical modelling of tracer transport in the area east of Japan and implemented a cruise to cross major streams and eddies in the area on the board of the R/V *Professor Gagarinskiy* of the Far Eastern Branch of Russian Academy of Sciences. The cruise has been conducted from June 12 to July 10, 2012, 15 months after the accident.

We focus in this section on comparing the experimental results, obtained with water samples at stations in the centers of some selected ACEs in the region, with the results of altimetry-based numerical simulation of spatial distribution of Fukushima-derived radionuclides [5]. That simulation helps to explain why we have detected comparatively high cesium concentrations in the cores of some eddies and lower ones inside other eddies.

7.2.1   R/V Professor Gagarinskiy Cruise in June–July 2012

The standard methods have been used to collect water and biota followed by laboratory processing and detection of ^{134}Cs and ^{137}Cs with a high purity germanium spectrometer. During the cruise surface water samples were collected along the cruise track by a submerged pump at 54 stations. Water samples from subsurface and deep layers (100–3621 m) were taken using a CTD/Rosette sampling system at 15 stations. A volume of each water sample was 95–120 l. Locations of stations are shown in Fig. 7.7. The water samples pre-treatment for measurement of ^{134}Cs and ^{137}Cs isotopes was performed on board by concentrating the cesium isotopes on the selective sorbent ANFEZH [3, 36, 37] with a preliminary separation of suspended matter. Water was pumped through a filter and then through the sorbent with a flow rate around 2 l/min.

Further processing of the samples and measurements of gamma activity have been continued at the land-based laboratory at the Pacific Oceanological Institute in Vladivostok. Each sorbent has been dried in the oven for 3–5 h at temperatures of 70–80 °C and then burned in a muffle furnace at temperatures of 430–450 °C during 10–15 h. Initial sorbent of each water sample had a volume of 320 cm^3 and a weight of 80 g, while the ash in the measuring vessel for gamma spectrometry, remained after combustion of the sorbent, had a volume of 4.5–4.7 cm^3. Minimizing the sample volume is essential for gamma-spectrometric analysis as it allows decreasing a minimum detective activity. Recovery efficiency of cesium and ^{60}Co

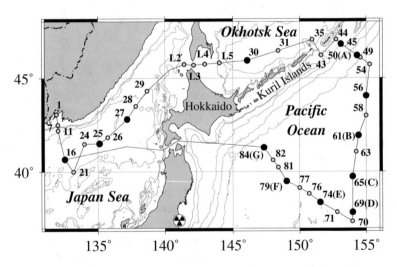

Fig. 7.7 Locations and numbers of stations where surface (the *open circles*) and deep seawater samples (the *full circles*) were collected during the cruise (June 12–July 10, 2012). Letters A, B, C, D, E, F, and G mark elliptic points in the centers of the corresponding mesoscale eddies to be studied. Radioactivity sign is location of the FNPP

was 0.98 ± 0.02 and 0.67 ± 0.1, respectively, where errors ± 0.02 and ± 0.1 are equal to 2σ of ten ^{137}Cs and ^{60}Co labeled seawater samples processing.

Gamma activity and radioisotope composition have been determined with a gamma spectrometer with a high purity germanium detector GEM150 and a digital multi-channel analyzer DSPEC jr 2.0 (ORTEC). To reduce the background activity, the detector was placed in a lead shield with 10 cm thickness of the wall and cover and inner walls covered with a copper layer of 1 mm thickness. Integral background count rate of the detector in the shield within the energy range of 50–2990 keV is 6.6 counts per second. We have obtained the following minimum detective activities for the measurement of 80 g ANFEZH blank sample during 160,000 s: ^{134}Cs— 0.004 Bq, ^{137}Cs—0.005 Bq, and ^{60}Co—0.014 Bq. For the spectrometer calibration we used an 80 g sorbent soaked with solution of known concentration of ^{134}Cs, ^{137}Cs, and ^{60}Co after its ashing. The gamma ray spectrum was analyzed using the software package Gamma Vision-32, version 6.09, ORTEC. True-coincidence correction factor for ^{134}Cs in our case was determined as 1.64. The ^{134}Cs to ^{137}Cs ratio was found to be 0.62 (Fig. 7.10), whereas at the time of accident this value was equal 1. This decrease corresponds to a partial disintegration of ^{134}Cs for 15 months after the accident. All results were corrected to the sampling date.

7.2.2 Observed and Simulated Horizontal Distribution of Cesium Isotopes and Identification of Mesoscale Eddies in the Area

Concentrations of ^{137}Cs in the Japan and Okhotsk seas have been found to be 1.4–2.3 and 1.5–1.9 Bq m^{-3}, accordingly (Table 7.1 and Fig. 7.8), and did not exceed much pre-accident level. A slightly increased concentration of ^{134}Cs (2.4 Bq m^{-3}) was found only at one station L2 located in the northeastern Japan Sea off northern tip of the Hokkaido Island. This station was sampled in the area of the Tsushima Current which transports water contaminated by river runoff from the Honshu Island that could explain a higher concentration of ^{137}Cs. This is in accordance with observations in [12] which showed an increase of ^{134}Cs and ^{137}Cs concentrations in surface water transported by the Tsushima Current along the west coast of Japan to Hokkaido.

All surface water samples, collected in the Pacific Ocean, contain an increased concentration of ^{134}Cs, 0.2–11.9 Bq m^{-3}, with except of Station 76 (Table 7.1 and Fig. 7.8). This station was taken in the warm streamer extended northward from the Kuroshio Extension jet and thus transported relatively clean water. ^{137}Cs concentrations in all samples in the western subarctic Pacific and Kuroshio–Oyashio frontal zone were in the range of 1.8–21 Bq m^{-3}. The surface water maximal ^{134}Cs and ^{137}Cs concentrations were registered in the frontal zone between stations 56–65 and stations 81–84 (Fig. 7.8). This confirms accumulation of surface water in the frontal zone with particular increased content in mesoscale ACEs. Below we will discuss a role of those eddies in more details.

Table 7.1 Concentrations of ^{134}Cs and ^{137}Cs in seawater collected in the North Western Pacific including the Japan and Okhotsk Seas

Station	Time	Longitude	Latitude	Depth, m	Temp, °C	Sal, psu	^{134}Cs, Bq/m^3	^{137}Cs, Bq/m^3
1	12.06.12 13:50	131.877° E	43.067° N	0	12.401	32.117	<0.09	1.9± 0.2
7	14.06.12 01:56	131.995° E	42.514° N	0	9.700	33.860	<0.08	1.7± 0.2
11	14.06.12 11:55	132.004° E	42.220° N	0	10.390	33.850	<0.08	1.9± 0.3
16	15.06.12 14:18	132.517° E	40.668° N	0	15.198	33.972	<0.05	1.4± 0.2
				201	1.454	33.978	<0.07	1.7± 0.3
				751	0.533	34.069	<0.07	1.7± 0.3
				1500	0.195	34.066	<0.08	0.7± 0.1
				2500	0.110	34.065	<0.06	0.4± 0.1
				3361	0.090	34.065	<0.09	1.0± 0.3
21	16.06.12 21:21	133.167° E	40.003° N	0	17.607	34.000	<0.07	2.2± 0.3
24	17.06.12 19:09	133.997° E	41.504° N	0	14.609	33.950	<0.08	1.6± 0.2
25	18.06.12 05:01	135.121° E	41.520° N	0	14.370	33.880	<0.05	2.0± 0.2
				2900	0.087	34.064	<0.06	0.3± 0.1
26	18.06.12 12:00	135.722° E	41.847° N	0	14.657	33.852	<0.08	1.6± 0.2
27	19.06.12 07:08	137.183° E	42.818° N	0	15.212	33.908	<0.07	2.2± 0.2
				3621	0.088	34.065	<0.06	1.1± 0.1
28	19.06.12 15:15	137.842° E	43.501° N	0	15.690	33.967	<0.09	2.3± 0.3
29	20.06.12 01:21	138.696° E	44.311° N	0	15.306	34.030	<0.08	2.0± 0.2
L2	21.06.12 01:30	141.408° E	45.721° N	0	11.980	33.806	0.3±0.2	2.4± 0.3
L3	21.06.12 08:00	142.110° E	45.686° N	0	9.045	33.540	<0.08	1.8± 0.2
L4	21.06.12 14:00	142.935° E	45.737° N	0	8.918	32.203	<0.06	1.5± 0.2

(continued)

Table 7.1 (continued)

Station	Time	Longitude	Latitude	Depth, m	Temp, °C	Sal, psu	^{134}Cs, Bq/m^3	^{137}Cs, Bq/m^3
L5	21.06.12 20:30	144.024° E	45.805° N	0	8.355	32.242	<0.06	1.5± 0.2
30	22.06.12 08:16	146.076° E	45.934° N	0	8.206	32.281	<0.03	1.5± 0.1
				150	−0.340	33.180	<0.07	1.4± 0.1
				500	1.560	33.640	<0.03	1.1± 0.1
				1005	2.220	34.240	<0.07	0.9± 0.2
				2200	1.785	34.506	<0.03	0.7± 0.1
				3236	1.626	34.604	<0.07	0.4± 0.1
31	23.06.12 07:43	148.393° E	46.456° N	0	8.591	32.272	<0.05	1.6± 0.1
35	24.06.12 05:36	150.932° E	47.043° N	0	5.810	32.640	<0.06	1.5± 0.1
43	24.06.12 23:11	151.609° E	46.204° N	0	3.843	32.976	0.4±0.1	1.9± 0.5
44	25.06.12 10:57	152.732° E	47.050° N	0	6.010	32.910	0.9±0.2	2.6± 0.2
45	25.06.12 15:16	153.095° E	46.803° N	0			1.2±0.2	3.7± 0.4
				200			0.4±0.1	1.9± 0.2
				500			<0.07	0.9± 0.1
				1000			<0.08	0.7± 0.1
49	26.06.12 12:10	154.577° E	46.083° N	0	6.700	33.030	0.4±0.1	1.8± 0.2
50	26.06.12 14:54	154.334° E	46.197° N	200	1.320	33.360	0.5±0.2	1.7± 0.3
				500	1.680	33.460	0.2±0.1	1.5± 0.1
				1000	3.240	34.150	<0.03	0.6± 0.1
54	27.06.12 08:03	155.279° E	45.717° N	0	6.501	32.932	1.2±0.1	3.0± 0.3

(continued)

Table 7.1 (continued)

Station	Time	Longitude	Latitude	Depth, m	Temp, °C	Sal, psu	^{134}Cs, Bq/m^3	^{137}Cs, Bq/m^3
56	27.06.12 23:05	155.000° E	44.050° N	0	7.256	33.011	3.0±0.3	5.9± 0.5
				200	2.932	33.573	0.5±0.2	2.0± 0.2
				500	3.334	34.122	<0.09	0.9± 0.1
				1000	2.554	34.404	<0.07	0.5± 0.1
				2002	1.705	34.601	<0.06	0.4± 0.1
				3002	1.297	34.661	<0.09	0.5± 0.3
58	28.06.12 17:27	155.005° E	42.999° N	0	12.800	34.020	2.4±0.2	5.2± 0.4
61	29.06.12 06:30	154.438° E	41.936° N	0	11.580	33.985	11.9 ± 0.6	21.0± 1.1
				203	6.817	33.904	12.5 ± 0.5	21.6± 0.9
				500	4.135	33.907	0.2±0.1	1.6± 0.3
				1000	3.064	34.347	0.8±0.2	1.9± 0.2
63	29.06.12 18:15	154.262° E	41.068° N	0	12.100	33.940	9.7±0.5	16.6± 0.9
65	30.06.12 06:42	154.003° E	39.750° N	0	17.139	34.444	2.1±0.2	4.8± 0.3
				200	9.218	34.155	13.5 ± 0.9	22.7± 1.5
				500	5.017	33.926	<0.07	0.9± 0.2
				1000	3.173	34.354	<0.11	0.7± 0.2
69	01.07.12 08:44	154.001° E	37.842° N	0	18.650	34.690	2.8±0.2	6.3± 0.4
				205	15.990	34.630	4.1±0.3	8.0± 0.5
				500	7.770	34.030	3.5±0.3	7.2± 0.6
				1000			<0.07	1.1± 0.2
70	01.07.12 20:25	153.981° E	37.366° N	0	22.368	34.390	0.2±0.1	2.3± 0.2
71	02.07.12 06:15	152.813° E	37.863° N	0	18.304	34.313	1.6±0.2	4.5± 0.3

(continued)

Table 7.1 (continued)

Station	Time	Longitude	Latitude	Depth, m	Temp, °C	Sal, psu	^{134}Cs, Bq/m^3	^{137}Cs, Bq/m^3
74	02.07.12 19:24	151.536° E	38.385° N	0	19.120	34.518	0.6±0.1	2.5± 0.2
				100	14.860	34.530	1.1±0.2	4.0± 0.4
				307	5.893	33.885	5.6±0.6	12.3± 0.8
				500	4.720	34.021	<0.07	1.4± 0.1
76	03.07.12 09:56	150.701° E	38.836° N	0	18.650	34.235	<0.06	0.4± 0.1
77	03.07.12 16:37	150.002° E	39.151° N	0	18.290	34.370	1.5±0.2	4.1± 0.3
79	04.07.12 02:41	149.003° E	39.495° N	0	18.174	34.319	1.7±0.2	4.3± 0.3
				100	10.643	34.236	2.9±0.4	7.3± 0.7
				280	3.906	33.489	3.5±0.6	9.0± 0.9
				500	4.229	33.964	<0.05	1.3± 0.2
81	04.07.12 14:54	148.387° E	40.234° N	0	15.950	33.530	3.1±0.3	6.7± 0.5
82	04.07.12 20:20	147.996° E	40.631° N	0	13.625	33.242	1.0±0.1	2.5± 0.2
84	05.07.12 05:29	147.329° E	41.299° N	0	14.750	33.846	6.1±0.4	11.0± 0.6
				100	8.352	33.979	10.4 ± 0.7	18.0± 1.3
				350	4.672	33.601	6.9±0.4	11.3± 0.6
				500	1.895	33.432	0.4±0.1	1.7± 0.2

Solving the advection equations (4.1) backward in time, we have computed FTLE and drift Lagrangian maps on each cruise day. Those maps have been sent electronically to the board and used to plan the cruise track in order to cross prominent eddies in the area. Here we compare simulated lateral distribution of radionuclides with cruise observations.

The initial distribution of radionuclides in simulation is supposed to be a patch with tracer concentration decreasing logarithmically with distance from the FNPP location. We advect particles by the AVISO velocity field, starting not from the date of tsunami but from March 25, 2011, in order to take into account not only a direct

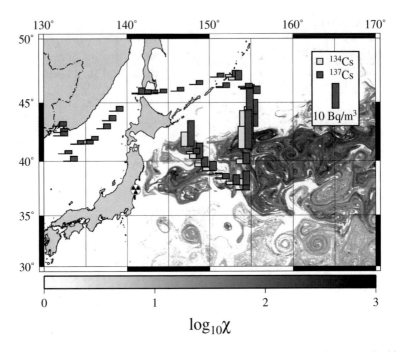

Fig. 7.8 Simulated distribution of radionuclides concentration in the North Western Pacific to the end of June 2012 with measured concentrations of ^{134}Cs and ^{137}Cs in Bq m^{-3} imposed. The relative simulated concentration, χ, is in a logarithmic scale

release of radioactive material from the FNPP but, as well, a subsequent atmospheric deposition on the ocean surface just after the tsunami on March 11. Variations in the initial date do not change significantly the simulation results. The concentration distribution, χ, in the end of June 2012 is shown in Fig. 7.8 in a logarithmic scale with measured concentrations of ^{134}Cs and ^{137}Cs imposed.

As expected, particles in the surface layer were transported mainly along the Kuroshio Extension to the east. The concentration is larger on the north flank of the parent jet because the larger part of the initial radioactive patch was situated to the north from the eastward jet. Transport of radionuclides to the southern flank of the Kuroshio Extension may be explained partly by tracer advection from the southern part of the initial patch. Moreover, we have shown in the preceding section a possibility of CJT across the Kuroshio Extension due to pinching off rings with contaminated water from the southern flank of the jet.

The cruise track, shown on the FTLE map in Fig. 7.9a, b, and c, was chosen to cross the eddies A, B, C, D, E, F, and G marked on that figure. All they are mesoscale ACEs of the Subarctic Front but of different origin and history. The eddy A is a long-lived quasistationary Kuril anticyclone. It is the Bussol' eddy A to be studied in the preceding chapter in detail. The eddy B appeared in the northern subarctic

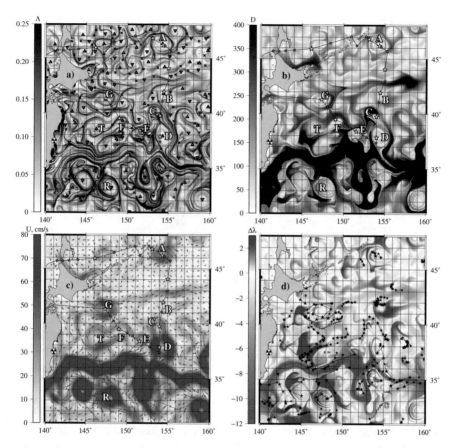

Fig. 7.9 Anticyclonic eddies of the Subarctic Front A, B, C, D, E, F, and G: (**a**) on the FTLE map (Λ is in days^{-1}), (**b**) on the drift map (the absolute displacement of particles D is in km), and (**c**) in the AVISO velocity field (the speed U is in cm/s) with elliptic (*triangles*) and hyperbolic (*crosses*) "instantaneous" stationary points imposed on June 28, 2012. The ship's track and some sampling stations are shown. (**d**) Backward-in-time zonal drift map on June 28, 2012 with "*red*" and "*blue*" waters passing for 15 days large distances in the west–east and the east–west directions, respectively. The values of the zonal displacements of particles, $\Delta\lambda$, are given in geographic degrees. Tracks of the ARGO floats (*stars*) and drifters (*squares*) are shown in the area for the same period of time

frontal area on the southern flank of a zonal eastward jet transporting waters along 42° N–43° N from the eastern coast of Japan (see Fig. 7.9d). The eddy C forms a pair along with the warm-core Kuroshio ring D that was pinched off from a Kuroshio Extension meander in the end of May 2012. Both are clearly seen on the velocity map in Fig. 7.9b as a vortex pair. The warm-core Kuroshio ring E was pinched off from a meander of the Kuroshio Extension jet on June 10–12, 2012 and disappeared in the middle of July. The eddy F has not been identified as a ring pinched off from the Kuroshio Extension. The eddy G, located southeast of Hokkaido, is a typical

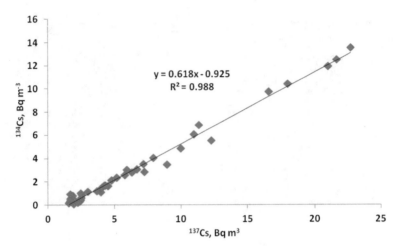

Fig. 7.10 Relation between ^{134}Cs and ^{137}Cs concentrations in Pacific water samples

warm-core ring. It has been found to be closely related with the warm mesoscale eddy located to the southwest off the Tohoku area (the Tohoku eddy T) (Fig. 7.9).

The frontal Kuroshio–Oyashio zone is populated with mesoscale eddies of different sizes and lifetimes. They can be visualized on Lagrangian synoptic maps of the region computed backward in time. We seed on a fixed day 1000×1000 particles distributed homogeneously over the region shown on the maps and integrate them backward in time for 15 days starting from June 28, 2012. The mesoscale eddies are delineated in Fig. 7.9a by black ridges of the FTLE field which approximate unstable manifolds of the hyperbolic trajectories to be present in the area during the integration period, 15 days in our case. The backward-in-time drift map in Fig. 7.9b on June 28, 2012 shows by shadows of grey color the absolute displacements of synthetic particles, D, in km for 15 days before the date indicated. The mesoscale eddies look like patches of different color than surrounding waters. The most prominent eddies with elliptic points in their centers are also visible in the AVISO velocity field (Fig. 7.9c).

In the following numerical experiment tracers, distributed on March 18, 2011 over the box around the FNPP [140° E–144° E, 36.5° N–38.5° N], have been advected forward in time in the AVISO velocity field. In Fig. 7.11 we show the simulated density of particles during the cruise period in the North Western Pacific region, from June 24 to July 6, 2012. The area with the Kuril mesoscale eddy A, centered at 154.33° E, 46.19° N, has not been visited during that period by potentially radioactive tracers. The measured concentrations of ^{134}Cs and ^{137}Cs at station 50 in the center of that eddy did not exceed the background level at 0, 200, 500, and 1000 m depth. As to the other eddies of interest, the maximal radionuclide concentrations have been detected inside the eddies B, centered at 154.4° E, 41.9° N, and G, centered at 147.3° E, 41.3° N, where the simulated density of tracers is really higher in Fig. 7.11. The Kuroshio rings D, centered at 154° E, 37.84° N, and E, centered at 151.5° E, 38.38° N, look like white patches on the map.

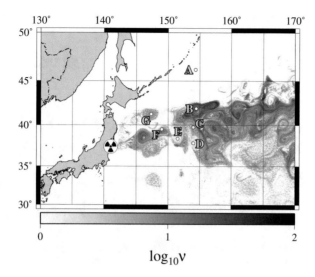

Fig. 7.11 Simulated forward-in-time tracking map for the "radioactive" particles distributed initially inside the box around the FNPP with the coordinates 140° E–144° E, 36.5° N–38.5° N. The map shows where those particles have been from June 24 to July 6, 2012. Letters *A, B, C, D, E, F, and G* mark elliptic points (*circles*) in the centers of the corresponding mesoscale eddies to be studied. The density of tracks, v, is in a logarithmic scale

It means that potentially radioactive tracers were able to visit during the simulation period their periphery but not the core.

Besides of the eddies of interest, the increased density of tracers are clearly seen in Fig. 7.11 at the places of the Tohoku eddy T with the center in the simulation period at 146.5° E, 38° N and a cold-core cyclonic Kuroshio ring R, centered at 147.5° E, 33° N. As to the Tohoku eddy, we recall that it was found [15] to be strongly contaminated by ^{134}Cs and ^{137}Cs just after the accident. The increased simulated density of tracers in the cold cyclonic ring R to the south of the Kuroshio Extension means that some tracers from the initial patch could be transported southward to the parent jet and then be trapped by that eddy. We have found that the ring R was pinched off from a meander of the Kuroshio Extension jet and moved slowly to the west along its southern flank. In fact, it is a demonstration of transport of radionuclides across the strong Kuroshio Extension jet documented and studied in the preceding section.

Concluding this section, the reader is referred to Fig. 7.12 where we show positions of those simulated tracers on July 2, 2012 that have visited just after the accident (from March 11 to April 10, 2011) two selected areas around the FNPP. The density of points in the centers of the eddies B, C, and G, where the highest cesium concentrations have been detected in the cruise, is really comparatively large in this figure.

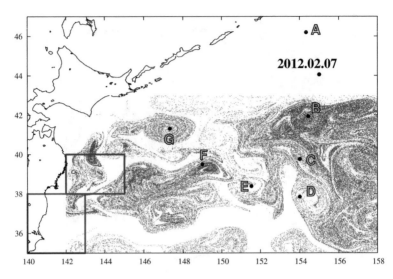

Fig. 7.12 Computed positions (on July 2, 2012) of those tracers that have visited just after the accident (from March 11 to April 10, 2011) the southern [140° E–143° E, 35° N–38° N] (*red*) and northern [142° E–145° E, 38° N–40° N] (*blue*) rectangles around the FNPP. The Kuroshio Extension jet looks like a white meandered band with a few branches. Locations of sampling stations in the centers of the eddies to be studied are shown by *black dots*

7.2.3 Vertical Structure of Eddies and Vertical Distribution of ^{134}Cs and ^{137}Cs

Results of CTD observations in the cruise demonstrate a very good correspondence with the modeled mesoscale eddy locations. Using the modeled maps of the eddy field (Fig. 7.9a–d) transmitted operationally on ship's board during the cruise, we managed to cross the eddies very close to their centers. The only exception was the mesoscale eddy D that was crossed along its western edge, not exactly through its center. Figure 7.13a shows a vertical distribution of potential density anomaly along the cruise track (see Fig. 7.7) from the central Kuril Islands in the north (station 44) and down to the Kuroshio Extension in the south (station 70) and then to the northwest towards Hokkaido (station 84). As ACEs contain water of lower density in their cores, their locations are clearly seen by a downward deflection of isopycnal lines. All of the 7 sampled eddies can be traced down to 1000 m depth and five of them (A, B, D, F, and G) can be seen down to the maximal depth of the CTD observations (2000 m). Considering the magnitude of isopycnal lines deflection, the Kuril mesoscale ACE A seems to be the most intense dynamic feature among the sampled eddies. A history of that eddy has been studied in detail in Sect. 6.2. The isopycnes of 26.8–27.6 kg/m^3 are deepened in its center by 450–550 m as compared with surrounding water. Other energetic eddies are the Kuroshio warm-core rings D and G.

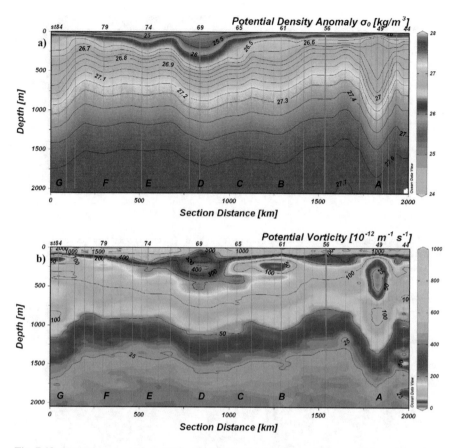

Fig. 7.13 Vertical cross-section of (**a**) potential density anomaly and (**b**) potential vorticity along the cruise track from the central Kuril Islands (*right*) to Hokkaido (*left*). Station numbers are indicated along the top axis, locations of ACE centers are marked as *A–G*

Distribution of potential vorticity (Fig. 7.13b) indicates an existence of low potential vorticity layers in the centers of the eddies A, B, C and G in the depth range of 50–750 m. This corresponds to well-mixed vertically uniform cores of the eddies formed during eddy evolution [14, 21, 24]. Vertical mixing, driven by winter convection, contributes to the formation of eddy cores. Thus, the older eddies have larger and less stratified cores. In opposite, relatively young eddies do not have a uniform layer in their centers. An absence of any noticeable well-mixed layer in the eddies D, E, and F confirms their relatively young age detected by satellite altimetry and our modelling results. The relatively old Kuroshio warm-core ring G demonstrated a more complicated structure having two low potential vorticity layers. The upper one, corresponding to warm and higher salinity core of the eddy, is located between 55 and 205 m (Figs. 7.14 and 7.15). While the secondary core of the eddy, formed by lower temperature and salinity waters subducted into the eddy

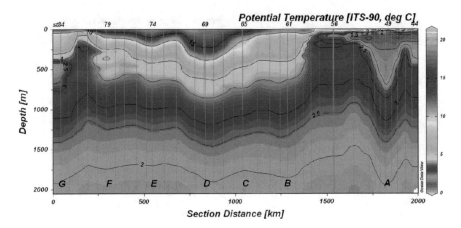

Fig. 7.14 Vertical cross-section of water temperature along the cruise track from the central Kuril Islands (*right*) to Hokkaido (*left*). Station numbers are indicated along the top axis, locations of the ACE centers are marked as *A–G*

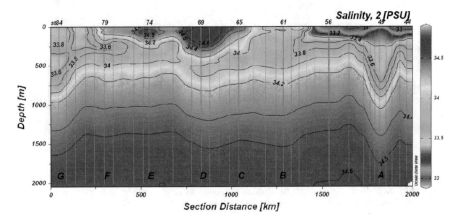

Fig. 7.15 Vertical cross-section of salinity along the cruise track from the central Kuril Islands (*right*) to Hokkaido (*left*). Station numbers are indicated along the top axis, locations of the ACE centers are marked as *A–G*

center, is located at 410–750 m. Low potential vorticity cores of the eddies A, B, C, and G also have a high content of dissolved oxygen (Fig. 7.16) which indicates recent ventilation of waters. Considering these features of the eddy structure, we may expect accumulation of surface water with radionuclides in the centers of ACEs and vertical transport of this water downward in the eddy cores.

Distribution of ^{134}Cs and ^{137}Cs concentrations in surface waters along the cruise track in the Pacific (Fig. 7.17) shows increased cesium content in the areas of stations 56–65 and stations 81–84 corresponding to the northern subarctic frontal area along 155° E and to the southeast off Hokkaido. Maximal concentrations have been observed in the mesoscale eddy B located at the northern Subarctic Front with

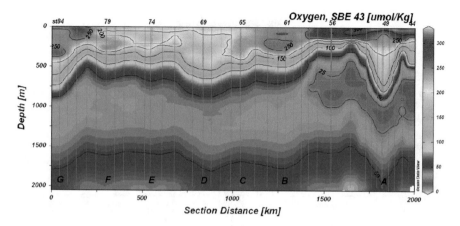

Fig. 7.16 Vertical cross-section of dissolved oxygen along the cruise track from the central Kuril Islands (*right*) to Hokkaido (*left*). Station numbers are indicated along the top axis, locations of the ACE centers are marked as *A–G*

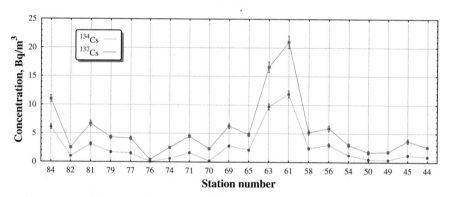

Fig. 7.17 Distribution of cesium isotopes ^{134}Cs (*dotted line*) and ^{137}Cs in surface water along the cruise track in the Pacific. Locations of stations, centers of the mesoscale eddies to be studied, and cruise track are shown in Fig. 7.7

^{134}Cs and ^{137}Cs contents at station 61 up to 11.9 and 21.0 Bq m^{-3}, respectively (Table 7.1). High concentration of both the cesium isotopes were also observed at the southern periphery of the mesoscale eddy B at station 63 (9.7 and 16.6 Bq m^{-3}) and in the area of the mesoscale eddy G at station 84 (6.1 and 11.8 Bq m^{-3}). This suggests a direct transport of water enriched by radionuclides from the Fukushima area by streamers and its trapping in those eddies that corresponds well with our modelling results.

The observed concentrations of cesium were low at the southern part of our transect (stations 70–79) within the surface layers of the eddies D, E, and F, influenced by relatively clean water transported by the Kuroshio Extension. In the area to the north of the Subarctic Front and the central Kuril Islands (the eddy A),

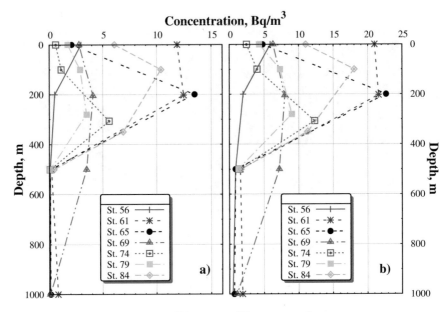

Fig. 7.18 Vertical distribution of (**a**) ^{134}Cs and (**b**) ^{137}Cs in Bq m^{-3} for some selected stations

again we have not found a high content of radiocesium. Concentrations of ^{134}Cs and ^{137}Cs at stations 45–50 in the area of the eddy A were around 0.4–1.2 and 1.7–3.7 Bq m^{-3} which was higher than in the Japan and Okhotsk Seas. However, this is rather a result of atmospheric deposition then direct advection by water flow.

Vertical distribution of radiocesium, sampled at some stations, is shown in Fig. 7.18. Inside most of the eddies, a higher concentration was observed not at the surface but within subsurface layers deepened down to 200–500 m. The maximal content of cesium isotopes in the eddies B, C, and G (stations 61, 65, and 84) was observed in the low vorticity cores of these eddies located around 200, 200, and 100 m, correspondingly (Fig. 7.13b). This proves our preliminary hypothesis that water with Fukushima-derived radionuclides was subducted and trapped in the cores of ACEs. Lower content of cesium at surface layer of the eddies may be explained by a faster ventilation of surface layer by mesoscale streamers and advection of relatively clean water originated from the Kuroshio Extension.

Our measurements show that maximal concentrations of cesium in the subarctic frontal area in June–July 2012 have been registered not at the surface but within subsurface and intermediate water layers in the potential density range of 26.5–26.7 kg/m^3. Deepening of these isopycnal surfaces in the ACEs resulted in deeper locations of radionuclide maxima. Thus, high concentrations of cesium were observed down to 300–400 m in the eddies F, E, and G (stations 75, 79, and 84) and down to 500 m in the mesoscale eddy D (station 69). Even deeper penetration of ^{134}Cs and ^{137}Cs, down to 1000 m, was observed in the mesoscale eddy B (station 61) of 0.8 ± 0.2 and 1.9 ± 0.2 Bq m^{-3}.

The first measurements of vertical distribution of Fukushima-derived cesium, taken in summer of 2011 (see, e.g., [6]), found increased content of cesium isotope at the upper surface layer down to 50–100 m. Then, starting from spring 2012 and later, a higher concentration of cesium was observed subducted down to subsurface and intermediate layers [16, 22]. This coincides well with our results which show that subduction in the frontal zone forms a cesium enriched intermediate water which then spreads southward at the depth of 200–370 m between potential density anomalies surfaces of 25.20–25.33 kg/m^3 as the North Pacific Subtropical Mode Water [22]. A simulation performed in [38] also demonstrates deep penetration of radiocesium rich water into the North Pacific Mode Water in the eastern areas of the North Pacific. We demonstrate that on the background of this large-scale subduction and advection mesoscale ACEs could be considered as an effective mechanism of downward transport and advection of cesium-rich water. Our measurements prove that high radiocesium concentrations are observed exactly inside the uniform core of the eddies located in the density range 26.5–26.7 kg/m^3 which corresponds to the North Pacific Intermediate Water forming in the subarctic frontal area. Thus, we expect that the eddies could trap this water with the maximal content of radiocesium and transport it horizontally and vertically. Considering the northward translation of the eddies, we may suggest that Fukushima-derived cesium should also be transported by the eddies to the north at the intermediate depth.

7.2.4 Tracking Maps for Samples Collected in Centers of the Eddies of the Subarctic Front

Our main goal in simulation is to trace out the origin and pathways of water samples that were collected 15 months and later after the accident to measure the cesium concentration. To do that we seed a square 2 × 2 km around the location of a given sampling station with artificial tracers and advect them backward in time in the AVISO velocity field starting from the date of sampling to the day of the accident. Fixing the places on a regional map, where the corresponding tracers have been found for 1 month after the accident, we can estimate by the tracer density the probability to detect higher concentrations of Fukushima-derived radionuclides in surface seawater samples at sampling stations. The corresponding backward-in-time tracking maps are shown in Figs. 7.19, 7.20, and 7.21. The simulated results are compared with the results of measurements of ^{134}Cs and ^{137}Cs. In this section we present the simulation results for the tracers distributed in the centers of the eddies A, B, C, D, E, F, and G where seawater samples were collected at some cruise stations.

Station 50 (154.33° E, 46.19° N) was located near the elliptic point of the mesoscale eddy A (the Bussol' eddy in Sect. 6.2) with the size $\simeq$ 3° × 1.5° located approximately at the same place from the day of the accident (and even earlier) to the end of the cruise (and even later). The observed concentrations at different depths (see Table 7.1) ranged for ^{137}Cs from 0.6 ± 0.1 to 1.7 ± 0.3 Bq m^{-3} and did not exceed the background level. It is seen in Fig. 7.19a that the tracers of station

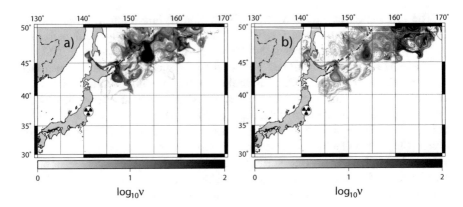

Fig. 7.19 Backward-in-time tracking maps for tracers distributed (**a**) in the center of the mesoscale eddy A at station 50 and (**b**) at station 56, outside any eddy. The maps show where the corresponding tracers have been just after the accident, from March 11 to April 10, 2011. The density of tracks, v, is in a logarithmic scale

50 have not visited the latitudes to the south off $40°$ N where one would expect a significant contamination due to the accident. We may conclude that the probability to detect an increased cesium concentration in the Bussol' eddy A is small. For comparison, we computed in Fig. 7.19b a tracking map for tracers of station 56 ($155°$ E, $44.05°$ N) located outside any eddy. The observed concentration of ^{137}Cs in surface water samples at that station, 5.9 ± 0.5 Bq m^{-3}, exceeded the background level more than in three times. The traces of simulated particles in Fig. 7.19b have been found to be closer to the FNPP location than the tracers in Fig. 7.19a.

Station 61 ($154.4°$ E, $41.9°$ N) was located near the elliptic point of the mesoscale eddy B with the size $\simeq 1.5° \times 1.5°$. The highest cesium concentrations, 21.1 ± 1.1 at surface and 21.6 ± 0.9 Bq m^{-3} at 203 m depth, have been observed in seawater samples at that station. Our observations are agreed with measurements to be carried out approximately at the same place and in the same period [15]. The authors [15] detected the concentration of ^{137}Cs in surface seawater samples, 18 ± 0.7 Bq m^{-3}, at their station B38 located nearby our station 61 and 17 ± 0.7 Bq m^{-3} and 13 ± 0.7 Bq m^{-3} at stations B37 and B39 located inside the eddy B. The tracking map in Fig. 7.20a shows that tracers of the eddy B have visited for the month after the accident the area with presumably high level of contamination. In particular, they have visited frequently the location of the Tohoku eddy T in March and April 2011 with the highest levels of cesium concentration (besides the FNPP discharge channels) to be detected just after the accident [15]. We traced out the history of the mesoscale eddy B and found that it was born on the southern flank of a zonal eastward jet transporting waters from the eastern coast of Japan. That jet is seen on the computed zonal drift Lagrangian map in Fig. 7.9d as a wave-like red band extending approximately along $42°$ N–$43°$ N. Red (blue) colors mean that the corresponding tracers passed for the integration time (15 days) large distances in the west–east and the east–west directions, respectively.

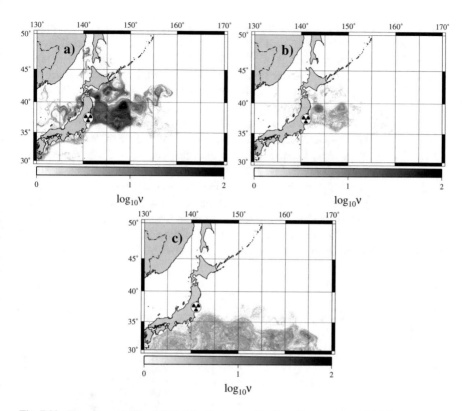

Fig. 7.20 The same as in Fig. 7.19 but for the tracers distributed in the center of (**a**) the mesoscale eddy B at station 61, (**b**) the mesoscale eddy C at station 65, and (**c**) the mesoscale eddy D at station 69

Station 65 (154° E, 39.75° N) was located near the elliptic point of the mesoscale eddy C with the size $\simeq 1° \times 1°$, which was born in January 2012 as a companion of the warm-core Kuroshio Extension ring D. The high concentration of ^{137}Cs, 22.7 ± 1.5 Bq m^{-3}, has been detected at 200 m depth, whereas it was relatively low at the surface and at 500 and 1000 m depths (see Table 7.1). The concentrations of ^{137}Cs in surface seawater samples at Japanese stations A34 and A33, located nearby our station 65, and at station A35 located in the core of the eddy C were found to be relatively high, 9.2 ± 0.5, 11 ± 0.6 and 13 ± 0.7 Bq m^{-3} [15]. The probability that waters in the core of the eddy C could contain a large amount of Fukushima-derived radionuclides is comparatively high because the tracers in Fig. 7.20b have frequently visited the contaminated area and, in particular, the Tohoku eddy T.

The Kuroshio ring D with the size $\simeq 2.5° \times 3°$ was pinched off from a meander of the Kuroshio Extension jet in the end of May 2012. Until the middle of August, it was a free ring sometimes to be connected with the parent jet by an arch. The probability to detect higher concentrations of cesium in its surface water is estimated to be low (see Fig. 7.20c) because it contains mainly Kuroshio waters. We have detected in surface water samples the concentration of ^{137}Cs to be

6.3 ± 0.4 Bq m^{-3}. It is greater than the background level that may be explained by water exchange with its companion, the mesoscale eddy C with a high level of radioactivity. The concentrations of ^{137}Cs in surface seawater samples at Japanese station B30, located closely to our station 69, and at station B29, located nearby our station 70, were found [15] to be slightly greater the background level, 3.6 ± 0.5 and 3.4 ± 0.4 Bq m^{-3}, respectively.

The Kuroshio ring E with the size $\simeq 1.5° \times 1°$ was pinched off from a meander of the jet on June 10–12, 2012 and disappeared in the middle of July. Station 74 (151.5° E, 38.38° N) was located near the elliptic point of that eddy where the increased concentration of ^{137}Cs, 12.3 ± 0.8 Bq m^{-3}, has been detected at 307 m depth. A comparatively small number of tracers over the whole broad area in Fig. 7.21a is explained by the history of core waters in the mesoscale eddy E which have been transported mainly by the Kuroshio from the south and then directed to the east by the Kuroshio Extension. The genesis of the eddy E shows a presence of the Tohoku eddy waters in its core (see the patch in Fig. 7.21a centered at 143° E, 39° N).

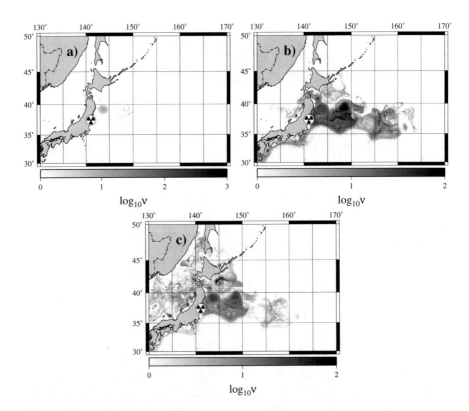

Fig. 7.21 The same as in Fig. 7.19 but for the tracers distributed in the centers of (**a**) the mesoscale eddy E at station 74, (**b**) the mesoscale eddy F at station 79, and (**c**) the mesoscale eddy G at station 84

The mesoscale eddy F with the size $\simeq 1° \times 1°$ has not been identified as a ring pinched off from the Kuroshio Extension. Station 79 (149.5° E, 39.5° N) was located near the elliptic point of that eddy where the increased concentrations of ^{137}Cs, ranged from 7.3 ± 0.7 to 9 ± 0.9 Bq m^{-3}, have been detected from the surface to 280 m depth. The tracking map in Fig. 7.21b demonstrates that water from its core really has visited potentially contaminated area around the FNPP location during the first month after the accident.

Station 84 (147.3° E, 41.3° N) was located near the elliptic point of the mesoscale eddy G with the size $\simeq 2° \times 1.5°$ situated at the traverse of the Tsugaru Strait. The tracking map for that station in Fig. 7.21c reveals its close connection with the Tohoku eddy T, and, therefore, the probability to detect increased cesium concentrations was expected to be comparatively large. In fact, we detected the concentration of ^{137}Cs at 100 m depth to be as large as 18 ± 1.3 Bq m^{-3}.

7.2.5 Concluding Remarks

Results of direct observations of Fukushima-derived ^{134}Cs and ^{137}Cs isotopes in seawater samples, collected at surface and different depths in the western North Pacific in June and July 2012 in the cruise of R/V *Professor Gagarinskiy*, have shown a background or slightly increased level of radiocesium in the Japan and Okhotsk seas and increased concentrations in the area of the Subarctic Front east off Japan. The highest concentrations of ^{134}Cs and ^{137}Cs (13.5 ± 0.9 and 22.7 ± 1.5 Bq m^{-3}) have been found to exceed ten times the background levels before the accident. Maximal content of radiocesium was observed inside the mesoscale ACEs. Among the sampled eddies, the anticyclonic ring of the northern Subarctic Front B and the warm-core ring G (located southeast of Hokkaido) presented the highest concentration of cesium.

The maximal concentrations of radionuclides were observed not at the surface but within subsurface and intermediate water layers (100–500 m) in the potential density range of 26.5–26.7 kg/m^3 with extremely high values inside the low potential vorticity cores of the eddies B, C, and G. This suggests that convergence and subduction of surface water inside eddies were main mechanisms of downward transport of radionuclides. In particular, in the eddies B and D a slightly increased content of radiocesium was observed even at the depth of 1000 m. Concentrations of radiocesium in the eddies, located closer to the Kuroshio Extension (D, E, and F), have been found to be very low at surface layer. It may be explained by a faster ventilation of surface layer by mesoscale streamers and advection of relatively clean water originated from the Kuroshio Extension. A higher content of radiocesium observed at 300–500 m depth was a result of subduction in the subarctic frontal zone and the following advection of intermediate water by the eddies.

The direct observations were compared with simulation of advection of these radioisotopes by the AVISO velocity field. Synthetic tracers, released at the locations of a number of stations inside the eddies, have been advected backward in time till the day of the accident. Fixing their traces for the month after the accident, we computed tracking maps for each of those stations which were used to reconstruct the history and origin of the tracers imitating measured seawater samples. Those maps allowed to explain why measured activities of [134]Cs and ^{137}Cs differed strongly in different samples. It has been shown that tracers with increased radioactivity really have visited the areas with presumably high level of contamination just after the accident. In particular, it has been found a close connection of the samples with increased radioactivity in the eddies B, C, E, F, and G with the Tohoku eddy T known to be strongly contaminated just after the accident [6, 10, 15].

Thus, the mesoscale eddies play the important role in the transport of water with radionuclides released from the FNPP. Considering subduction and accumulation of high-cesium-content water in the anticyclonic eddies, we may suggest that Fukushima-derived cesium should also be transported by the eddies northward at the intermediate depth.

References

1. Andrade-Canto, F., Sheinbaum, J., Zavala Sansón, L.: A Lagrangian approach to the loop current eddy separation.   Nonlinear Process. Geophys. **20**(1), 85–96 (2013). 10.5194/npg-20-85-2013
2. Aoyama, M., Uematsu, M., Tsumune, D., Hamajima, Y.: Surface pathway of radioactive plume of TEPCO Fukushima NPP1 released [134]Cs and [137]Cs. Biogeosciences **10**(5), 3067–3078 (2013). 10.5194/bg-10-3067-2013
3. Bandong, B.B., Volpe, A.M., Esser, B.K., Bianchini, G.M.: Pre-concentration and measurement of low levels of gamma-ray emitting radioisotopes in coastal waters. Appl. Radiat. Isot. **55**(5), 653–665 (2001). 10.1016/s0969-8043(01)00081-1
4. Behrens, E., Schwarzkopf, F.U., Lübbecke, J.F., Büning, C.W.: Model simulations on the long-term dispersal of ^{137}Cs released into the Pacific Ocean off Fukushima. Environ. Res. Lett. **7**(3), 034004 (2012). 10.1088/1748-9326/7/3/034004
5. Budyansky, M.V., Goryachev, V.A., Kaplunenko, D.D., Lobanov, V.B., Prants, S.V., Sergeev, A.F., Shlyk, N.V., Uleysky, M.Y.: Role of mesoscale eddies in transport of Fukushima-derived cesium isotopes in the ocean.   Deep-Sea Res. I Oceanogr. Res. Pap. **96**, 15–27 (2015). 10.1016/j.dsr.2014.09.007
6. Buesseler, K.O., Jayne, S.R., Fisher, N.S., Rypina, I.I., Baumann, H., Baumann, Z., Breier, C.F., Douglass, E.M., George, J., Macdonald, A.M., Miyamoto, H., Nishikawa, J., Pike, S.M., Yoshida, S.: Fukushima-derived radionuclides in the ocean and biota off Japan. Proc. Natl. Acad. Sci. **109**(16), 5984–5988 (2012). 10.1073/pnas.1120794109
7. Bulatov, N.V., Lobanov, V.B.: Influence of the Kuroshio warm-core rings on hydrographic and fishery conditions off Southern Kuril Islands. In: Proceedings of the PORSEC-92 in Okinawa, pp. 1127–1131 (1992)
8. Dietze, H., Kriest, I.: [137]Cs off Fukushima Dai-ichi, Japan — model based estimates of dilution and fate. Ocean Sci. **8**(3), 319–332 (2012). 10.5194/os-8-319-2012
9. Ebuchi, N., Hanawa, K.: Trajectory of mesoscale eddies in the Kuroshio recirculation region. J. Oceanogr. **57**(4), 471–480 (2001). 10.1023/A:1021293822277
10. Honda, M.C., Aono, T., Aoyama, M., Hamajima, Y., Kawakami, H., Kitamura, M., Masumoto, Y., Miyazawa, Y., Takigawa, M., Saino, T.: Dispersion of artificial caesium-134 and -137 in the western North Pacific one month after the Fukushima accident. Geochem. J. **46**(1), e1–e9 (2012)
11. Inoue, M., Kofuji, H., Hamajima, Y., Nagao, S., Yoshida, K., Yamamoto, M.: [134]Cs and ^{137}Cs activities in coastal seawater along Northern Sanriku and Tsugaru Strait, northeastern Japan, after Fukushima Dai-ichi nuclear power plant accident. J. Environ. Radioact. **111**, 116–119 (2012). 10.1016/j.jenvrad.2011.09.012

12. Inoue, M., Kofuji, H., Nagao, S., Yamamoto, M., Hamajima, Y., Yoshida, K., Fujimoto, K., Takada, T., Isoda, Y.: Lateral variation of [134]Cs and ^{137}Cs concentrations in surface seawater in and around the Japan Sea after the Fukushima Dai-ichi nuclear power plant accident. J. Environ. Radioact. **109**, 45–51 (2012). 10.1016/j.jenvrad.2012.01.004
13. Itoh, S., Yasuda, I.: Characteristics of mesoscale eddies in the Kuroshio–Oyashio Extension region detected from the distribution of the sea surface height anomaly. J. Phys. Oceanogr. **40**(5), 1018–1034 (2010). 10.1175/2009JPO4265.1
14. Itoh, S., Yasuda, I.: Water mass structure of warm and cold anticyclonic eddies in the western boundary region of the Subarctic North Pacific. J. Phys. Oceanogr. **40**(12), 2624–2642 (2010). 10.1175/2010jpo4475.1
15. Kaeriyama, H., Ambe, D., Shimizu, Y., Fujimoto, K., Ono, T., Yonezaki, S., Kato, Y., Matsunaga, H., Minami, H., Nakatsuka, S., Watanabe, T.: Direct observation of [134]Cs and ^{137}Cs in surface seawater in the western and central North Pacific after the Fukushima Dai-ichi nuclear power plant accident. Biogeosciences **10**(2), 4287–4295 (2013). 10.5194/bg-10-4287-2013
16. Kaeriyama, H., Shimizu, Y., Ambe, D., Masujima, M., Shigenobu, Y., Fujimoto, K., Ono, T., Nishiuchi, K., Taneda, T., Kurogi, H., Setou, T., Sugisaki, H., Ichikawa, T., Hidaka, K., Hiroe, Y., Kusaka, A., Kodama, T., Kuriyama, M., Morita, H., Nakata, K., Morinaga, K., Morita, T., Watanabe, T.: Southwest intrusion of [134]Cs and ^{137}Cs derived from the Fukushima Dai-ichi nuclear power plant accident in the western North Pacific. Environ. Sci. Technol. **48**(6), 3120–3127 (2014). 10.1021/es403686v
17. Kameník, J., Dulaiova, H., Buesseler, K.O., Pike, S.M., Šťastná, K.: Cesium-134 and 137 activities in the central North Pacific Ocean after the Fukushima Dai-ichi nuclear power plant accident. Biogeosciences **10**(9), 6045–6052 (2013). 10.5194/bg-10-6045-2013
18. Kanda, J.: Continuing ^{137}Cs release to the sea from the Fukushima Dai-ichi nuclear power plant through 2012. Biogeosciences **10**(9), 6107–6113 (2013). 10.5194/bg-10-6107-2013
19. Karasev, E.V.: Monitoring of ecological conditions of the Far East seas. In: Proceedings of the 2nd International Meeting of Amur-Okhotsk consortium, pp. 75–80. Amur-Okhotsk Consortium (2012). http://amurokhotsk.com/wp-content/uploads/2012/04/Proceedings.pdf
20. Kawai, H.: Hydrography of the Kuroshio Extension. In: Stommel, H.M., Yoshida, K. (eds.) Kuroshio: physical aspects of the Japan current, pp. 235–352. University of Washington Press, Seattle (1972)
21. Kitano, K.: Some properties of the warm eddies generated in the confluence zone of the Kuroshio and Oyashio currents. J. Phys. Oceanogr. **5**(2), 245–352 (1975). 10.1175/1520-0485(1975)005<0245:SPOTWE>2.0.CO;2
22. Kumamoto, Y., Aoyama, M., Hamajima, Y., Aono, T., Kouketsu, S., Murata, A., Kawano, T.: Southward spreading of the Fukushima-derived radiocesium across the Kuroshio extension in the North Pacific. Sci. Rep. **4**, 1–9 (2014). 10.1038/srep04276
23. Kuznetsov, L., Toner, M., Kirwan Jr., A.D., Jones, C.K.R.T., Kantha, L.H., Choi, J.: The loop current and adjacent rings delineated by Lagrangian analysis of the near-surface flow. J. Mar. Res. **60**(3), 405–429 (2002). 10.1357/002224002762231151
24. Lobanov, V.B., Rogachev, K.A., Bulatov, N.V., Lomakin, A.F., Tolmachev, K.P.: Long-term evolution of the Kuroshio warm eddy. Dokl. USSR Akad. Sci. **317**(4), 984–988 (1991). [in Russian]
25. Maderich, V., Bezhenar, R., Heling, R., de With, G., Jung, K.T., Myoung, J.G., Cho, Y.K., Qiao, F., Robertson, L.: Regional long-term model of radioactivity dispersion and fate in the Northwestern Pacific and adjacent seas: application to the Fukushima Dai-ichi accident. J. Environ. Radioact. **131**, 4–18 (2014). 10.1016/j.jenvrad.2013.09.009
26. Mendoza, C., Mancho, A.M.: The Lagrangian description of aperiodic flows: a case study of the Kuroshio current. Nonlinear Process. Geophys. **19**(4), 449–472 (2012). 10.5194/npg-19-449-2012
27. Mendoza, C., Mancho, A.M., Rio, M.H.: The turnstile mechanism across the Kuroshio current: analysis of dynamics in altimeter velocity fields. Nonlinear Process. Geophys. **17**(2), 103–111 (2010). 10.5194/npg-17-103-2010

28. Miyazawa, Y., Masumoto, Y., Varlamov, S.M., Miyama, T., Takigawa, M., Honda, M., Saino, T.: Inverse estimation of source parameters of oceanic radioactivity dispersion models associated with the Fukushima accident. Biogeosciences 10(4), 2349–2363 (2013). 10.5194/bg-10-2349-2013
29. Nakano, M., Povinec, P.P.: Long-term simulations of the ^{137}Cs dispersion from the Fukushima accident in the world ocean. J. Environ. Radioact. 111, 109–115 (2012). 10.1016/j.jenvrad.2011.12.001
30. Nakata, K., Sugisaki, H. (eds.): Impacts of the Fukushima Nuclear Accident on Fish and Fishing Grounds. Springer, Tokyo (2015). 10.1007/978-4-431-55537-7
31. Oikawa, S., Takata, H., Watabe, T., Misonoo, J., Kusakabe, M.: Distribution of the Fukushima-derived radionuclides in seawater in the Pacific off the coast of Miyagi, Fukushima, and Ibaraki prefectures, Japan. Biogeosciences 10(7), 5031–5047 (2013). 10.5194/bg-10-5031-2013
32. Povinec, P.P., Hirose, K., Aoyama, M.: Fukushima Accident: Radioactivity Impact on the Environment. Elsevier, Amsterdam (2013). 10.1016/B978-0-12-408132-1.01001-9
33. Prants, S.V., Budyansky, M.V., Ponomarev, V.I., Uleysky, M.Y.: Lagrangian study of transport and mixing in a mesoscale eddy street. Ocean Model. 38(1–2), 114–125 (2011). 10.1016/j.ocemod.2011.02.008
34. Prants, S.V., Uleysky, M.Y., Budyansky, M.V.: Numerical simulation of propagation of radioactive pollution in the ocean from the Fukushima Dai-ichi nuclear power plant. Dokl. Earth Sci. 439(2), 1179–1182 (2011). 10.1134/S1028334X11080277
35. Qiu, B., Chen, S.: Variability of the Kuroshio extension jet, recirculation gyre, and mesoscale eddies on decadal time scales. J. Phys. Oceanogr. 35(11), 2090–2103 (2005). 10.1175/JPO2807.1
36. Remez, V.P.: The application of caesium selective sorbents in the remediation and restoration of radioactive contaminated sites. In: Luykx, F., Frissel, M. (eds.) Radioecology and the Restoration of Radioactive-Contaminated Sites. NATO ASI Series, vol. 13, pp. 217–224. Springer, Dordrecht (1996). 10.1007/978-94-009-0301-2_17
37. Remez, V.P., Sapozhnikov, Y.A.: The rapid determination of caesium radionuclides in water systems using composite sorbents. Appl. Radiat. Isot. 47(9–10), 885–886 (1996). 10.1016/s0969-8043(96)00080-2
38. Rossi, V., Sebille, E.V., Gupta, A.S., Garçon, V., England, M.H.: Multi-decadal projections of surface and interior pathways of the Fukushima Cesium-137 radioactive plume. Deep-Sea Res. I Oceanogr. Res. Pap. 80, 37–46 (2013). 10.1016/j.dsr.2013.05.015
39. Rypina, I.I., Jayne, S.R., Yoshida, S., Macdonald, A.M., Douglass, E., Buesseler, K.: Short-term dispersal of Fukushima-derived radionuclides off Japan: modeling efforts and model-data intercomparison. Biogeosciences 10(7), 4973–4990 (2013). 10.5194/bg-10-4973-2013
40. Sugimoto, S., Hanawa, K.: Relationship between the path of the Kuroshio in the south of Japan and the path of the Kuroshio extension in the east. J. Oceanogr. 68(1), 219–225 (2012). 10.1007/s10872-011-0089-1
41. Tomosada, A.: Generation and decay of Kuroshio warm-core rings. Deep Sea Res. Part A 33(11–12), 1475–1486 (1986). 10.1016/0198-0149(86)90063-4
42. Tsumune, D., Tsubono, T., Aoyama, M., Hirose, K.: Distribution of oceanic ^{137}Cs from the Fukushima Dai-ichi nuclear power plant simulated numerically by a regional ocean model. J. Environ. Radioact. 111, 100–108 (2012). 10.1016/j.jenvrad.2011.10.007
43. Tsumune, D., Tsubono, T., Aoyama, M., Uematsu, M., Misumi, K., Maeda, Y., Yoshida, Y., Hayami, H.: One-year, regional-scale simulation of ^{137}Cs radioactivity in the ocean following the Fukushima Dai-ichi nuclear power plant accident. Biogeosciences 10(8), 5601–5617 (2013). 10.5194/bg-10-5601-2013
44. Waseda, T.: On the Eddy-Kuroshio interaction: Meander formation process. J. Geophys. Res. Oceans 108(C7), 3220 (2003). 10.1029/2002JC001583
45. Yasuda, I., Okuda, K., Hirai, M.: Evolution of a Kuroshio warm-core ring — variability of the hydrographic structure. Deep Sea Res. Part A 39(Supplement 1), S131–S161 (1992). 10.1016/s0198-0149(11)80009-9

List of Internet Resources

[GDP] The Global Drifter Program.
http://www.aoml.noaa.gov/phod/dac

Chapter 8
Lagrangian Fronts and Coherent Structures Favorable for Fishery and Foraging Strategy of Top Marine Predators

8.1 Hydrological and Lagrangian Fronts

Fronts and frontal zones are widespread features at different spatial and temporal scales in the ocean. In a frontal zone, the spatial gradients of main physical and chemical characteristics of water are much larger than their background distribution. A hydrological front is the wake of frontal partitioning with the free surface of the ocean [10, 18], i.e., the geometric locus of points with a maximal horizontal gradient of a hydrological characteristic (see, e.g., Fig. 3.1). One may distinguish between planetary, mesoscale, and local fronts. Regardless of the scale, the most important properties are their complex structure and spatiotemporal variability. It is difficult to overestimate the significance of oceanic fronts, which in addition are areas with increased biological productivity and intensive fishery.

Each elementary volume of water can be attributed to physicochemical properties (temperature, salinity, density, radioactivity, etc.) that characterize this volume as it moves. In addition, each parcel of water can be attributed to more specific characteristics as trajectory's functions that carry key data but they are not physicochemical properties. For example, adjacent waters can be indistinguishable by temperature, and the corresponding SST images indicate no thermal hydrological front. However, they may differ by values of some Lagrangian indicators containing information about the origin and history of water masses. The boundary of partitioning of waters with strongly different Lagrangian indicators was called a "Lagrangian front" (LF) in [23, 24, 26]. It may be a physical property, such as SST, salinity, and density, or concentration of chlorophyll a (Fig. 6.21a). Large lateral gradients of those properties would indicate on common hydrological fronts, thermal, salinity, density, and chlorophyll ones, which are often connected with each other. However, one may consider more specific Lagrangian indicators which are functions of a particle's trajectory (see Sect. 4.1).

S.V. Prants et al., *Lagrangian Oceanography*, Physics of Earth and Space Environments, DOI 10.1007/978-3-319-53022-2_8

Hydrological fronts are manifestations of the current state of a water medium that can be detected by direct measurements. SST and ocean color fronts are often visible on satellite images (see Chap. 3). The LFs of some indicators reflect history and origin of water masses and can be computed and plotted on geographical Lagrangian maps. They can be, in principle, measured by launching a large number of drifters in appropriate places. The LFs may coincide with hydrological fronts but may not. It is possible to compute the LF of such a Lagrangian indicator that would not manifest itself as a hydrological front but would give a useful information on water motion. Even if a specific LF does not coincide with a hydrological front, it does not mean that it would be useless. Such the LF may reflect other properties of convergent water masses. In general, not every hydrological front is a LF. For example, an ephemeral local front may appear due to some reasons without significant convergence of water masses, and one could not identify that as an LF. They are connected with kinematic characteristics of water masses, both scalar and vector ones, such as their coordinates, current time, and velocity vectors.

The relationship between LCSs (see Sect. 4.4) and LFs is not trivial. Any surface LF by definition is a curve with the local maximal gradient of a Lagrangian indicator which varies significantly on both sides of the LF, whereas the FTLE values are almost the same on both sides of any ridge in the 2D FTLE field. Local extrema of that field approximate the corresponding LCSs. By definition, any Lagrangian indicator is a function of trajectory, whereas in order to compute the FTLE it is necessary in addition to know the dynamical system as well. Lagrangian coherent structures contain information about evolution of some medium segments. Displacement and other Lagrangian indicators are attributes of a given fluid particle whereas the Lyapunov exponent is a characteristic of the medium surrounding that particle. For example, it is possible to calculated Lagrangian indicators for the trajectory of a buoy. We would like to stress the important role of LFs because, in difference from rather abstract geometric objects of an associated dynamical system, like stable and unstable invariant manifolds, they are fronts of real physical quantities that are, in principle, measurable.

8.2   Lagrangian Fronts Favorable for Saury Fishing

8.2.1   Identifying Lagrangian Fronts

The importance of hydrological fronts to ecosystems may be explained by the fact that they are often associated with a convergent flow with an intensified flux of nutrients due to vertical flows. If the front is sufficiently long-lived, populations of phyto- and zooplankton will increase attracting other higher level organisms in the trophical chain which are able to detect the front. By the common opinion [2, 3, 20, 21], good fishing areas are often found at the boundaries of warm and cold currents and around warm-core eddies where the energy of the physical system is transferred in some way to biological processes. This strong physical–biological interaction provides favorable conditions for marine organisms. Surface convergent fronts of considerable physical and biological activity occur in zones where different water masses impinge. The reasons leading to aggregation of tuna, saury, and some other pelagic fish at oceanic fronts are unknown in detail, but there are in the literature some speculations about physical mechanisms providing a transfer of energy of the physical system to biological processes [2, 3, 20, 21]. They include transport of nutrients, phytoplankton blooms, and aggregation of other marine organisms at fronts and eddy edges. Thus, oceanic fronts work as aggregating mechanisms for zooplankton, the main food for pelagic fish [2, 3, 20, 21, 34, 37]. Sea surface temperature gradients have long been the main indicators used to find places in the ocean with rich marine resources.

In the North Western Pacific, the cold Oyashio Current flows out of the Arctic along the Kamchatka Peninsula and the Kuril Islands and converges with the warmer Kuroshio Current off the eastern shore of Japan (Fig. 8.1) [29]. This frontal zone is known to be one of the richest fisheries in the world due to the large nutrient content

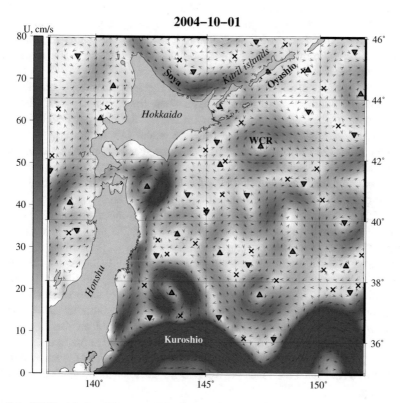

Fig. 8.1 AVISO velocity field in the Kuroshio–Oyashio frontal zone on October 1, 2004, in the season with the Second Oyashio Intrusion. Hyperbolic (*crosses*) and elliptic (*triangles*) stagnation points are shown. WCR is a Hokkaido warm-core ring

in the Oyashio waters (for a review see [31]). The Kuroshio–Oyashio frontal zone has long been recognized by Japanese fishermen to attract squids, fish, and mammals [37]. It was found that fishing grounds depend on the location of the Oyashio fronts and vary from year to year [11, 12, 30, 34, 41]. The fishing grounds were located near shore if there existed the First Oyashio Intrusion along the eastern coast of the Hokkaido Island [41]. In other years, the fishing grounds were located offshore where the Second Oyashio Intrusion was formed due to the presence there of a large-scale warm-core ACE often called the Hokkaido eddy [30, 34]. It has been shown that locations of the fishing grounds depend not only on instantaneous and local oceanographic conditions nearby the fishing grounds but also on the conditions over the whole region [11, 12, 30, 34, 41].

We focus in this section on fishing grounds of Pacific saury (*Cololabis saira*), one of the most commercial pelagic fish in the region (Fig. 8.2). The Pacific saury is a migratory fish moving in schools [15]. They used to pass for 5–6 months distances of the order of 2500 miles from spawning grounds to feeding grounds. They spend much of the time near the surface in nights and in deeper waters in daily time. Pacific saury, in general, migrate seasonally from the south to the north. In winter and spring, spawning grounds are formed in the south, off the eastern coast of Honshu, Japan. In spring and summer, juvenile and young saury migrate northward to the Oyashio area. After feeding in those productive waters, adult saury migrate to the south in the late summer [11, 12, 31, 41]. Commercial fishery begins in August and ends in December.

We restrict our analysis in this section by the region to the east off the Hokkaido Island and the Southern Kuril Islands coasts where the daily saury catch data from the database of the Federal Agency for Fishery of the Russian Federation were collected for six fishery seasons. In order to detect LFs separating waters with different origin and histories, we advect a large number of synthetic tracers in the AVISO velocity field backward in time for 2 weeks and compute zonal, $\Delta\lambda$, meridional, $\Delta\phi$, and absolute, D (4.2), displacements from their positions on the initial day. Two weeks have been found empirically to be an optimal period to detect strong LFs in the region. If the time of integration is too short, some fragments of an LF may not have enough time to be formed. If it is too long, we may miss comparatively short-lived but strong fronts.

In some seasons, there existed the First Oyashio Intrusion along the eastern coast of Hokkaido which is visible on the SST satellite image in Fig. 8.3a averaged for September 23–25, 2002 as a blue "tongue" of cold Oyashio waters along the southern Kuril Islands. The AVISO velocity field and "instantaneous" hyperbolic and elliptic stationary points on September 24 are imposed on that SST image. The data of saury catch for September 23–25, 2002 overlaid allow to conclude that the fishing grounds were concentrated in the waters of the First Oyashio Intrusion (the radius of the black circles in the figure is proportional to the catch in tons per a given ship).

Backward-in-time zonal and meridional drift maps allow to visualize clearly the LFs separating water masses with different origin and histories. Nuances of the colors in Fig. 8.3b, c code the distance passed by the corresponding tracers in

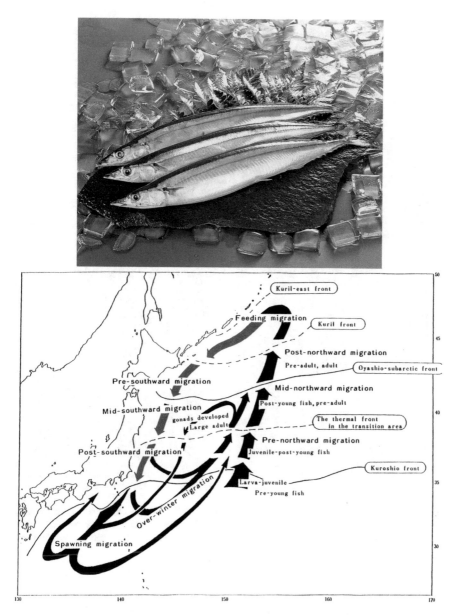

Fig. 8.2 Pacific saury (*Cololabis saira*) and the scheme of the saury seasonal migration in the area (by Kosaka [15])

the zonal or meridional directions in geographic degrees. The "red" ("green") colors in Fig. 8.3b code the tracers drifted zonally from the west to east (from the east to west) for 2 weeks in the past, whereas the "red" ("green") colors in Fig. 8.3c code the

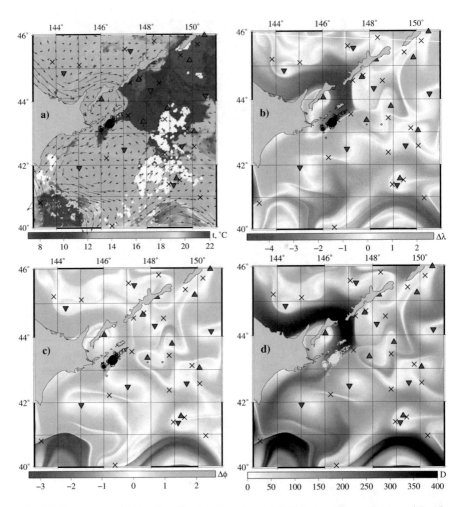

Fig. 8.3 The season with the First Oyashio Intrusion. (**a**) SST image, (**b**) zonal, (**c**) meridional, and (**d**) absolute drift maps on September 24, 2002 with locations of saury catches imposed. Zonal, $\Delta\lambda$, and meridional, $\Delta\phi$, displacements are in geographic degrees, and the absolute displacements, D, is in km

tracers drifted meridionally from the south to north (from the north to south) for the same period of time. It is seen on both the maps that productive cold waters of the Oyashio Current and warmer waters of the southern branch of the Soya Current, flowing through the straights between the southern Kuril Islands, converge at the LF. This LF demarcates the boundary between those "green" and "red" waters. The absolute drift map in Fig. 8.3d confirms this conclusion visualizing the same LF as a sharp boundary between the "dark" and "grey" waters which is strictly defined as a local maximal gradient of D. Here nuances of the grey color code the absolute displacement of the corresponding particles in km.

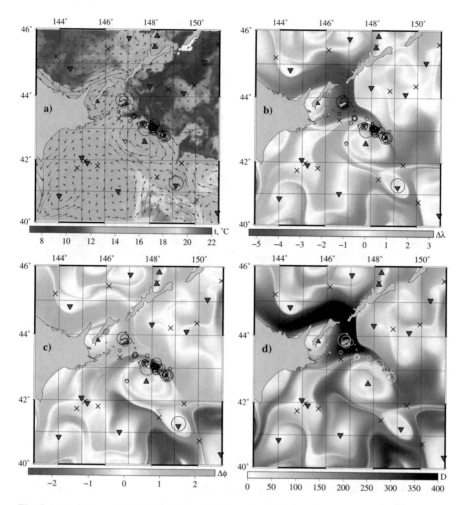

Fig. 8.4 The season with the Second Oyashio Intrusion. (**a**) SST image, (**b**) zonal, (**c**) meridional, and (**d**) absolute drift maps on October 17, 2004 with locations of maximal saury catches imposed

   The oceanographic situation in the region cardinally differed in the years with the Second Oyashio Intrusion when a large warm-core Kuroshio anticyclonic ring (WCR in Fig. 8.1) approached the Hokkaido eastern coast and forced the Oyashio Current to shift to the east rounding the eddy. The SST image in Fig. 8.4a averaged for October 16–18, 2004 illustrates that the saury catch locations were concentrated mainly at the convergence front between the Oyashio and Soya waters and the periphery of the Kuroshio ring with the center around 147°30′ E, 42°30′ N. The zonal, meridional, and absolute drift maps on the same date in Fig. 8.4b–d with the circles of saury catch locations overlaid show clearly that the fishing grounds

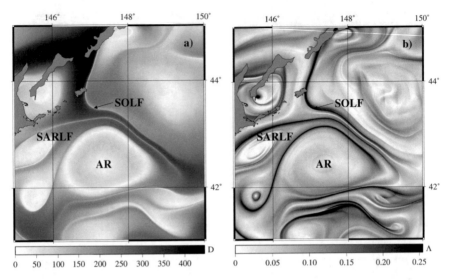

Fig. 8.5 Backward-in-time Lagrangian maps for (**a**) particle's displacement D and (**b**) FTLE in the season with the Second Oyashio Intrusion (October 1, 2004). AR is for the anticyclonic ring, SARLF—the LF between Soya and AR waters, SOLF—the LF between Soya and Oyashio waters. *Grayscale* gradations encode the values of D in km and FTLE in day^{-1}, respectively

with maximal catches were located along the LFs where waters of the Oyashio Current and of the southern branch of the Soya Current converged with waters of the Kuroshio ring.

Let us look more carefully at the LFs around the warm-core Kuroshio ring in autumn of 2004. This is a place where waters with different properties converge: cold and nutrient rich waters of the Oyashio Current from the north, warm and salty waters of the Kuroshio from the south, and the Soya Current waters flowing from the south along the west coast of Hokkaido (part of these waters enters the region through the straits between the islands). Figure 8.5a shows the drift map on October 1, 2004, for the displacement D of particles with given initial conditions for 15 days in the past. The lines of local maxima of gradients of D visualize the main LFs in the region including the SOLF, which separates the waters flowing through the straits between the islands: the waters of a branch of the Soya (S) current and the Oyashio (O) waters and the SARLF, which separates the waters of the Soya Current from subtropical waters of the anticyclonic Kuroshio ring (AR).

Other LFs are associated with eddies and intrusions in the region. Each of these LFs can be identified also by a narrow white strip adjacent to the line of the maximal gradient of D. The white color means that the corresponding particles experienced very small displacements for the integration period, 15 days in our case. To find the cause of this effect, we calculated the FTLEs backward in time. Figure 8.5b shows clearly the Kuroshio anticyclonic ring (AR), which is surrounded by the black ridges with local FTLE maxima which are located along white curves of local minima of

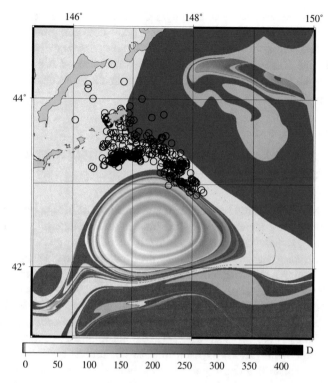

Fig. 8.6 Backward-in-time drift map in the season with the Second Oyashio Intrusion shows where the waters of the converging LFs came from for half a year in the past: *yellow*—from the west, *green*—from the east, *blue*—from the north, *red*—from the south, *grey*—the waters that have been in the box during all that period in the past. Locations of maximal saury catches for 1 week before and 1 week after October 1, 2004 are imposed

D in Fig. 8.5a, because the motion of advecting particles near unstable manifolds of hyperbolic trajectories slows down due to the effect of dynamical saddle traps (see Sect. 2.2.3).

To visualize where the waters of the converging LFs came from for a given period in time, it is instructive to compute specific backward-in-time drift maps. A large number of artificial tracers are uniformly distributed in the box shown in Fig. 8.6 and are integrated backward in time in the season with the Second Oyashio Intrusion from October 10, 2004 to April 10, 2003 in the AVISO velocity field. Coding by different colors the particles which entered the area through its geographical borders for that period of time, we plot in Fig. 8.6 the drift Lagrangian map that shows where water masses of the converging LFs came from for half a year in the past. The yellow color means that the corresponding particles entered the box through its western boundary, green, blue, and red ones—through its eastern, northern, and southern boundaries, respectively. The "grey" particles are those ones which were present in the box for the whole integration period. In this figure the yellow color marks waters of the southern branch of the Soya Current and waters from the southern Okhotsk Sea, green one—Pacific waters, blue one—cold and productive subarctic Oyashio waters, red one—warm and less productive subtropical waters. The grey color marks the Kuroshio ring waters and streamers around it and some Pacific waters to the northeast which experienced rather small displacement during half a year in the past.

8.2.2   Accumulation of Saury Catches at Strong Lagrangian Fronts

We illustrated a relationship between LFs and locations of fishing boats with saury catches in two typical oceanographic situations. In order to determine quantitatively whether saury catches were indeed associated with LFs or not, we compute here for available fishery seasons the frequency distribution of the distances, r, between locations of fishing boats with saury catches and strong LFs. The maximal gradient of the absolute displacement, $\nabla D = \sqrt{(\partial D/\partial x)^2 + (\partial D/\partial y)^2}$, is supposed to be an identificator of the presence of an LF. Such gradients delineate boundaries between waters that passed distances that may differ in two orders of magnitude (see Figs. 8.3d and 8.4d). It is shown in Fig. 8.6 that contrast boundaries are strong LFs separating waters of the Oyashio and Soya currents and waters of warm-core Kuroshio rings. In order to get rid of ephemeral LFs, we choose a threshold, $\nabla D_{th} = 60$, and only the LFs with $\nabla D_{th} \geq 60$ are supposed to be strong. We have empirically found that such gradient values correspond to permanent LFs in the Kuroshio–Oyashio frontal area. Then we compute on each day the distance r between the location of each boat with saury catch and the nearest geographical point where $\nabla D_{th} \geq 60$. The corresponding probability distribution function (PDF) for each season is compared with the random PDF which is computed by the same way but with 10,000 points randomly distributed over the same region.

The statistical results are shown in Fig. 8.7 for all available fishery seasons. The number of events, i.e., the number of locations of the boats with saury catches, varied from season to season and in average was about 1000 per season. As expected, the random PDFs (thin curves) are rather smooth curves with long tails. The real PDFs (bold curves) have a tooth-like structure that can be explained partly by congregation of boats near strong LFs with large D gradients and partly by a fishery strategy. The vertical solid and dashed lines represent the medians for real and random PDFs, respectively. The median is a more robust statistical indicator than the mean value and can be used as a measure of location if a distribution is skewed having, for example, a heavy tail. In all the seasons, the medians were closer to strong LFs for the real PDFs than for the corresponding random ones. Moreover, the random PDFs have more longer tails than the corresponding real ones proving that fishing boats really tend to be closer to strong LFs then be randomly distributed over the region.

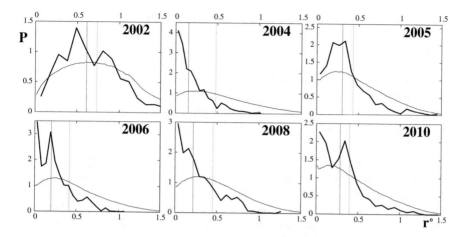

Fig. 8.7 PDFs for six fishery seasons with real locations of the fishing boats with saury catches (*solid curves*) and the points randomly distributed over the same region (*dashed curves*)

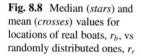

Fig. 8.8 Median (*stars*) and mean (*crosses*) values for locations of real boats, r_b, vs randomly distributed ones, r_r

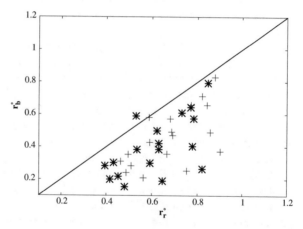

Plot in Fig. 8.8 illustrates relations between the medians (stars) for real and random PDFs and between the mean values (crosses) for real and random PDFs. In order to take into account the effect of choice of the threshold value for the displacement gradients, ∇D_{th}, we computed the median and mean values for all the seasons at $\nabla D_{th} = 60$, 100, and 130. The points below the slope line in Fig. 8.8 mean that the corresponding median and mean values are closer to strong LFs for real fishing boats than for randomly distributed ones.

The oceanographic situation in 2002 was not typical being the single fishery season, among the studied ones, with the First Oyashio Intrusion. Moreover, the oceanographic situation in the region has changed significantly during that season. Animation of daily Lagrangian maps in 2002 with the saury catch locations overlaid [ANIM] illustrates clearly that in addition to an intrusion of Soya Current waters along the northeastern coast of Hokkaido Island there appeared in October–November an intrusion of cold Oyashio waters from the north into the fishery region which cardinally changed the oceanographic situation. The saury catch locations moved to the east and south in that time.

The most stable oceanographic situation among the years with available fishery data was in the year 2004 with the Second Oyashio Intrusion, when a quasistationary warm Kuroshio anticyclonic ring was located in the region for the whole fishery period. Strong LFs have been found to be stable in that year. The real PDF in Fig. 8.7 in 2004 exceeds significantly the random one and decays rapidly. The tail of the random PDF extends significantly over the distance r as compared with the real one. All these facts demonstrate that saury catches really were located mainly along the strong LFs in that year. In the other years with available fishery data, the oceanographic situations resemble the 2004 case with the Second Oyashio Intrusion. Probability distribution functions for those fishery seasons in Fig. 8.7 provide a statistical evidence that saury fishing locations were not randomly distributed over the region but were concentrated near the strong LFs around the Kuroshio ring (see Fig. 8.4). Based on statistical results, we may conclude that the more stable the oceanographic situation in the fishery region is, the closer fishing boats to LFs tend to be.

Oceanic fronts are areas with strong horizontal and vertical mixing. Highly turbid waters, however, are unsuitable for saury because it is a visual predator hunting in comparatively clear waters outside the exact locations of fronts [11, 12, 31, 41]. Saury avoid highly turbid waters and waters with large phytoplankton concentration, more than $5 \, g/m^3$, which are turbid due to organic matter [11]. On the other hand, extremely oligotrophic waters contain little food. Food abundance and water clarity are known to be two factors affecting the rate of food encounter [12, 30, 34, 41].

As to physical and biological reasons that may cause saury aggregation near strong LFs, we suggest the following ones. Strong LFs in the Kuroshio–Oyashio frontal area demarcate convergence of water masses with different productivity. They are zones with increased lateral and vertical mixing and often with increased primary and secondary production. Intrusion of nutrient rich Oyashio waters into more oligotrophic Kuroshio ones creates LFs there and provides higher phyto- and zooplankton concentrations along them with a net effect of aggregation of saury to forage on the lower trophic level organisms. Strong stretching of material lines in the vicinity of hyperbolic objects in the ocean, a hallmark of chaotic advection (see Chap. 1), is one of the possible effective mechanisms providing such intrusions. Those filament-like intrusions may expand over hundreds of kilometers and are easily captured by the Lagrangian diagnostics but may be not visible on SST or chlorophyll a images.

We illustrate that kind of transport of nutrients in Fig. 8.9a where evolution of patches with synthetic particles seeded on September 15, 2004 near five hyperbolic trajectories in the region is shown. It is a fishery season with the Second Oyashio Intrusion and a prominent quasistationary Kuroshio warm anticyclonic ring in the region (Figs. 8.1 and 8.4). Let us suppose that some of the patches are rich in food and trace their evolution. Computing the FTLE backward in time for 2 weeks,

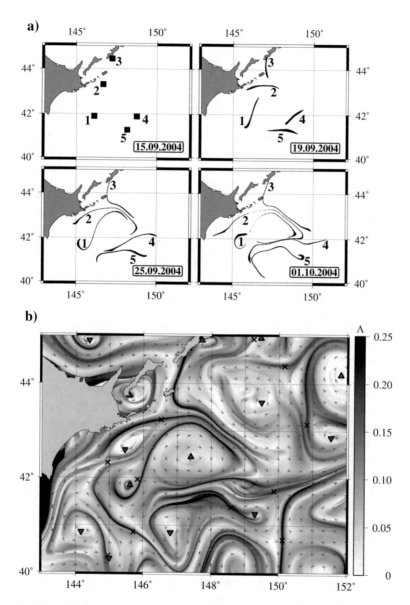

Fig. 8.9 (**a**) Evolution of synthetic food patches selected near five hyperbolic trajectories in the 2004 fishery season with the Second Oyashio Intrusion. (**b**) Backward-in-time FTLE map on October 1, 2004 with the values of Λ in units of day^{-1}. For 2 weeks the patches delineate rapidly the corresponding unstable manifolds around the Kuroshio warm-core ring approximated by the *black ridges* on the FTLE map

we have checked that all the patches in Fig. 8.9a in the course of time delineated the corresponding ridges on the FTLE map in Fig. 8.9b which approximated the unstable manifolds of selected five hyperbolic trajectories. The passive marine organisms in those fluid patches have been advected along with them and might attract saury for feeding.

Comparing D map in Fig. 8.4d with Fig. 8.9a, it is clear that the stretched patches in Fig. 8.9a delineate the corresponding LFs in Fig. 8.4d with maximal gradients of the absolute displacement looking in that figure as white stripes. The perimeter of some patches increased more than in 100 times for only 2 weeks increasing significantly, in turn, the chance for saury to find food. Such a mechanism of export of nutrient rich waters into more poor ones is supposed to be typical because of a large number of hyperbolic trajectories in the frontal oceanic zones with increased mixing activity and rich in eddies (see Fig. 8.1).

Local minima of the displacement field, the white stripes with very low values in Figs. 8.3b–d and 8.4b–d, coincide with the corresponding local maxima of the FTLE field. It is clear when comparing Figs. 8.4d and 8.9b. It is known from theory of dynamical systems that moving particles slow down when approaching a hyperbolic saddle point along an incoming separatrice or a hyperbolic trajectory along its stable manifold [27, 38] (see Sect. 2.2.3). Therefore, one gets low D values while computing displacements of those particles. That is why in vicinity and along of any unstable manifold there should be trajectories of particles approaching the corresponding hyperbolic trajectory at exponentially slow rate whose initial positions are coded by white color on the drift maps.

Generally speaking, LFs are 3D features. Altimetry-based Lagrangian maps allow to visualize their manifestations at the ocean surface. They may expand over hundreds of kilometers, but fish is expected to aggregate only near comparatively small LF segments. Saury, as a predator of zooplankton, prefer the places with an aggregation of forage zooplankton. Favorable fishing grounds may differ from the adjacent waters by a complex of conditions including hydrological situation in the upper ocean layer. It has been empirically found in [11] that there should be a strong seasonal thermocline (with the vertical gradient more than 0.19 °C/m), coinciding in that region with a seasonal pycnocline. Moreover, the width of the upper mixed layer should be more than 6 m with the forage zooplankton concentration more than 0.4 g/m^3 but the phytoplankton concentration less than 5.6 g/m^3. Such conditions are expected to be formed at LFs where water masses of different densities meet. The heavy cold water tends to flow under the light warm water resulting in formation of a strong pycnocline which in turn forces passive phytoplankton to concentrate all time near the surface but not to be distributed over the depth. Zooplankton is able to move vertically concentrating near the surface in the nighttime. The absolute values of SST in the locations with saury aggregation may vary from 5 to 20 °C [11]. The important factors of favorable fishery conditions are thermostructural peculiarities and the abundance in forage zooplankton but not SST, salinity, and other hydrographic factors.

In the end of this section we would like to discuss the connection between potential fishing grounds, LFs, and thermal fronts visible on SST images.

The correspondence between the SST images and the drift maps, visualizing LFs, is illustrated by Figs. 8.3 and 8.4. Intrusion of cold Oyashio waters in September 2002 (the blue "tongue" in Fig. 8.3a with black circles) is manifested in Fig. 8.3b–d as the corresponding "tongues" of waters contrasting with adjacent waters of different origin. As to the season with the Second Oyashio Intrusion, it is seen on all the drift maps in Fig. 8.4b–d that "tongues" of the cold Oyashio water, visible in Fig. 8.4a, encircle the anticyclonic warm-core Kuroshio ring contrasting with its core water.

SST fronts have long been the main indicators used to find places in the ocean with rich marine resources. So, in order to find potential fishing grounds, it is instructive to use as the first guess the strong thermal gradients visible on satellite SST images if they are available. Unfortunately, SST images are not available in cloudy and rainy days which often occur in the fishery period in the region studied. In some fishery seasons, up to half of days have been found to be cloudy or rainy in the Kuroshio–Oyashio frontal area. Typically, the saury fishing grounds with maximal catches have been found not exactly at the SST fronts but in a comparatively large area around that front where a few LFs may be detected. Computation of LFs is a simple way to visualize a fine structure of the frontal zone which is a problem when using SST and/or chlorophyll a images. Any strong large-scale LFs can be accurately detected in a given altimetric velocity field or in a numerically generated velocity field by computing drift maps for the absolute displacement D. Moreover, by computing meridional and zonal drift maps, one gets an information on the origin and pathways of convergent waters that may be useful to determine by which waters this or that LF has been formed.

8.3 Lagrangian Fronts Favorable for Fishery of Neon Flying Squid

Neon flying squid (*Ommastrephes bartramii*), shown in Fig. 8.10a, is a large oceanic squid distributed in temperate and subtropical waters of the Pacific, Indian, and Atlantic oceans. In the North Western Pacific, this species plays an important role in the pelagic ecosystem and is an international fishery resource with high commercial value. We use in this section commercial fisheries data provided by the TINRO Centre (Vladivostok, Russia) for the South Kuril region [40° N–46° N, 145° E–155° E] which is the main target stock for the squid-jigging vessels of Russia. The fishery usually starts there in August and ends in December. The fishing methods are jigging with using of attracting lamps. The schools of neon flying squid are rather small in size and density as compared to other pelagic species, and it is difficult to detect them by sound radars.

The North Pacific population comprises an autumn cohort and a winter–spring cohort (see, e.g., [1, 19] and the references therein). The winter–spring cohort makes a clockwise, annual round-trip migration in the region shown in Fig. 8.10b between its spawning grounds (subtropical region) and its northern feeding grounds. It feeds

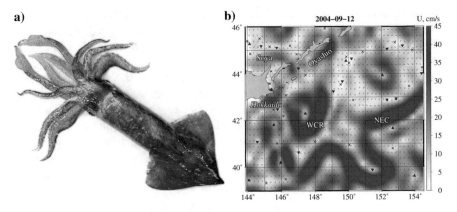

Fig. 8.10 (a) Neon flying squid (*Ommastrephes bartramii*). (b) AVISO velocity field in the Kuroshio–Oyashio frontal zone on September 12, 2004 with imposed elliptic (*triangles*) and hyperbolic (*crosses*) stationary points. NEC is for the Northeastern Current that is a branch of the Kuroshio Current, and WCR is a warm-core Kuroshio ring. The southward Oyashio Current along the Kuril Islands and the Soya Current between the Kuril Islands and the Hokkaido Island are shown

in the spawning grounds until 2–3 months old and then takes a northward migration. Most of them, besides those that have been caught off the coast of northeast Japan and the Kuril Islands, return to the spawning grounds in the following fall and winter. Their lifespans are thought to be 1 year.

A number of authors have studied influence of oceanographic environment on fishing grounds of neon flying squid in the area to the east off the Kuril Islands and the Hokkaido Island. Most of them have studied SST and chlorophyll *a* distributions in the area as easy available environmental parameters which have been used by fishermen for years. It was concluded in [39] that around 40% of catch in the area 140° E–150° E has been recorded nearby SST gradients and around 50%—nearby chlorophyll *a* gradients. The results show that the favorable temperature for neon flying squid living lies in the wide range from 10 to 22 °C and the chlorophyll *a* concentration is 0.15–3 mg/m^3 [9]. Salinity has not been found to affect catches. Connection between SSHs and favorable fishing grounds has also been studied. The authors [32] reported that areas with small negative SSH anomalies up to −10 cm have been found to be optimal for good fishery. However, the authors [9] reported that the maximal catches have been found at positive SSH anomalies up to 30 cm. Some authors [4] argue that the probability of forming fishing ground becomes higher at the locations with the vertical temperature gradient in the upper mixed layer to be in the range 0.1–0.25 °C/m.

In this section, the relationship between squid catch locations and the LFs in the South Kuril area is studied. We compute daily altimetry-based Lagrangian maps in the fishery seasons of 1998, 1999, 2001–2005, impose on them available fishery data, and try to find those Lagrangian structures at which the majority of squid catch locations have been accumulated.

We calculate absolute displacements (4.2) of tracers distributed in the area and code by different colors the tracers which entered the area through all its geographical borders. 250,000 tracers are uniformly distributed on each day for each fishery season in the box shown in Fig. 8.10b. Their trajectories are integrated backward in time for 150 days in the AVISO velocity field. Locations of squid catches for a few days before and after the date of the D map are imposed. Nuances of the grey color code the tracers with corresponding values of D in km, whereas yellow color means that the corresponding particles entered the box for 150 days in the past through its western boundary, green, blue, and red ones—through its eastern, northern, and southern boundaries, respectively.

Typical oceanographic situations in the South Kuril Islands area (see Fig. 8.10b) have been described in the preceding section. The cold Oyashio Current transports productive subarctic Pacific waters from the north. The branch of the Soya Current transports modified subtropic waters of the Japan Sea which enter to the Pacific through the Kuril Straits from the west. The Northeastern Current (NEC Fig. 8.10b) is a branch of the Kuroshio Current. The part of the NEC from about 39° N, 148° E–43° N, 155° E is called sometimes by the Isoguchi jet [13] which is a topographically constrained jet intensified in fall.

The South Kuril region is known for its eddy activity. Large mesoscale eddies appear to the east off the Hokkaido almost each year. Some of them are large-scale warm-core Kuroshio rings which split off from large meanders of the Kuroshio Extension (see Sects. 7.1 and 8.2). Mesoscale cyclones and anticyclones appear regularly in the southern part of the region due to meandering of the NEC. Moving to the north, they may reach the Bussol' Strait to form large-scale Bussol' eddies (see Sect. 6.2). Moreover, smaller scale and comparatively short lived single eddies, dipoles, tripoles, and even quadrupoles appear in the central part of the region. Thus, it is a challenge to find the oceanographic features in this turbulent media which are favorable for squid catching.

Anticipating our Lagrangian analysis for seven fishery seasons, the results can be resumed as follows. Locations of squid catch have been concentrated mainly at or nearby the specific LFs which are boundaries of large-scale Lagrangian intrusions. The boundaries of such intrusions are LFs by definition [25, 26] because they separate waters with different values of Lagrangian indicators (see Sect. 8.1).

They are tongues of water with some values of Lagrangian indicators which penetrate into the area with another values of those indicators. By large scale, we mean here horizontal features with a linear size of about 100–300 km in one direction. Their size in the other direction may vary from 10 to 100–300 km. Moreover, squid schools prefer aggregate at those places where waters with different values of some Lagrangian indicators do not just converge but often mix producing filaments, swirls and tendrils typical for chaotic advection.

8.3.1 Fishery at Lagrangian Intrusions of the Subarctic Front

In this section we illustrate the episodes when squid catch locations were distributed at the Subarctic Front. Only 1998, 1999, and 2002 fishery seasons are discussed. In 1998 good fishery occurred at the Subarctic Front from September 27 to November 17. Maximal catches were recorded at the boundaries of the Lagrangian intrusions of subtropical water into subarctic one to the north off the front. To the end of September, a large-scale intrusion of "yellow" water was formed with the coordinates of its "head" at about 41° N, 148.5° E. One cluster of catch locations was mainly at the very "head" (Fig. 8.11a). Another cluster was along the LF separating "yellow" and "grey" waters in the area around 42° N, 147° E. This intrusion deformed greatly to October 10 with its "head" moved to the west and catch locations followed it (Fig. 8.11b). Moreover, a few catch locations accumulated at the "head" of the Lagrangian intrusion of the "red" subtropical water into the "grey" subarctic one in the area around [41° N–42° N, 149° E–150° E]. As to chlorophyll *a* distributions, satellite MODIS data were not available for most days in this fishery season because of cloudiness. When it was possible to get ocean color images, we have not found significant correlations of chlorophyll *a* concentration with squid catch locations.

From August to the end of September 1999, a fishery with maximal catches was at the boundary of the intrusion of "yellow" subtropical water into the "blue" subarctic one and at the northeastern periphery of the Hokkaido anticyclone with the center at around 43° N, 148° E (Fig. 8.12a). From the end of September to the end of October, catch locations were at the boundary of the Lagrangian intrusion of the "red" subtropical water into the "blue" subarctic one (Fig. 8.12b).

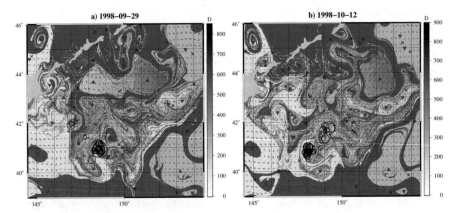

Fig. 8.11 The drift maps show squid catch locations in September–October 1998 concentrated near the Subarctic Front, mainly, at the "heads" of the Lagrangian intrusions of *"yellow"* and *"red"* modified subtropical waters entered the area through its western and southern boundaries, respectively. Catch locations for 6 days before and 6 days after the date indicated on the maps are shown by the *circles* with the radius proportional to the catch value

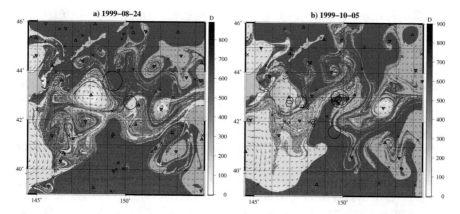

Fig. 8.12 The drift maps show squid catch locations in August and October 1999 concentrated at the LF separating the *"yellow"* (*"red"*) subtropical and *"blue"* subarctic waters and at the northeastern periphery of the Hokkaido anticyclone with the center at around 43° N, 148° E. Catch locations for 8 days before and 8 days after the date indicated on the maps are shown by *circles*

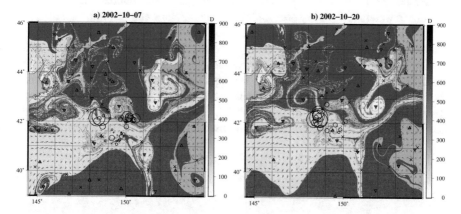

Fig. 8.13 The drift maps show squid catch locations in October 2002 concentrated at the LF separating the *"yellow"* subtropical and *"blue"* subarctic waters. Catch locations for 8 days before and 8 days after the date indicated on the maps are shown by *circles*

In 2002, a good fishery was at the Subarctic Front from the end of September to the end of October without any catches recorded after that. The drift maps in Fig. 8.13 show squid catch locations in October 2002 concentrated at the LF separating the "yellow" subtropical and "blue" subarctic waters. Due to lack of space, we do not show drift maps for a number of other episodes with maximal catches that occurred during fishery seasons of 2001–2004 at prominent Lagrangian intrusions of the Subarctic Front.

8.3.2   Fishery Inside and Around Hokkaido Mesoscale Eddies

In fall 2004, the oceanographic situation in the region was called in Sect. 8.2 as a regime with the Second Oyashio Intrusion when a large warm-core Kuroshio anticyclonic ring (WCR in Fig. 8.10b) approached the Hokkaido eastern coast and forced the Oyashio Current to shift to the east rounding the eddy. Cold and nutrient rich "blue" waters of the Oyashio from the north, warm and salty "grey" Kuroshio waters from the south, and "yellow" Soya waters converged there forming strong LFs. As we know from the preceding section, these mesoscale features have been found to be favorable for saury fishing in fall 2004.

In the middle of September, the Kuroshio ring catched modified subtropical "yellow" Soya waters and subarctic "blue" Oyashio waters and wound them around itself starting from the northern periphery (Fig. 8.14). The catch locations were concentrated in the beginning of October at the eastern side of the ring, at the boundary separating Kuroshio, Soya, and Oyashio waters and partly inside the ring (Fig. 8.14a). In the second part of October, they were recorded inside the ring and at its western periphery (Fig. 8.14b). The eddy periphery is a boundary between "old" subtropic waters in the eddy core and "fresh" subarctic waters around the eddy.

In fall 2005, a mesoscale anticyclone was located to the east off Hokkaido. An intensification of the Oyashio and Soya flows was recorded from September 30 to October 14 in the studied area. Rich in nutrients "blue" Oyashio waters wound onto the eddy creating a strong LF. The squid catch locations were concentrated in this fishery season mainly around and inside that eddy (Fig. 8.15). A mesoscale anticyclone with the center at around 43° N, 148° E also was observed in fall 1999 to the east off Hokkaido, and catch locations were recorded around and inside it.

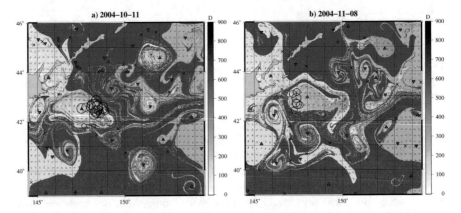

Fig. 8.14 The drift maps show squid catch locations in October–November 2004 around and inside the Kuroshio ring with the center at around 43° N, 148° E. Catch locations for 8 days before and 8 days after the date indicated on the maps are shown by the *circles*

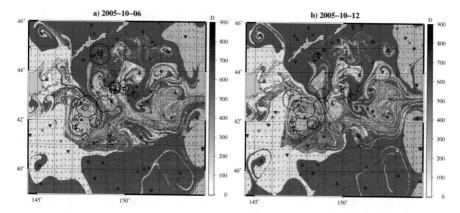

Fig. 8.15 The drift maps show squid catch locations in October 2005 around and inside the anticyclonic Hokkaido eddy with the center at around 42° N, 148° E. Catch locations for 8 days before and 8 days after the date indicated on the maps are shown by the *circles*

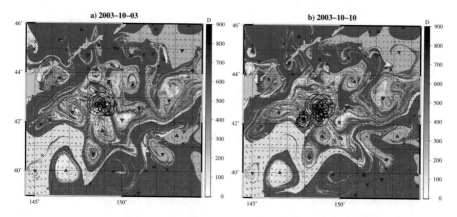

Fig. 8.16 The drift maps show squid catch locations in October 2003 concentrated in the central part of the region [41° N–42.5° N, 148° E–149.5° E], mainly, at the LF separating *"yellow"* subtropical water and *"blue"* subarctic water

8.3.3 Fishery at Lagrangian Intrusions in the Central Part of the Studied Area

Mesoscale and submesoscale eddies of the Subarctic Front are able to catch and retain subtropic water and transport it to the north off the front. The streamers of subtropic water around those eddies are specific narrow Lagrangian intrusions which may penetrate far to the north in the central part of the region (Fig. 8.16). Sometimes the streamers of subtropic water split off and create patches of the water with properties that differ strongly from the surroundings. Boundaries of such features may be favorable for squid catching. A situation with such patches of "yellow" subtropical water in the fishery season 2003 is shown in Fig. 8.16. In the beginning of October, catch locations were concentrated mainly in the region about [41° N–42.5° N, 148° E–149.5° E], at the LF between subtropical and subarctic waters.

We have computed in this section altimetry-based daily drift maps in the South Kuril area for a few fishery seasons with available data on neon flying squid catching. The majority of catch locations were concentrated in vicinity of specific LFs, namely near large-scale Lagrangian intrusions, i.e., the "tongues" of water wedging into surrounding water with other Lagrangian properties. It has been shown that the catch locations were accumulated in those places where waters with different Lagrangian indicators not just converge but mix producing filaments, tendrils, and swirls typical for chaotic advection. They have been mainly found in a vicinity of (1) Lagrangian intrusions at the Subarctic Front, (2) filament-like intrusions of Soya and Oyashio waters encircling mesoscale anticyclones to the east off the Hokkaido Island with subsequent penetration inside the eddies, and (3) intrusions of subtropical waters into the central part of the area due to eddy–eddy and eddy–jet interactions.

We have compared Lagrangian data with Terra and Aqua MODIS satellite data on SST and chlorophyll a concentrations in the area   [OC] and have not found significant correlations of squid catch locations with definite ranges of the chlorophyll a concentration. As to SST, the boundaries of the Subarctic Front and filament-like intrusions of Soya and Oyashio waters coincide rather good on satellite SST images and computed D maps. However, the Lagrangian maps demonstrate, as compared to the SST images, much more fine features with filaments, eddies of different size, streamers, swirls, etc. Backward-in-time Lagrangian maps contain information not only on a given day, as Eulerian snapshots, but they are also able to "record" a history of water motion.

Some biophysical reasons for aggregation of zooplankton at strong LFs have been discussed in the end of the preceding section. It was underlined there that LFs are 3D features. Vertical structure, the width of the mixed layer, and the thermocline depth are important factors in forming potential fishing grounds. Large horizontal and vertical temperature gradients typically arise at fronts. Wind intensification in fall in the South Kuril area causes intensive mixing of warm surface water with a cold depth one that decreases SST and increases the width of the mixed layer containing a nutrient rich water. It, in turn, intensifies photosynthesis creating more favorable conditions for squid food supply. Lagrangian maps with identified LFs could help in finding "horizontal" conditions favorable for fishery. In order to find the LFs with favorable "vertical" conditions and to identify potential fishing grounds more exactly, complementary 3D hydrological information is necessary.

Lagrangian fronts, including boundaries of large-scale intrusions and of mesoscale eddies, are convergence zones where zooplankton and fish could aggregate. As to potential fishing grounds at the Subarctic Front, it is the NEC that is known as a transport corridor for larvae and small pelagic fish to the feeding places in the South Kuril area [13, 14]. The Lagrangian intrusions of subtropical water into the central part of the region also are transport corridors for larvae and small pelagic fish, the main food for squid. Filament-like intrusions of productive Soya and Oyashio waters around anticyclonic Hokkaido eddies and their convergence with more oligotrophic waters at eddy's cores provide aggregation of zooplankton and fish there [25, 28] creating a food base for neon flying squid.

8.4 Foraging Strategy of Top Marine Predators and Lagrangian (Sub)Mesoscale Features

Currents, eddies, streamers, fronts, and other mesoscale and submesoscale structures present in the ocean impact marine ecosystem dynamics in a variety of ways (see, e.g., [2, 3, 20, 21] and the references therein). We have shown in the preceding sections that strong LFs could be oceanographic structures favorable for saury and squid fishing. In this section we briefly review recent studies of other authors on foraging behavior of some top marine predators and its relationship with Lagrangian sub- and mesoscale features such as LCSs, eddies, and filaments.

8.4.1 Foraging Strategy of Great Frigatebirds and Lagrangian Coherent Structures

Rapid advancement in electronic tagging and tracking of top marine predators such as sharks, seals, tuna, turtles, some seabirds, and other migratory marine species provides a new source of information on foraging behavior and its relationship with the marine environment. Some authors have studied foraging strategy of different species of top predators and observed the potential role of sub- and mesoscale features. However, it is still unclear why and how top predators choose these or those mesoscale structures to forage and feed. Lagrangian approach provides tools to develop quantitative methods and progress in understanding how marine species exploit dynamic environment and which oceanographic processes drive their foraging strategy.

Foraging strategy of great frigatebirds (*Fregata minor*) in Fig. 8.17 has been investigated in connection with LCSs in [7, 35, 36]. The frigatebirds are tropical seabirds living in an environment characterized by an oligotrophic marine ecosystem. They feed at the sea surface, because they cannot wet their feathers or dive into the water to feed. Frigatebirds forage mainly through association with subsurface predators such as tuna or cetaceans, which bring prey to the surface. Their diet is composed mainly of flying-fish and *Ommastrephids* squids.

The foraging strategy at sea of great frigatebirds, breeding on Europa Island in the Mozambique Channel, has been studied in [40] by using satellite transmitters and altimeters. The Europa Island is located in the central part of the Channel (22.3° S, 40.3° E) with the frigatebird population of about 700–1100 pairs. The Mozambique Channel is a deep strait between the African continent and Madagascar with the central part to be dominated by mesoscale cyclonic and anticyclonic eddies of 100–300 km in diameter and with a lifetime of several

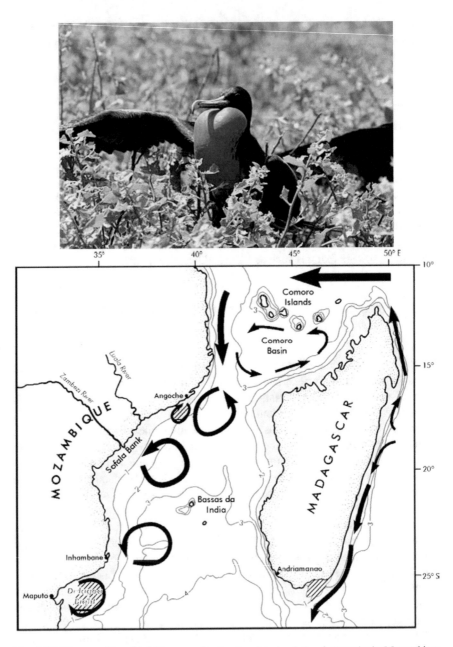

Fig. 8.17 The great frigatebird (*Fregata minor*) and major circulation features in the Mozambique Channel. *Hatched areas* denote upwelling. The train of mesoscale eddies is moving to the south along the Mozambique coast (from [17])

months (Fig. 8.17). Four to seven eddies per year are known to transit through the Channel from the north to south [33]. The higher productivity in the western part of the Channel is due to the presence of anticyclonic eddies and upwelling.

Eight birds were tracked through the Argos system between August 18 and September 30, 2003, resulting in 1864 positions [40]. Frigatebirds foraged over extensive distances up to 600 km from the island. When foraging, they move at slow speeds continuously climbing and descending and rarely come close to the surface to feed. Birds did not return to the same area. Flight speed and periods with reduced flight speed, derived from satellite telemetry data, have been used as indicators of foraging activity. A foraging patch has been defined as the area where flight speed between at least three successive Argos locations was lower than 10 km per hour. Feeding opportunities were derived from altimetry measurements when birds were close to the surface (below 5 m).

To characterize habitat at the mesoscale and to find whether birds selected some areas with enhanced productivity or specific SSH, the authors [40] used the distributions of chlorophyll concentration from the SeaWiFS database and of SSH from the AVISO database. They compared those distributions to that of waters over which birds crossed and foraged. The composite images of the Mozambique Channel in Fig. 8.18a, b show maximum chlorophyll concentrations between September 5 and 28, 2003 and SSH anomalies between September 13 and 18, 2003, respectively. Apart from the coastal waters off Africa, the productivity is low, but the western oceanic part, where frigatebirds spent most of their time, has a higher overall production than the eastern part (Fig. 8.18a). Frigatebirds foraged

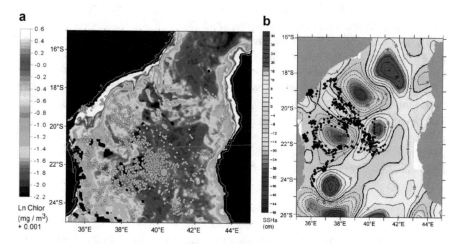

Fig. 8.18 (**a**) Chlorophyll concentration in the Mozambique Channel derived from the SeaWiFS database between September 5 and 28, 2003 with the bird locations during that period (*grey dots*). (**b**) Sea level anomalies (SLA) between September 13 and 18, 2003 with the bird locations during this time (*black dots*). Positive SLAs, associated with warm anticyclonic eddies, are in *red* and negative ones, associated with cold cyclonic eddies, are in *blue. Large green dot* is Europa Island. Courtesy of H. Weimerskirch

mainly over oceanic waters of the Mozambique Channel. The western part of the Channel is populated with mesoscale cyclonic and anticyclonic eddies (Fig. 8.18b) which slowly move to the south. Birds tended to avoid the centers of the eddies and remained at the periphery of the warm and cold eddies.

The foraging strategy of this tropical species differs strongly from that of the seabirds living in more cold regions. In colder waters, birds generally tend to concentrate around predictable features related to bathymetry or tidal forcing at a small scale or like fronts at a larger scale where productivity is enhanced. Most species tend to return to feed to the same area displaying a high degree of predictability. Frigatebirds do not cluster at specific predictable oceanographic features. Instead, they focus on areas with higher productivity on a regional scale, but the prey patches they feed on are clearly not predictable in space and time. From the energetic point of view, low cost and long range foraging strategy is preferable in tropical waters. At a mesoscale level, frigatebirds appear to favor large areas with slightly enhanced productivity such as frontal zones between mesoscale eddies [40].

Because of close association of frigatebirds with other subsurface predators like tuna and dolphins, the results obtained can be used to study the distribution of marine resources in the poorly known tropical oceanic waters. Using the same data on frigatebirds tracking, tuna catch statistics, SST, SSH, and chlorophyll a concentration data, the authors [35] have studied the role of mesoscale eddies in the Mozambique Channel in foraging strategy of frigatebirds and tuna. Their results have shown that seabirds are more closely tied to mesoscale eddies compared to tuna. The role of eddy boundaries on the response of frigatebirds and tuna has been especially underlined. Good foraging conditions along the edge of eddies have been considered as a result of the interplay of the maturation process from cyclonic eddies and the concentration process by eddy interactions.

The authors [36] attempted to find a connection between foraging strategy of great frigatebirds in the Mozambique Channel and LCSs there. To characterize the mixing activity in the region and to identify LCSs (for LCSs, see Sect. 4.4), that control transport at specific scales, they computed the so-called finite-size Lyapunov exponent, Λ_s, which is an alternative to the FTLE extensively used in the book. The quantity Λ_s is calculated through computing the time, τ, at which two tracers initially separated at a distance, δ_0, reach a final separation distance δ_f, following their trajectories in the AVISO velocity field. At position x and time t the finite-size Lyapunov exponent is given by

$$\Lambda_s(x, t, \delta_0, \delta_f) = \frac{\ln(\delta_f/\delta_0)}{\tau}. \tag{8.1}$$

The trajectories have been followed for 200 days. The authors [36] calculated Λ_s on all of the points of a grid with a spacing $\delta_0 = 0.025°$, and δ_f has been chosen to be approximately $1°$ to correspond to mesoscale. In this respect, the finite-size Lyapunov exponent represents the inverse time scale for mixing up fluid parcels between the grid and the characteristic scales of the mesoscale eddies.

Using the finite-size Lyapunov exponents, the authors [36] identified LCSs present in the surface flow in the Channel over a 2-month observation period (August and September 2003). Maps of Λ_s have been calculated weekly. The main aim was to test whether measured seabird positions during their foraging trips are related to the ridges of the Lyapunov exponent field and, therefore, to the corresponding LCSs in the area. The authors [36] specified a threshold defining a significant presence of LCSs to be $\Lambda_s > 0.1$ day^{-1} corresponding to mixing times smaller than 1 month.

The distributions of Λ_s in the whole Channel, its central part, and in areas crossed by seabirds were tested for conformity to the normal distribution using the Kolmogorov–Smirnov sample test, and they all have been found to be non-normal. In the whole channel and its central part, LCSs detected by $\Lambda_s > 0.1$ day^{-1} represented a minority of locations, occupying $\leq 30\%$ of the total area. However, in areas crossed by frigatebirds, $>60\%$ of the bird positions have been found on LCSs (see Fig. 3 in [36]). The sample tests confirmed that distributions of the finite-size Lyapunov exponents in areas crossed by seabirds are different from those found over the whole area and the central part.

The authors [36] concluded that the great frigatebirds during the observation period were not randomly distributed throughout the Λ_s range, and that seabirds moved over specific areas rich in LCSs, despite the area occupied by LCSs being comparatively small. Nearly two-thirds of the bird's positions were on LCSs, even though only $\leq 30\%$ of the whole area or the central part contained high Λ_s values. It seems that this top predator is able to track somehow LCSs to locate food patches.

Some possible reasons to understand how frigatebirds can follow these LCSs have been discussed in [36]. The birds might use atmospheric current changes over (sub)mesoscale structures to move along them. SST fronts generated by eddies and upwelling could influence on the perturbation of summertime wind stress curl. Mesoscale eddies and their interaction would force the atmosphere and generate air currents favorable to birds, which might take advantage of the wind to save energy in flight. The other hypothesis is that birds may follow visual and olfactory cues. Foraging behavior of seabirds is complex and results from a number of behavioral parameters and environmental parameters such as chlorophyll concentration, wind speed, and direction. Seabirds may use olfaction to track high concentrations of odor compounds, such as dimethyl sulfide, and sight when they locate prey patches. Dynamic LCSs may induce vertical mixing favorable to phytoplankton enhancement and their patchy distribution. The grazing of phytoplankton by zooplankton induces the production of dimethyl sulfide which is very attractive for different species of seabirds (for the corresponding references see [36]).

Because the LCSs are mobile, a simple memory is not sufficient for frigatebirds to return to a productive prey area. They could undertake long flights to find area where eddies are likely to be found (see Fig. 8.17) and then move until they cross a Lyapunov exponent ridge, which they will follow until they encounter a prey patch. Frigates are often in association with tuna schools to forage. So, the presence of other subsurface predators rather than directly of their prey could help them to feed.

Three-dimensional GPS tracking of the great frigatebirds has been used in the Mozambique Channel in September–October 2008, when 16 birds were equipped with GPS recorders on the Europa Island [7]. A total of 21 complete trajectories were obtained, nine of which were long-distance flights that lasted between 16 and 59 h, spanning a region several hundreds of km wide around the Europa Island. The birds had the occasion to repeatedly sample the (sub)mesoscale structures in the area. These data provided high-resolution vertical and horizontal bird's positions. Behavioral patterns were classified based on the bird's vertical displacement (e.g., fast/slow ascents and descents). Overlaying these data on SST and Lyapunov maps, the authors [7] related the behavior of frigatebirds to the physical environment at the sub- and mesoscale (10–100 km, days–weeks).

It was argued that frigatebirds modify their behavior concurrently to transport and thermal fronts. The transport fronts (LCSs, in fact) have been identified with the help of finite-size Lyapunov exponents in altimetry-derived velocity field. As in [36], the significant transport fronts were defined as a region where $\Lambda_s > 0.1$ day^{-1}. The bird's co-occurrence with these structures has been supposed to be a consequence of their search not only for food (preferentially searched over thermal fronts) but also for upward vertical wind. Frigatebirds may exploit vertical winds not only for sustaining their flight, but also for locating fronts even when these structures are not visible. Following favorable winds may constitute a good searching strategy for locating prey-enriched regions.

8.4.2 Preference of Southern Elephant Seals and Mediterranean Whales for Distinct Mesoscale Features

The authors of papers [6, 8] analyzed post-moulting trips of elephant seals (*Mirounga leonina*) who left the Kerguelen Islands (49° S, 69° E) between 2005 and 2011. The patterns of elephant seal distribution overlaid on bathymetry are shown in Fig. 8.19 for the area in the southern Indian Ocean. The circumpolar frontal system consists of the Southern Subtropical Front, sub-Antarctic Front, and Polar Front. The last two are related to the jets of the dynamic Antarctic Circumpolar Current with an intensive meandering creating mesoscale and submesoscale activity.

A total of 42 adult females were equipped with Argos tags. Their at-sea distribution and behavior were monitored using satellite devices. Tracking data consist of locations in a 3D-space (longitude, latitude, time). The authors [6] use the terminology "intensive foraging" vs "travelling" to refer to two distinct seal behavioral states with slow movement of animals and fast and directed movement, respectively. Adult elephant seals performed two foraging trips during their year cycle. After breeding on land in September–October, seals performed a 2–3 months post-breed foraging trip and they return to land to moult in December–January. After the moult they remained at sea for an extended 7–8 month foraging trip (for details see [6] and the references therein).

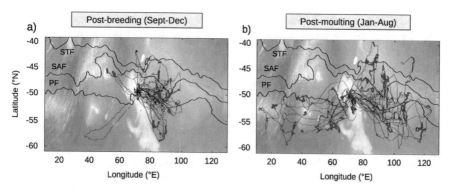

Fig. 8.19 The patterns of elephant seal distribution overlaid on bathymetry. Foraging trips of elephant seals equipped between 2005 and 2011 from Kerguelen Islands (49° S, 69° E) during (**a**) post-breeding period (September–December) and (**b**) post-moulting period (January–August). Travelling (extensive behavior) and foraging (intensive behavior) bouts of trips are, respectively, in *black and red*. Southern Subtropical Front (STF), sub-Antarctic Front (SAF), and Polar Front (PF) are shown. Courtesy of C. Cotté

The two periods have been considered: post-breeding (October–December) and post-moult (January–August) that corresponded to different conditions of biological activity in the Kerguelen Plateau area, namely the phytoplankton bloom period and more oligotrophic period, respectively. Elephant seals, and particularly females, forage in the open ocean, where they alternate hundreds of meters deep diving in search for fish or squid, and short breathing intervals at the surface. After moulting on land, females leave their colony to engage in solitary trips, lasting several months, where they typically head towards a zone of strong mesoscale activity between the Subtropical and the Subpolar fronts (see Fig. 8.19).

In order to describe the surface dynamic environment and identify habitat preference of seals, multi-satellite data have been used in [6] including SST, SSH, and chlorophyll *a* concentration distributions. Moreover, altimetry-based Lagrangian diagnostic has been applied to quantify mesoscale mixing and transport in the area and to examine the preference of seals for LCSs referred in [6] as (sub)mesoscale transport fronts. Using the multi-year tracking data set on elephant seal movements, those authors investigated the relationships between seal behavior (travelling vs intense foraging) and physical environmental properties. They extracted each satellite-derived physical environmental property at the seal location in space and time and compared it with the value in the surrounding mesoscale environment to highlight a possible difference. The observed differences were interpreted as a preference for a given physical parameter for seals. To estimate whether a seal preference can be inferred or not, some statistical tests were performed.

The results obtained in paper [6] can be summarized briefly as follows. Relationships between elephant seal behavior and (sub)mesoscale features have been found for the post-moulting period (January–August): travelling along thermal fronts and intensive foraging in cold and long-lived mesoscale water patches. These patches corresponded to waters which have supported the bloom during spring. In contrast, no clear preference at the (sub)mesoscale has been found for the post-breeding period (October–December), although seals were distributed within the chlorophyll-rich water plume detaching from the plateau. These contrasted foraging strategies were supposed in [6] to be connected to seasonally contrasted biological environment. Two contrasted trophic conditions for primary production during the bloom period and from the bloom onwards induced a fundamental difference in foraging strategies of elephant seals.

However, post-moulting elephant seals were still influenced by the spring bloom that had occurred several months earlier and that had progressively drifted eastward. Those authors hypothesized that while the marine ecosystem develops up to higher trophic levels within these waters where fish congregate, elephant seals actively tracked post-bloom waters. These productive waters were increasingly stirred by the mesoscale activity with other waters with lower biological activity. This process was likely to induce an increasingly heterogeneous pattern in the prey field.

Stirring, in turn, creates a strong filamentary field induced by numerous eddy–eddy interactions. These filaments are elongated structures reaching hundreds of km in length and widths of around 10 km. The authors [6] argued that elephant seals were associated with submesoscale fronts not only for feeding but while travelling as well. They formulated two hypotheses: "(1) seals may use these filaments of cold water as an environmental tracer to reach cold patches which may offer favorable foraging conditions; and/or (2) seal's trajectories could be stretched by advection during their displacements along frontal structures when they swim in the vicinity of a filament" (cited from [6]).

A strong SST gradient associated with cold filaments could serve as a local environmental cue to reach favorable mesoscale features. Moreover, filaments were reported to carry high zooplankton densities [16, 22]. The second hypothesis relies on the effect of LCSs (transport fronts) that are often areas of high current velocities which can be used by seals to move. Favorable foraging eddies targeted by elephant seals are rather stable and long-lived features populated by large communities of zooplankton and small pelagic fish. The centers of cyclonic eddies and the edges of anticyclonic eddies are well known to be enriched in organisms of different trophic levels [2, 3, 8, 20, 21, 23–25]. So elephants seals could benefit from the enhanced local biological production and aggregation of prey there.

We conclude this section by mentioning the study of influence of the ocean circulation on the seasonal distribution of eight Mediterranean fin whale equipped with Argos tracking devices from August 2003 to June 2004 [5]. Using altimetry, SST, and chlorophyll a satellite data, statistical analysis, and the field of the finite-scale Lyapunov exponent as a Lagrangian descriptor, the authors tested two hypotheses: (1) whales were preferentially associated with eddies and (2) they were mainly found in regions of strong filament activity. They came to the conclusion that fin whale distribution in the western Mediterranean Sea is linked to their foraging strategy by two physical mechanisms: at the basin scale, a southern boundary of the krill habitat constituted by the North Balearic Front, and at mesoscale and submesoscale, zooplankton aggregation by the high filamentary activity at eddy peripheries [5]. At the basin scale, fin whales were associated with the anticlockwise gyre in the northern part of the western Mediterranean Sea, which defines the habitat of krill, the whale's main prey. At mesoscale and submesoscale, and only during the seasonal phytoplankton biomass minimum in summer, whales prefer the periphery of eddies and associated filaments to feed on. Whales were not associated with productive areas, probably due to the spatiotemporal lag between phytoplankton and krill.

We presented in this chapter some recent results on relationship between potential fishing grounds and foraging behavior of some top marine predators and mesoscale and submesoscale LFs and LCSs. Both the LFs and LCSs can be identified with the help of different Lagrangian descriptors computed in altimetry-based near-surface velocity field which is now available globally and in near real time with the spatiotemporal resolution increasing year after year. Thus, it is routinely now to get daily Lagrangian maps of mixing and transport of surface water practically for any basin in the ocean (except for the ones covered by ice). On the other hand, satellites provide now continuous and global data at high space resolution for many oceanic and atmospheric parameters: SST and salinity, SSH and its anomalies, ocean color and concentration of chlorophyll, winds at the ocean surface, sea ice, wave heights, surface roughness, and others (see Sect. 3). These data become used only recently to study a connection of the LFs, LCSs, and some of the oceanic and atmospheric parameters with fish catch locations and tracks of tagged sea birds and animals.

The results obtained up to now should be considered as preliminary ones. It has been shown in this chapter and in the corresponding papers that fish, marine animals, and seabirds are able to track somehow mesoscale and submesoscale Lagrangian structures use them in the foraging strategy and to feed on. As to physical and biological reasons of that, a number of hypotheses have been suggested and discussed in this chapter. The problem is, of course, very difficult because one deals with living creatures whose foraging strategy is defined not only by changing dynamic environment but also their reflexes and even by free will. Moreover, the Lagrangian structures are 3D features, but satellite images and altimetry-based Lyapunov and other Lagrangian maps allow to identify them near the ocean surface only. The vertical motion, especially at submesoscale, is expected to be important factor in congregation of phytoplankton, forage zooplankton, and other food in a trophic chain at those Lagrangian structures.

Continuous improvement of numerical models of regional and global ocean circulation with assimilation of real data allows to hope that in the near future they will be able to reproduce adequately many of mesoscale and submesoscale characteristic features of circulation at different depths. If so, we could get a 3D picture of submesoscale circulation and would know what happens below the surface LFs and LCSs and how they are structured in 3D. This, together with further progress in satellite products and in tagging program, will open new possibilities for understanding physico–biological interactions between marine organisms and ocean circulation.

Climate change, pollution, and overfishing threaten the existence of pelagic ecosystems and make their conservation challenging. Understanding how fish, marine animals, and seabirds exploit their dynamic environment and which oceano-graphic processes drive their foraging strategy might help us to recognize areas where they prefer to congregate and create protectable marine reservations there.

References

1. Alabia, I.D., Saitoh, S.I., Mugo, R., Igarashi, H., Ishikawa, Y., Usui, N., Kamachi, M., Awaji, T., Seito, M.: Seasonal potential fishing ground prediction of neon flying squid (*Ommastrephes bartramii*) in the western and central North Pacific. Fish. Oceanogr. **24**(2), 190–203 (2015). 10.1111/fog.12102
2. Bakun, A.: Fronts and eddies as key structures in the habitat of marine fish larvae: opportunity, adaptive response and competitive advantage. Sci. Mar. **70**(S2), 105–122 (2006). 10.3989/scimar.2006.70s2105
3. Belkin, I.M., Cornillon, P.C., Sherman, K.: Fronts in large marine ecosystems. Prog. Oceanogr. **81**(1–4), 223–236 (2009). 10.1016/j.pocean.2009.04.015
4. Chen, X., Tian, S., Guan, W.: Variations of oceanic fronts and their influence on the fishing grounds of *Ommastrephes bartramii* in the Northwest Pacific. Acta Oceanol. Sin. **33**(4), 45–54 (2014). 10.1007/s13131-014-0452-3
5. Cotte, C., d'Ovidio, F., Chaigneau, A., Levy, M., Taupier-Letage, I., Mate, B., Guinet, C.: Scale-dependent interactions of Mediterranean whales with marine dynamics. Limnol. Oceanogr. **56**(1), 219–232 (2010). 10.4319/lo.2011.56.1.0219
6. Cotté, C., d'Ovidio, F., Dragon, A.C., Guinet, C., Lévy, M.: Flexible preference of southern elephant seals for distinct mesoscale features within the Antarctic Circumpolar Current. Prog. Oceanogr. **131**, 46–58 (2015). 10.1016/j.pocean.2014.11.011
7. De Monte, S., Cotte, C., d'Ovidio, F., Levy, M., Le Corre, M., Weimerskirch, H.: Frigatebird behaviour at the ocean-atmosphere interface: integrating animal behaviour with multi-satellite data. J. R. Soc. Interface **9**(77), 3351–3358 (2012). 10.1098/rsif.2012.0509
8. d'Ovidio, F., De Monte, S., Penna, A.D., Cotté, C., Guinet, C.: Ecological implications of eddy retention in the open ocean: a Lagrangian approach. J. Phys. A Math. Theor. **46**(25), 254023 (2013). 10.1088/1751-8113/46/25/254023
9. Fan, W., Wu, Y., Cui, X.: The study on fishing ground of neon flying squid, *Ommastrephes bartrami*, and ocean environment based on remote sensing data in the Northwest Pacific Ocean. Chin. J. Oceanol. Limnol. **27**(2), 408–414 (2009). 10.1007/s00343-009-9107-1
10. Fedorov, K.N.: The Physical Nature and Structure of Oceanic Fronts. Lecture Notes on Coastal and Estuarine Studies, vol. 19. Springer, Berlin (1986). 10.1029/ln019
11. Filatov, V.N.: Pacific saury migrations in the areas of the Kuril Islands and the Sea of Okhotsk. In: Proceedings of the 20th International Symposium on Okhotsk Sea and Sea Ice, pp. 257–260 (2005)
12. Fukushima, S.: Synoptic analysis of migration and fishing conditions of saury in the northwest Pacific Ocean. Bull. Tohoku Reg. Fish. Res. Lab. **4**, 1–70 (1979) [in Japanese, English abstract]
13. Isoguchi, O., Kawamura, H., Oka, E.: Quasi-stationary jets transporting surface warm waters across the transition zone between the subtropical and the subarctic gyres in the North Pacific. J. Geophys. Res. Oceans **111**(C10), C10003 (2006). 10.1029/2005jc003402
14. Ito, S.I., Wagawa, T., Kakehi, S., Okunishi, T., Hasegawa, D.: Importance of advection to form a climate and ecological hotspot in the western North Pacific. In: Proceedings of 3rd International Symposium on Effects of Climate Change on the World's Oceans, March 23–27, 2015, Santos City, Brazil (2015)
15. Kosaka, S.: Relation of the migration of pacific sauries to oceanic fronts in the northwest pacific ocean. INPFC Bull. **47**, 229–246 (1986). http://www.npafc.org/new/inpfc/INPFC%20Bulletin/Bull%20No.47/Bulletin47.pdf

16. Labat, J.P., Gasparini, S., Mousseau, L., Prieur, L., Boutoute, M., Mayzaud, P.: Mesoscale distribution of zooplankton biomass in the northeast Atlantic Ocean determined with an optical plankton counter: relationships with environmental structures. Deep-Sea Res. I Oceanogr. Res. Pap. **56**(10), 1742–1756 (2009). 10.1016/j.dsr.2009.05.013

17. Lutjeharms, J.R.E.: The coastal oceans of South-Eastern Africa. The Sea: Ideas and Observations on Progress in the Study of Seas. Interdisciplinary Regional Studies and Syntheses, vol. 14, pp. 783–834. Harvard University Press, Cambridge (2006)

18. Monin, A.S., Krasitsky, V.P.: Phenomena on the Ocean Surface. Gidrometeoizdat, Leningrad (1985)

19. Nishikawa, H., Toyoda, T., Masuda, S., Ishikawa, Y., Sasaki, Y., Igarashi, H., Sakai, M., Seito, M., Awaji, T.: Wind-induced stock variation of the neon flying squid (*Ommastrephes bartramii*) winter-spring cohort in the subtropical North Pacific Ocean. Fish. Oceanogr. **24**(3), 229–241 (2015). 10.1111/fog.12106

20. Olson, D., Hitchcock, G., Mariano, A., Ashjian, C., Peng, G., Nero, R., Podesta, G.: Life on the edge: marine life and fronts. Oceanography **7**(2), 52–60 (1994). 10.5670/oceanog.1994.03

21. Owen, R.W.: Fronts and eddies in the sea: mechanisms, interactions and biological effects. In: Longhurst, A.R. (ed.) Analysis of Marine Ecosystems, pp. 197–233. Academic Press, London (1981)

22. Perruche, C., Rivière, P., Lapeyre, G., Carton, X., Pondaven, P.: Effects of surface quasi-geostrophic turbulence on phytoplankton competition and coexistence. J. Mar. Res. **69**(1), 105–135 (2011). 10.1357/002224011798147606

23. Prants, S.V.: Dynamical systems theory methods to study mixing and transport in the ocean. Phys. Scr. **87**(3), 038115 (2013). 10.1088/0031-8949/87/03/038115

24. Prants, S.V.: Chaotic Lagrangian transport and mixing in the ocean. Eur. Phys. J. Spec. Top. **223**(13), 2723–2743 (2014). 10.1140/epjst/e2014-02288-5

25. Prants, S.V., Budyansky, M.V., Uleysky, M.Y.: Identifying Lagrangian fronts with favourable fishery conditions. Deep-Sea Res. I Oceanogr. Res. Pap. **90**, 27–35 (2014). 10.1016/j.dsr.2014.04.012

26. Prants, S.V., Budyansky, M.V., Uleysky, M.Y.: Lagrangian fronts in the ocean. Izv. Atmos. Oceanic Phys. **50**(3), 284–291 (2014). 10.1134/s0001433814030116

27. Prants, S.V., Budyansky, M.V., Uleysky, M.Y., Zaslavsky, G.M.: Chaotic mixing and transport in a meandering jet flow. Chaos **16**(3), 033117 (2006). 10.1063/1.2229263

28. Prants, S.V., Uleysky, M.Y., Budyansky, M.V.: Lagrangian coherent structures in the ocean favorable for fishery. Dokl. Earth Sci. **447**(1), 1269–1272 (2012). 10.1134/S1028334X12110062

29. Qiu, B.: Kuroshio and Oyashio currents. In: Steele, J.H. (ed.) Encyclopedia of Ocean Sciences, 2nd edn., pp. 358–369. Academic Press, Oxford (2001). 10.1016/B978-012374473-9.00350-7

30. Saitoh, S., Kosaka, S., Iisaka, J.: Satellite infrared observations of Kuroshio warm-core rings and their application to study of Pacific saury migration. Deep Sea Res. Part A **33**(11–12), 1601–1615 (1986). 10.1016/0198-0149(86)90069-5

31. Sakurai, Y.: An overview of the Oyashio ecosystem. Deep-Sea Res. II Top. Stud. Oceanogr. **54**(23–26), 2526–2542 (2007). 10.1016/j.dsr2.2007.02.007

32. Samko, E.V., Bulatov, N.V., Nikitin, A.A., Muktepavel, L.S., Kapshiter, A.V.: Results of remote sensing data using for maintenance of fishery in the Far-Eastern Seas. Sovremennye problemy distantsionnogo zondirovaniya Zemli iz kosmosa **7**(2), 209–220 (2010)

33. Schouten, M.W., de Ruijter, W.P., van Leeuwen, P.J., Ridderinkhof, H.: Eddies and variability in the Mozambique channel. Deep-Sea Res. II Top. Stud. Oceanogr. **50**(12–13), 1987–2003 (2003). 10.1016/s0967-0645(03)00042-0

34. Sugimoto, T., Tameishi, H.: Warm-core rings, streamers and their role on the fishing ground formation around Japan. Deep Sea Res. Part A **39**, S183–S201 (1992). 10.1016/S0198-0149(11)80011-7

35. Tew Kai, E., Marsac, F.: Influence of mesoscale eddies on spatial structuring of top predators' communities in the Mozambique channel. Prog. Oceanogr. **86**(1–2), 214–223 (2010). 10.1016/j.pocean.2010.04.010

36. Tew Kai, E., Rossi, V., Sudre, J., Weimerskirch, H., Lopez, C., Hernandez-Garcia, E., Marsac, F., Garçon, V.: Top marine predators track Lagrangian coherent structures. Proc. Natl. Acad. Sci. **106**(20), 8245–8250 (2009). 10.1073/pnas.0811034106
37. Uda, M.: Researches on "Siome" or current rip in the seas and oceans. Geophys. Mag. **11**, 307–372 (1938)
38. Uleysky, M.Y., Budyansky, M.V., Prants, S.V.: Effect of dynamical traps on chaotic transport in a meandering jet flow. Chaos **17**(4), 043105 (2007). 10.1063/1.2783258
39. Wang, W., Shao, Q.: The spatial relationship between the distribution of *Ommastrephes bartrami* and marine environment in the western North Pacific Ocean. In: ISPRS Workshop on Service and Application of Spatial Data Infrastructure, XXXVI(4/W6), October 14–16, 2005, Hangzhou, China, pp. 177–182 (2005)
40. Weimerskirch, H., Le Corre, M., Jaquemet, S., Potier, M., Marsac, F.: Foraging strategy of a top predator in tropical waters: great frigatebirds in the Mozambique channel. Mar. Ecol. Prog. Ser. **275**, 297–308 (2004). 10.3354/meps275297
41. Yasuda, I., Kitagawa, D.: Locations of early fishing grounds of saury in the northwestern Pacific. Fish. Oceanogr. **5**(1), 63–69 (1996). 10.1111/j.1365-2419.1996.tb00018.x

List of Internet Resources

[ANIM] Animation of daily Lagrangian maps in 2002 with the saury catch locations.
  http://dynalab.poi.dvo.ru/data/GRL12/2002
[OC] NASA Ocean Color.
  http://oceancolor.gsfc.nasa.gov/cms/

Chapter 2
Chaotic Transport and Mixing in Idealized Models of Oceanic Currents

© Springer International Publishing AG 2017
S.V. Prants et al., *Lagrangian Oceanography*, Physics of Earth and Space
Environments, DOI 10.1007/978-3-319-53022-2

DOI 10.1007/978-3-319-53022-2_9

In the original version of this chapter the citation on page 20 line 11 "Using that concept, chaotic advection has been studied for a variety of dynamically consistent models in the barotropic and baroclinic ocean [21]" and the corresponding reference was erroneously mentioned as:

21. Kozlov, V.F.: Background currents in geophysical hydrodynamics. Izv. Atmos. Oceanic Phys. **31**(2), 229–234 (1995)

Correct citation and corresponding references below have now been updated as "Using that concept, chaotic advection has been studied for a variety of dynamically consistent models in the barotropic and baroclinic ocean [48–54]"

48. Kozlov, V.F., Koshel, K.V.: Barotropic model of chaotic advection in background ows. Izv. Atmos. Ocean. Phys. **35**(1), 123–130 (1999)
49. Kozlov, V.F., Koshel, K.V.: Some features of chaos development in an oscillatory barotropic ow over an axisymmetric submerged obstacle. Izv. Atmos. Ocean. Phys. **37**(1), 351–361 (2001)
50. Izrailsky, Y.G., Kozlov, V.F., Koshel, K.V.: Some specific features of chaotization of the pulsating barotropic ow over elliptic and axisymmetric seamounts. Phys. Fluids **16**(8), 3173–3190 (2004). 10.1063/1.1767095
51. Ryzhov, E., Koshel, K., Stepanov, D.: Background current concept and chaotic advection in an oceanic vortex ow. Theor. Comput. Fluid Dyn. **24**(1–4), 59–64 (2010). 10.1007/s00162-009-0170-1
52. Ryzhov, E.A., Koshel, K.V.: Estimating the size of the regular region of a topographically trapped vortex. Geophys. Astrophys. Fluid Dyn. **105**(4–5), 536–551 (2010). 10.1080/03091929.2010.511205

The updated original online version for this chapter can be found at
DOI 10.1007/978-3-319-53022-2_2

© Springer International Publishing AG 2017
S.V. Prants et al., *Lagrangian Oceanography*, Physics of Earth and Space
Environments, DOI 10.1007/978-3-319-53022-2_9

53. Ryzhov, E.A., Koshel, K.V.: Interaction of a monopole vortex with an isolated topographic feature in a three-layer geophysical ow. Nonlinear Process. Geophys. **20**(1), 107–119 (2013). 10.5194/npg-20-107-2013

54. Ryzhov, E.A., Sokolovskiy, M.A.: Interaction of a two-layer vortex pair with a submerged cylindrical obstacle in a two layer rotating fluid. Phys. Fluids 28(5), 056,602 (2016). 10.1063/1.4947248

Glossary of Some Terms in Dynamical Systems Theory

A brief and simple description of basic terms in dynamical systems theory with illustrations is given in the alphabetic order. Only those terms are described which are used actively in the book. Rigorous results and their proofs can be found in many textbooks and monographs on dynamical systems theory and Hamiltonian chaos (see, e.g., [1, 6, 15]).

Bifurcations

Bifurcation means a qualitative change in the topology in the **phase space** under varying control parameters of a dynamical system under consideration. The number of **stationary points** and/or their stability may change when varying the parameters. Those values of the parameters, under which bifurcations occur, are called *critical* or *bifurcation values*. There are also bifurcations without changing the number of stationary points but with topology change in the phase space. One of the examples is a separatrix reconnection when a heteroclinic connection changes to a homoclinic one or vice versa.

Cantori

Some **invariant tori** in typical unperturbed Hamiltonian systems break down under a perturbation. Suppose that an invariant torus with the frequency f breaks down at a critical value of the perturbation frequency ω. If f/ω is a rational number, then a chain of resonances or **islands of stability** appears at its place. If f/ω is an irrational number, then a *cantorus* appears at the place of the corresponding invariant torus. Cantorus is a Cantor-like invariant set [7, 12] the motion on which is unstable and quasiperiodic. Cantorus resembles a closed curve with an infinite number of gaps. Therefore, cantori are **fractal**. Since the motion on a cantorus is unstable, it has stable and unstable **manifolds**. All the points on a cantorus belong to the same quasiperiodic **trajectory** if its initial point belongs to it.

Cantori are singular objects that do not occupy a volume of a finite measure in the **phase space**. However, they form an infinite hierarchy around **islands of**

© Springer International Publishing AG 2017
S.V. Prants et al., *Lagrangian Oceanography*, Physics of Earth and Space Environments, DOI 10.1007/978-3-319-53022-2

stability. The closer a cantorus is to the island's boundary, the narrower are its gaps. Cantori influence essentially the transport in the phase space because it may take a long time for particles and their trajectories to percolate through cantori gaps. Since the smallest gaps appear in the cantori which are close to the very boundaries of the islands, one observes at those places an increased density of phase points on **Poincaré sections**. The island's boundaries are called **dynamical traps**, and the long stay of trajectories there is called a *stickiness*.

Cantori are not a single reason for stickiness and dynamical traps. Hyperbolic trajectories along with their stable and unstable **manifolds** produce such a complicated tangle where particles and their trajectories may be trapped for a long time.

Dynamical traps

Dynamical trap is a domain in the **phase space** where particles (and their **trajectories**) may spend an arbitrary long but finite time [14, 15], in spite of the fact that the corresponding trajectory is chaotic in any relevant sense. Strictly speaking, it is the definition of a quasitrap. Absolute traps, where particles could spend an infinite time, are not possible in Hamiltonian systems which have no attractors. However, there may exist separatrix-like trajectories with infinite time, but with zero measure of initial conditions. The dynamical traps are caused by a *stickiness* of trajectories, mainly, to the boundaries of **islands of stability** where **cantori** are situated. There are also traps of unstable periodic orbits, including saddle traps associated with unstable periodic trajectories. Up to now, there is no full classification and description of dynamical traps. Dynamical traps influence significantly transport in Hamiltonian systems specifying its anomalous statistical properties.

Fractals

The name *"fractal"* was coined by B. Mandelbrot [9] in order to describe irregular and self-similar structures, i.e., objects small parts of which are similar in a sense to big parts and those in turn are similar to the whole object. Such objects in mathematics as Cantor sets, Weierstrass functions, which are everywhere continuous but not differentiable anywhere, Julia sets, etc. [9] have been known for many years. However, they have been considered as exotic objects not existing in the real world. One of the definitions of fractality is the following: fractal is a set whose Hausdorff–Besikovich dimension is larger than its topological dimension [9].

Hamiltonian chaos produces different kinds of fractals. Stochastic layers, hierarchies of **islands of stability** and of **cantori** are fractal. Typical chaotic trajectories are fractal in a sense [15]. Chaotic scattering, exit-time functions [3, 10], and Poincaré recurrences [15] are fractal as well. Chaotic invariant sets in Hamiltonian systems are fractal and have a Cantor-like structure.

Fractal sets appear in dynamical systems in a natural way if there exists a mechanism removing phase points out off a given region in the phase space. In dissipative systems it is a dissipation which shrinks in the course of time an initial phase volume to a set called an *attractor*. In chaotic dissipative systems such attractors may be fractal sets (strange attractors). In **Hamiltonian systems** the phase volume is conserved (the Liouville theorem), and fractal sets may appear when a scattering problem is formulated [2, 3, 8].

The famous Cantor fractal is formed as follows [9]. Take the closed interval $[0, 1]$ and remove the open middle third interval $(1/2, 2/3)$, leaving the two intervals $[0, 1/3]$ and $[2/3, 1]$. Then remove the middle open thirds of each of these two intervals, leaving four closed intervals of length $1/9$ each, etc. The total length of remaining segments is $\lim_{n \to \infty} 2^n r = \lim_{n \to \infty} (2/3)^n \equiv \lim_{n \to \infty} e^{-n \ln(3/2)}$ (one gets $N = 2^n$ segments with the length $r = (1/3)^n$ each after n iterations). Though the set of remaining segments is infinite, its total length or the Lebesque measure is zero. Topological dimension of the classical Cantor fractal d_t is zero. Other measures have been introduced to characterize such dust-like objects called *Cantor sets*. The Hausdorff–Besikovich dimension is a common used one

$$d_{HB} = \lim_{r \to 0} \frac{\ln N}{\ln(1/r)}.$$

One gets in the case of the classical Cantor fractal: $d_{HB} = \ln 2 / \ln 3 = 0.63 \ldots$, i.e., the fractal dimension is not an integer. It is larger than the topological dimension of a point ($d_t = 0$), but smaller than the topological dimension of an interval ($d_t = 1$).

Typical chaotic Hamiltonian systems, having the mixed phase space with **islands of stability** and **dynamical traps**, produce, as a rule, fractals with PDFs having power-law "tails." Hyperbolic chaotic Hamiltonian systems, that do not have **KAM tori** and **cantori**, produce Cantor-like fractals with exponential PDFs.

Hamiltonian chaos

Hamiltonian chaos is a dynamical chaos in **Hamiltonian systems**. A deterministic dynamical system is called chaotic if it has at least one positive **Lyapunov exponent** and generates mixing. The *mixing* is defined as follows. Let B is a region with dye in a waterpool A with a circulation. The volume of B at $t = 0$ is $V(B_0)$. Let C is another region in A. The amount of dye in C is $V(B_t \cap C)$ at the moment of time t and its concentration in C is $V(B_t \cap C)/V(C)$. The definition of mixing is: $V(B_t \cap C)/V(C) - V(B_t)/V(A) \to 0$ at $t \to \infty$, i.e., the concentration of dye in any region C in the waterpool A is the same as in the entire waterpool. Recall that in Hamiltonian systems the phase fluid is incompressible, i.e., $V(B_t) = V(B_0)$.

Instability produces an exponential sensitivity of trajectories to small variations in initial conditions and/or control parameters. It is difficult to prove analytically existence of chaos, especially in nonhyperbolic systems. Dynamical chaos becomes evident after computing **Poincaré sections**, Melnikov integrals, intersections of stable and unstable **manifolds**, and maximal **Lyapunov exponents**.

Theory of Hamiltonian chaos is presented in a number of monographs and textbooks [1, 6, 15]. The **phase space** in a typical Hamiltonian system is mixed, i.e., the regions with regular motion coexist with chaotic ones. In Fig. G1 we show

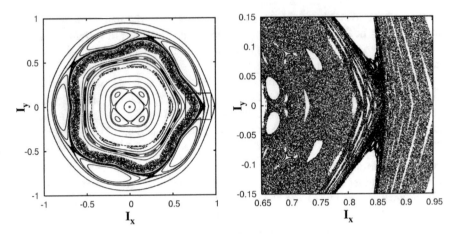

Fig. G1 *Left panel*—**Poincaré section** of a **Hamiltonian system** with **islands of stability** (*closed curves*) and a stochastic layer between confining **invariant tori**. *Right panel*—zoom of the small region in the stochastic layer indicated in the right panel

Poincaré section of a Hamiltonian system, simulating propagation of sound rays in the underwater sound channel in the ocean [8], with the mixed phase space with chains of **islands of stability** separated by stochastic layers (where motion is chaotic) and confined between **invariant KAM tori**. KAM tori are stable invariant **manifolds** with boundaries that are impenetrable to particle's transport. There are an infinite number of **cantori** around the nested islands of stability (not shown in the figure). They are Cantor-like unstable invariant sets with gaps, transport throw which is possible but difficult. Hamiltonian chaos is a special type of motion with properties both of regular motion (due to determinism of equations of motion) and stochastic motion (due to a local instability of trajectories).

Hamiltonian dynamics

Hamiltonian dynamics is a geometry in the **phase space** [1]. State of a Hamiltonian system with N degrees of freedom in the phase space is described by N generalized positions $(q_1, \ldots, q_N)$ and momenta $(p_1, \ldots, p_N)$ which are pairwise canonically conjugated variables. The equations of motion are specified with the help of a Hamiltonian function of the generalized positions and momenta

$$\dot{\mathbf{q}}_i = \frac{\partial H}{\partial \mathbf{p}_i}, \quad \dot{\mathbf{p}}_i = -\frac{\partial H}{\partial \mathbf{q}_i}. \tag{G.1}$$

If the Hamiltonian $H(p, q, t)$ depends on time, then the corresponding system can be studied in the enlarged $(2N+1)$-dimensional phase space $(q_1, \ldots, q_N; p_1, \ldots, p_N; t)$ where it has $N + 1/2$ degrees of freedom.

Equation (G.1) satisfy to the incompressibility condition

$$\sum_{i=1}^{N} \left(\frac{\partial \dot{q}_i}{\partial q_i} + \frac{\partial \dot{p}_i}{\partial p_i} \right) = 0. \tag{G.2}$$

If one specifies a volume of initial conditions, then this expression means that the phase fluid conserves its volume in the course of time (the Liouville theorem). A drop of phase fluid can be transformed during the evolution in a very complicated way.

The Hamilton equations (G.1) are not, in general, integrable. The Liouville–Arnold theorem states that a Hamiltonian system with N degrees of freedom is *fully integrable* if there exist N linearly independent first integrals of motion C_i in involution, i.e., with zero Poisson brackets $\{C_i, C_j\} \equiv 0$, $i, j = 1, 2, \ldots, N$. Equations of motion (G.1) for a fully integrable system can be always transformed to the following form:

$$\dot{I}_i = -\frac{\partial H}{\partial \theta_i} = 0, \quad \dot{\theta}_i = \frac{\partial H}{\partial I_i} \equiv \omega_i \quad (I_1, \ldots, I_N), \tag{G.3}$$

where I_i and θ_i are pairwise canonically conjugated variables known as *action* and *angle*, respectively. They are functions of positions and momenta.

Trajectories in a Hamiltonian system with N integrals of motion lie on N-dimensional invariant **manifolds** in the $2N$-dimensional phase space. These manifolds have torus topology and are called **invariant tori**. Any trajectory, starting on a given torus, stays on it all the time. If a Hamiltonian system is fully integrable, then the representation in terms of I_i and θ_i is global, i.e., the phase space is partitioned to invariant tori, and any trajectory is located on some torus. If a system is not integrable, then some trajectories do not lie on invariant tori. Up to now, there is no complete theory of behavior of nonintegrable Hamiltonian systems. However, there exists very important **Kolmogorov–Arnold–Moser theorem** about the behavior of Hamiltonian systems under weak perturbations.

Heteroclinic and homoclinic structures

Separatrices in an integrable 1D Hamiltonian system connect either two hyperbolic **stationary points** in such a way that a stable separatrix of one point $W_s^{(0)}(h_1)$ coincides with an unstable separatrix of the other point $W_u^{(0)}(h_2)$ and vice versa (see Fig. G2a), or $W_s^{(0)}$ and $W_u^{(0)}$ of the same hyperbolic point coincide (see Fig. G2b). In the former case, one gets a *heteroclinic connection*, whereas in the latter one—a *homoclinic connection*.

In Fig. G2 those connections are shown in the phase plane (x, y) of a Hamiltonian system with one degree of freedom, and in Fig. G3a, b they are shown in the enlarged phase space (x, y, t). Under a perturbation with a period T_0, hyperbolic (saddle) points of the unperturbed system become unstable periodic **trajectories** $\gamma(t)$ with stable and unstable separatrix branches which are called stable $W_s(\gamma)$ and unstable

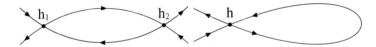

Fig. G2 (**a**) A heteroclinic connection of unperturbed separatrices of two hyperbolic points h_1 and h_2 and (**b**) a homoclinic connection with one hyperbolic point h

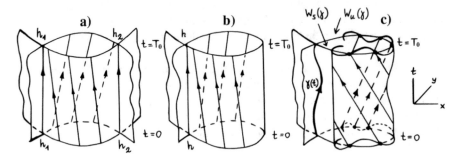

Fig. G3 Schematic representation of (**a**) heteroclinic and (**b**) homoclinic connections of a unperturbed one-degree-of-freedom system in the extended phase space (x, y, t). (**c**) Under a perturbation with the period T_0, a saddle point h becomes a periodic hyperbolic trajectory $\gamma(t)$ whose stable, $W_s(\gamma)$, and unstable, $W_u(\gamma)$, manifolds are surfaces intersecting in the extended phase space. The lines with arrows on those manifolds represent typical trajectories

$W_u(\gamma)$ **manifolds** of the corresponding hyperbolic trajectory $\gamma(t)$. In the enlarged phase space the manifolds $W_s(\gamma)$ and $W_u(\gamma)$ are two-dimensional surfaces which do not coincide but intersect (see Fig. G3c). The corresponding curves $W_s(\gamma)$ and $W_u(\gamma)$ intersect each other on a **Poincaré section surface** in *homoclinic points*. H. Poincaré proved that there are an infinite number of homoclinic points of intersections of stable and unstable manifolds of a hyperbolic trajectory [13].

Any point belonging to an invariant manifold maps, by definition, on the Poincaré section surface to another point on the same manifold. When moving away from a hyperbolic point, the amplitude of oscillations of the curve W_u increases. When approaching the same or another hyperbolic point, the "period" of oscillations decreases (the successive distances between the points of intersections of W_s and W_u decrease when approaching to h) because of slowing down of the motion nearly h. It results in a complicated *heteroclinic* or *homoclinic structure*.

Invariant tori

We described briefly Hamiltonian systems in the article **"Hamiltonian dynamics"**. The Liouville–Arnold theorem specifies that: (1) all the trajectories of a fully integrable Hamiltonian system with N degrees of freedom and N first integrals in involution C_i lie on N-dimensional invariant **manifolds** in the $2N$-dimensional **phase space** which are *invariant tori*; (2) the corresponding **trajectories** are quasiperiodic and specified by N incommensurate frequencies $\omega_i = \omega_i(C_1, \ldots, C_N)$, and (3) satisfy to the equations of motion (G.3). In fully integrable Hamiltonian systems, the action I_i is a constant on the corresponding invariant torus, and the angle variable has a simple solution $\theta_i = \omega_i t + \text{const}$ [1].

Invariant torus is a *resonant* one if its eigenfrequencies are commensurate, i.e., if $k_1\omega_1 + k_2\omega_2 + \cdots + k_N\omega_N = 0$ for nonzero integer k_i. If it is not the case, the torus is called a *nonresonant* one. In the former case, a trajectory is closed on the torus, and the motion is *(multi)periodic*. In the latter one, trajectories are not closed, and the corresponding motion is *quasiperiodic*. In a nondegenerate fully integrable system, i.e., if

$$\det \left| \frac{\partial \omega_i(\mathbf{I})}{\partial I_j} \right| = \det \left| \frac{\partial^2 H(\mathbf{I})}{\partial I_i \partial I_j} \right| \neq 0, \tag{G.4}$$

each invariant torus has its own frequencies. The set of nonresonant tori in a nondegenerate system is more powerful than that of resonant tori (however, the latter one is dense) because rational numbers constitute in the set of real numbers a subset of zero measure.

Islands of stability

Island of stability is a domain on a **Poincaré section surface** filled with regular trajectories. The islands of stability appear as a result of **nonlinear resonances** between natural frequencies of a nonlinear dynamical system under consideration and perturbation frequencies. Rotational islands correspond to finite regular motion in a bounded region in the **phase space**. Ballistic islands correspond to infinite regular motion, i.e., they all filled with ballistic regular trajectories.

Kolmogorov–Arnold–Moser theorem and KAM tori

The *Kolmogorov–Arnold–Moser theorem* (KAM theorem) states that under a sufficiently small conservative Hamiltonian perturbation a majority of **nonresonant invariant tori** of an integrable Hamiltonian system do not disappear, but they are slightly deformed in such a way that there appear invariant tori (called *KAM tori*) in the **phase space** filled up everywhere densely with (quasi)periodic **trajectories** [1]. The KAM theorem says nothing about the fate of **resonant tori**. It has been shown in numerous studies that they may break down with the onset of **Hamiltonian chaos**. Since resonant tori is a set of zero measure (the probability to find such a torus under a random choice of initial conditions is equal to zero), the KAM theorem can be reformulated more simply as follows: under a sufficiently small perturbation, almost all invariant tori of the integrable system under consideration are conserved. When proving the theorem, it is stated what is it "sufficiently small" and "almost all" [1].

Lyapunov exponents

Chaotic motion is characterized by an exponential sensitivity to small variations in initial conditions. It means that initially close trajectories may diverge exponentially fastly in time (not linearly as in the case with regular trajectories). *Lyapunov exponent* is a measure of mean velocity of exponential divergence (convergence) of initially close trajectories.

Equations of motion for a dynamical system are

$$\dot{x}_i = F_i(x_1, \ldots x_n), \quad i = 1, \ldots n. \tag{G.5}$$

Linearizing Eq. (G.5) nearby a given trajectory $x(t) = (x_1, x_2, \ldots x_n)$ with the initial condition $X(0)$, we get equations of motion for small deviations

$$\delta\dot{x}_i = \sum_{j=1}^{n} \delta x_j \left(\frac{\partial F_i}{\partial x_j} \right)_{X=X(t)}, \tag{G.6}$$

where $(\partial F_i/\partial x_j)_{X=X(t)}$ are elements of the Jacobian matrix. The norm

$$|\Delta(t)| = \sqrt{\sum_{i=1}^{n} \delta x_i^2(t)} \tag{G.7}$$

is a measure of divergence between the chosen trajectory X and a neighbor trajectory with close initial condition $x(0) + \delta x(0)$. Let us introduce the mean velocity of exponential divergence of trajectories

$$\Lambda(X(0)) = \lim_{t \to \infty} \frac{1}{t} \ln \frac{|\Delta(t)|}{|\Delta(0)|}, \tag{G.8}$$

where $|\Delta(0)| = \sqrt{\sum_{i=1}^{n} \delta x_i^2(0)}$.

A small initial phase volume stretches in the course of time mostly in the direction corresponding to a largest Lyapunov exponent. Computation with the expression (G.8) gives namely that value which is known as a *maximal Lyapunov exponent*. Generally speaking, values of Lyapunov exponents depend on the choice of a trajectory $x(t)$. It is not the case in hyperbolic chaotic systems, but the choice of a test trajectory, say, inside an **island of stability** gives obviously $\Lambda = 0$. The limit (G.8) can be achieved in chaotic systems with the bounded phase space for a reasonable computation time. In open systems $\Lambda \to 0$ at $t \to \infty$, and chaos in such systems is transient. The so-called *finite-time* and *finite-size Lyapunov exponents* may serve measures of transient chaos.

Manifolds

Manifold is a fundamental notion in topology. Its rigorous definition can be found in any textbook on this subject. It is sufficient here to define a manifold as a smooth subspace in the **phase space**. An infinite line and a circle are examples of one-dimensional manifolds, surfaces of a sphere and a torus are two-dimensional manifolds, the three-dimensional linear space R^3 is an example of a three-dimensional manifold. A segment with its limit points and the surface of a cone are not manifolds, because the limit points and the top of the cone do not satisfy to the smoothness criterion.

Nondegenerated hyperbolic **invariant tori** in Hamiltonian systems have *stable*, W_s, and *unstable*, W_u, *invariant manifolds* filled up with trajectories asymptotic to quasiperiodic trajectories on a hyperbolic torus at $t \to \infty$ (W_s) and $t \to -\infty$ (W_u). In integrable Hamiltonian systems, the manifolds W_s and W_u coincide, as a rule, pairwisely. In nonintegrable systems, they may intersect each other transversally forming a complicated **homoclinic** or **heteroclinic structure**.

To give a visual picture of these abstract objects, let us consider stable and unstable manifolds of a periodic saddle trajectory $\gamma(t)$ appearing in a plane flow of incompressible fluid under a periodic perturbation from a stationary saddle point of the corresponding integrable system. The manifolds $W_s(\gamma(t))$ and $W_u(\gamma(t))$ are collections of points through which pass at the moment of time t those trajectories of fluid particles which are asymptotic to the saddle trajectory $\gamma(t)$ at $t \to \infty$ and $t \to -\infty$, respectively. These manifolds evolve in time. In Fig. G4 geometry of stable and unstable manifolds nearby a periodic saddle trajectory $\gamma(t)$ is shown schematically at different time moments. Both $W_{s,u}(\gamma(t))$ and the corresponding linear invariant sets $E_{s,u}(t)$, which are specified with the help of a linearization of the velocity field nearby the corresponding saddle point, evolve in space and time. Under a periodic perturbation, stable and unstable manifolds are periodic functions of time. What happens with W_s and W_u far away from $\gamma(t)$ is discussed in the article **"Heteroclinic (homoclinic) structure"**.

The existence of these manifolds and their structural stability follow from the corresponding theorems which can be found, for example, in the textbook [4]. Unstable manifolds can be seen with a naked eye in laboratory experiments on chaotic advection in fluids with a dye [5, 11]. Theoretically, they are curves of

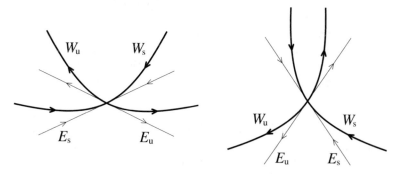

Fig. G4 Stable, $W_s(\gamma(t))$, and unstable, $W_u(\gamma(t))$, manifolds of a saddle trajectory $\gamma(t)$ are shown on the phase plane at t_1 (*left*) and $t_2 > t_1$ (*right*) along with the corresponding linear invariant submanifolds $E_s(t)$ and $E_u(t)$

an infinite length and complicated form. In real experiments with dyes, they are, of course, diffusive-like objects owing to molecular diffusion and a technical noise. Since the stable and unstable manifolds are material lines in two-dimensional flows, trajectories of fluid particles cannot cross them, i.e., W_s and W_u are transport barriers. Any material line in a fluid flow is, of course, a transport barrier. Exclusiveness of stable and unstable manifolds is in its partition a flow in topologically and dynamically distinct regions.

Nonlinear resonance

A resonance occurs in linear systems at the perturbation frequency close to a natural frequency of the system under consideration. In a nonlinear system with sufficiently strong nonlinearity, resonances may occur practically at any frequency of the excitation $\Omega = 2\pi T_0$. Since nonlinear systems possess, in general, infinitely many natural frequencies ω_i, the resonance condition $m\omega_i = n\Omega$ is satisfied with an infinite number of positive integers m and n. The corresponding resonance is denoted as $m : n$.

An isolated nonlinear resonance is represented on the **Poincaré section surface** by nested invariant curves, forming an **island of stability** or a resonant island, with an **elliptic point** in its center. Elliptic points on the Poincaré section surface are images of periodic trajectories in the **phase space**.

There are nonlinear resonances of different orders. A *primary nonlinear resonance*, $\omega_1 = \Omega$, in a system with one-and-half degrees of freedom is illustrated in Fig. G5 as it looks in the extended phase space (x, y, t), on the phase plane (x, y), and on the Poincaré section surface (x, y). In the extended phase space (Fig. G5a), a tube with quasiperiodic trajectories winds around the cylindrical surface that contains the periodic trajectory S_1 of that resonance. The quasiperiodic trajectories lie on the surfaces of the nested cylinders which are densely filled with those trajectories. The periodic, S_1, and one of the quasiperiodic trajectories, R_1, are shown in Fig. G5b by the dashed closed and solid open curves, respectively. The periodic trajectory is represented on the Poincaré section surface (Fig. G5c) by the point S_1, whereas a family of the quasiperiodic trajectories is mapped onto the corresponding nested resonant invariant curves.

The periodic trajectory of a *secondary nonlinear resonance*, S_2, winds the surface of the tube filled with quasiperiodic trajectories, R_1, of the corresponding primary resonance (see Fig. G6a). The tubes with the quasiperiodic trajectories of the secondary resonance R_2 (not shown in the figure) wind around S_2. This complicated motion is simplified on the Poincaré section surface (x, y) in Fig. G6b demonstrating schematically an island of the primary resonance with the elliptic point, S_1, surrounded by three islands of the secondary resonance with the elliptic points and invariant curves of the corresponding quasiperiodic trajectories R_2. The phase point on the periodic trajectory of the secondary resonance, S_2, turns around the elliptic point of the primary resonance S_1 and returns to its initial position for three perturbation periods.

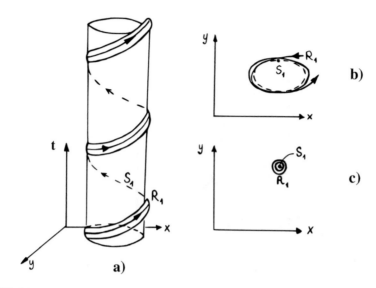

Fig. G5 Schematic illustration of a primary nonlinear resonance. (**a**) In the extended phase space (x, y, t) a tube, filled with quasiperiodic trajectories, R_1, of the primary resonance, winds the surface that contains the periodic trajectory S_1 of that resonance. (**b**) The periodic trajectory (*dashed closed curve* S_1) and one of the quasiperiodic trajectories of the primary resonance R_1 (*solid open curve*) are shown on the phase plane (x, y). (**c**) Stationary elliptic point S_1 and the invariant resonant curves of the quasiperiodic trajectories R_1 are shown on the Poincaré section surface (x, y)

There are infinitely many nonlinear resonances of different orders in typical chaotic Hamiltonian systems. They are represented on Poincaré section surfaces (see Fig. G1) by chains of islands of a different size. Islands of primary resonances are surrounded by chains of smaller islands of secondary resonances which, in turn, are surrounded by islands of higher-order resonances of smaller sizes, etc.

Phase space

The *phase space* is an n-dimensional abstract space with the coordinates being components, $x_i (i = 1, 2, \ldots, n)$, of a state vector of the dynamical system under consideration $\dot{x}_i = F_i(x_1, x_2, \ldots, x_n, t)$. In mechanical systems generalized positions and momenta are coordinates in the phase space. A state of a dynamical system at each time moment is a *phase point* in the phase space. A phase point moves in the course of time along a curve which is called a *phase trajectory* beginning at an initial point $[x_1(t = 0), x_2(t = 0), \ldots, x_n(t = 0)]$. A set of phase trajectories with all possible initial conditions constitutes a *phase portrait*. The *extended phase space* is a phase space with time as an additional coordinate.

Poincaré map

The Poincaré's idea [13] was to fix coordinates of a phase point at specified time moments or when it crosses a given surface in the **phase space**. If the dynamical

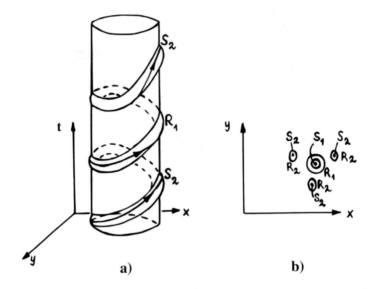

Fig. G6 Schematic illustration of a secondary nonlinear resonance. (**a**) In the extended phase space (x, y, t) the periodic trajectory of the secondary resonance, S_2, winds the surface of a tube filled with quasiperiodic trajectories of the primary resonance R_1. (**b**) On the Poincaré section surface, (x, y), an island of a primary resonance with the elliptic point S_1 is surrounded by three islands of a secondary resonance filled with invariant curves of the secondary resonance R_2

system under consideration is autonomous, then one chooses a surface in the phase space with the dimension which is smaller by one than the phase-space dimension and fixes the moments when the corresponding **trajectory** intersects it transversally. That surface is called a *Poincaré section surface*.

In difference from autonomous systems, nonautonomous ones are described by the additional variable, time, and their evolution should be considered in the extended phase space. The **Hamiltonian systems** with 3/2 degrees of freedom, which are studied in this book as simplified oceanographic models, have a three-dimensional phase space. If it is periodic with the period T_0, we can rid of time variable using a *Poincaré map*. An ordinary differential equation is replaced by a discrete mapping which associates coordinates of a trajectory $X(t_0)$ at the moment of time t_0 with its coordinates $X(t_0 + T_0)$ over the period T_0: $X(t_0 + T_0) = G_{T_0} X(t_0)$, where $G_{T_0} \equiv G(t_0, t_0 + T_0)$ is an evolution operator. Thus, one considers a discrete *orbit* consisting of the points $X_i = G_{T_0}^i X_0$, $i = 0, \pm 1, \pm 2, \ldots$ on the plane instead of the corresponding continuous trajectory in extended phase space. Geometrically, those points are intersections of a trajectory in the extended phase space by the planes $t = t_0 + i T_0$. At the moments of time corresponding to any section, own trajectory passes through each point of the corresponding orbit.

If a trajectory is periodic with the period kT_0, $k = 1, 2, \ldots$, then the corresponding orbit consists of k points. The periodic orbits can **bifurcate** under changing control parameters of the system under consideration as its **stationary**

points. Aperiodic trajectories are associated with orbits with an infinite number of points. The method of Poincaré sections is very convenient for systems with an inhomogeneous phase space because both the specific objects and inherent effects, such as **stability islands**, periodic trajectories, and sticking, are clearly manifested on the Poincaré sections. For example, **invariant nonresonant tori** (quasiperiodic trajectories) are associated with one or more closed curves with elliptic points at their centers. Chaotic orbits look like sets of points on the Poincaré sections filling some area with increased density at the borders of stability islands due to stickiness.

Separatrix

In the systems with one degree of freedom, a *separatrix* is a special, singular trajectory connecting hyperbolic **stationary points** and separating topologically different regions of motion. The period of the phase point motion along a separatrix is infinite because the velocity at stationary points is zero by definition. One gets a *homoclinic connection* if a separatrix connects the same hyperbolic point (see Fig. G3b). If a separatrix connects different saddle points, one gets a *heteroclinic connection* (Fig. G3a). Perturbed separatrices may arise in dynamical systems under a perturbation. They are stable and unstable **manifolds** of the corresponding hyperbolic points (trajectories).

Stable and unstable motion

Stability and *instability* are fundamental properties of motion which are manifested not only nearby **stationary points**. A **trajectory** $X(t)$ with the initial condition X_0 is called *stable by Lyapunov* if for any number ε there exists a number $\delta(\varepsilon)$ such that for all $\tilde{X}(t)$ the inequality $\|X(t) - \tilde{X}(t)\| < \varepsilon$ is satisfied for any trajectory $\tilde{X}(t)$ such that $\|X_0 - \tilde{X}_0\| < \delta$. It means that the diameter of a phase drop with the center at X_0 at $t = 0$ does not exceed in the course of time a given value ε, if it was smaller than δ at $t = 0$ (see Fig. G7). If a trajectory is stable by Lyapunov, then the corresponding phase drop is forever compact in a stream tube. An initially compact drop in a stream tube with an orbitally stable trajectory stays forever in that tube but spreads along that trajectory. In other words, two initially close points in the drop may diverge from each other in the course of time.

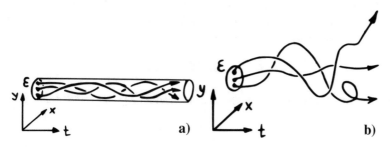

Fig. G7 Lyapunov (**a**) stable and (**b**) unstable motion

Stationary points

Analysis of the dynamical system under consideration begins with finding the **phase portrait** of an autonomous version of the system, finding its stationary points, and studying motion in a small neighborhood of each of them. Let us represent the equations of motion in the form of a set of the first-order differential equations

$$\dot{X} = F(X),$$

where X is a vector with components being state variables of the system. *Stationary points* (the other names: equilibrium, rest, special, singular or fixed points) are specified as $\dot{X} = 0$ or $F(X_s) = 0$. To study a character of motion nearby a stationary point, let us expand the function $F(X)$ in a Taylor series and analyze the corresponding linearized equations of motion. As an illustrative example, we consider a set with two equations

$$\dot{x} = f(x, y),$$
$$\dot{y} = g(x, y),$$

with the coordinates of their stationary points satisfying to the equations: $f(x_s, y_s) = 0$ and $g(x_s, y_s) = 0$. Let us introduce small deviations δx and δy nearby one of the points $x = x_s + \delta x$ and $y = y_s + \delta y$ and expand f and g in a series in powers of δx and δy:

$$\delta\dot{x} = f_x(x_s, y_s)\delta x + f_y(x_s, y_s)\delta y + f_{xy}(x_s, y_s)\delta x \delta y + \cdots,$$
$$\delta\dot{y} = g_x(x_s, y_s)\delta x + g_y(x_s, y_s)\delta y + g_{xy}(x_s, y_s)\delta x \delta y + \cdots.$$

Neglecting terms above the first order, one can represent the equations for small deviations as a set

$$\frac{d}{dt}\begin{pmatrix} \delta x \\ \delta y \end{pmatrix} = \begin{pmatrix} f_x(x_s, y_s) & f_y(x_s, y_s) \\ g_x(x_s, y_s) & g_y(x_s, y_s) \end{pmatrix}\begin{pmatrix} \delta x \\ \delta y \end{pmatrix},$$

which is a set of *linearized equations of motion*. Denoting the column vector $(\delta x, \delta y)^T$ by δX, 2×2 matrix by $\hat{F}$, two its eigenvectors by d_1 and d_2, and the corresponding eigenvalues by λ_1 and λ_2, a general solution of the linearized equations of motion can be represented in the form

$$\delta X = c_1 d_1 e^{\lambda_1 t} + c_2 d_2 e^{\lambda_2 t},$$

where $c_{1,2}$ are integration constants. The eigenvalues $\lambda_{1,2}$ are roots of the equation

$$\det[\hat{F} - \lambda\hat{I}] = 0,$$

where $\hat{I}$ is a unit matrix.

The full classification of stationary points can be found in any textbook on dynamical systems. Here, we reproduce general statements about Hamiltonian systems and area-preserving maps. There are the following three possibilities:

1. λ_1 and λ_2 is a complex-conjugated pair $\lambda_1 = e^{i\alpha}$ and $\lambda_2 = e^{-i\alpha}$ on a unit circle. Then small deviations δx and δy rotate around the corresponding point and the corresponding phase trajectories are ellipses. Such a stationary point is called *stable* or *elliptic*.
2. λ_1 and λ_2 are real numbers with the condition $\lambda_2 = \lambda_1^{-1}$. The motion nearby such a point is unstable, and it is called a *hyperbolic* or a *saddle* point.
3. There is a special case when $\lambda_1 = 1$ and $\lambda_2 = -1$. Such a stationary point is called *parabolic*.

Trajectories

We give below definitions of the types of trajectories in dynamical systems. The phase point, specifying a state of the dynamical system at a given time moment, changes its position in the phase space in the course of time. The corresponding curve is called a *phase trajectory*. Because of uniqueness of solutions of differential equations, phase trajectories cannot cross each other. If the motion is periodic, then the corresponding trajectory is called a *periodic* one. Circle is the simplest image of a periodic trajectory. Trajectories with a long period usually have more complicated forms. Multi-frequency motion can be periodic if the frequencies are commensurate, i.e., if there exists a set of nonzero integers (positive or negative) $k_1, k_2, \ldots$ such that $k_1\omega_1 + k_2\omega_2 + \cdots = 0$. If such a set does not exist, the motion is called *quasiperiodic*. In the case with the two frequencies, a quasiperiodic trajectory winds the surface of a torus without self-intersections and is not closed. The motions along and across the torus have different frequencies, and their ratio is an irrational number. *Aperiodic* or *chaotic* trajectories do not lie on the surfaces of tori in the phase space.

Periodic and quasiperiodic trajectories can be stable and unstable. The latter ones are called *hyperbolic trajectories*. Chaotic trajectories are, generally speaking, unstable, however, their fragments of an arbitrary but finite length may demonstrate a kind of stability. There are special types of chaotic trajectories. Let $\mathbf{X}(t)$ be an unstable (hyperbolic) trajectory. A trajectory which asymptotically approaches a hyperbolic trajectory at $t \to -\infty$ and $t \to \infty$ is called *homoclinic*. Let $\mathbf{X}(t)$ and $\tilde{\mathbf{X}}(t)$ be two hyperbolic trajectories. A trajectory which asymptotically approaches $\mathbf{X}(t)$ at $t \to -\infty$ and $\tilde{\mathbf{X}}(t)$ at $t \to \infty$ is called *heteroclinic*.

A periodic trajectory in the extended phase-space winds a surface of a (deformed) cylinder or torus, and its projection onto a phase plane is a smooth closed curve (perhaps, with self-intersections). The phase point along a periodic trajectory returns to its initial position for the time T, where T is a period of the trajectory. Figure G8a demonstrates a periodic trajectory on the surface of a straight cylinder. Projection of the quasiperiodic trajectory onto a phase plane is a smooth open curve winding in a comparatively narrow strip (Fig. G8b). Projection of an aperiodic trajectory onto the phase plane is a smooth open curve (Fig. G8c).

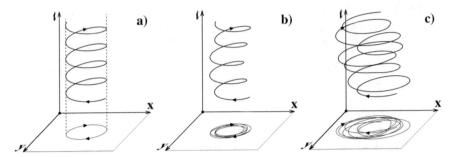

Fig. G8 (a) Periodic, (b) quasiperiodic, and (c) aperiodic trajectories in the extended phase space (x, y, t) and their projections onto the phase plane (x, y)

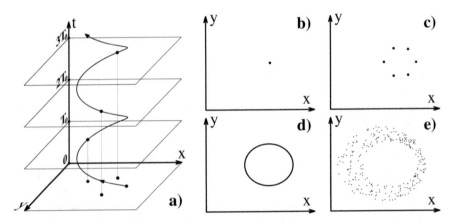

Fig. G9 (a) Trajectory in the extended phase space with the four first crossings at the time instants nT_0 $(n = 0, 1, 2, \dots)$. The Poincaré section surfaces (b) for a trajectory with the period T_0, (c) for a periodic trajectory with the period nT_0, (d) for a quasiperiodic trajectory, and (e) for aperiodic trajectory

Figure G9 illustrates **Poincaré section surfaces** of different kinds of trajectories in the phase space. A trajectory with the period equal to T_0 is represented by a point on the Poincaré section surface with the same period (Fig. G9b). A periodic trajectory with another value of the period is represented by a finite set of points on the Poincaré section surface with the period T_0 (Fig. G9c). A quasiperiodic trajectory never returns to its starting point. If we wait long enough, the quasiperiodic orbit will be as close as we want to returning at some point in time. Therefore, it covers a continuous invariant curve on the Poincaré section surface (Fig. G9d), whereas an aperiodic trajectory is represented by a cloud of points (Fig. G9e). It should be stressed that the pattern of motion in a multi-dimensional system can be different for different surfaces of sections.

References

1. Arnold, V.I., Kozlov, V.V., Neishtadt, A.I.: Mathematical Aspects of Classical and Celestial Mechanics. Encyclopaedia of Mathematical Sciences, vol. 3, 3rd edn. Springer, New York (2006). 10.1007/978-3-540-48926-9
2. Budyansky, M., Uleysky, M., Prants, S.: Chaotic scattering, transport, and fractals in a simple hydrodynamic flow. J. Exp. Theor. Phys. 99, 1018–1027 (2004). 10.1134/1.1842883
3. Gaspard, P.: Chaos, Scattering and Statistical Mechanics. Cambridge Nonlinear Science Series, vol. 9. Cambridge University Press, Cambridge (1998)
4. Guckenheimer, J., Holmes, P.: Nonlinear Oscillations, Dynamical Systems, and Bifurcations of Vector Fields. Applied Mathematical Sciences, vol. 42. Springer, New York (1983). 10.1007/978-1-4612-1140-2
5. Koshel, K.V., Prants, S.V.: Chaotic Advection in the Ocean. Institute for Computer Science, Moscow (2008) [in Russian]
6. Lichtenberg, A.J., Lieberman, M.A.: Regular and Chaotic Dynamics. Applied Mathematical Sciences, vol. 38. Springer, New York (1992). 10.1007/978-1-4757-2184-3
7. Mackay, R., Meiss, J., Percival, I.: Transport in Hamiltonian systems. Physica D 13(1–2), 55–81 (1984). 10.1016/0167-2789(84)90270-7
8. Makarov, D., Prants, S., Virovlyansky, A., Zaslavsky, G.: Ray and Wave Chaos in Ocean Acoustics: Chaos in Waveguide. Series on Complexity, Nonlinearity and Chaos, vol. 1. World Scientific, Singapore (2011). 10.1142/9789814273183_fmatter
9. Mandelbrot, B.B.: The Fractal Geometry of Nature. W.H. Freeman and Company, New York (1982)
10. Ott, E.: Chaos in dynamical systems, 2nd edn. Cambridge University Press, Cambridge (2002)
11. Ottino, J.M.: The Kinematics of Mixing: Stretching, Chaos, and Transport. Cambridge Texts in Applied Mathematics, vol. 3. Cambridge University Press, Cambridge (1989)
12. Percival, I.C.: Variational principles for invariant tori and cantori. AIP Conf. Proc. 57(1), 302–310 (1980). 10.1063/1.32113
13. Poincaré, H.: New methods of celestial mechanics. NASA Technical Translation, vol. 450–452. NASA, Springfield (1967)
14. Zaslavsky, G.: Dynamical traps. Physica D 168–169, 292–304 (2002). 10.1016/s0167-2789(02)00516-x
15. Zaslavsky, G.M.: Hamiltonian Chaos and Fractional Dynamics. Oxford University Press, New York (2005)

Printed in the United States
By Bookmasters